Palgrave Studies in Law, Neuroscience, and Human Behavior

Neuroscience is drawing increasing attention from lawyers, judges, and policy-makers because it both illuminates and questions the myriad assumptions that law makes about human thought and behavior. Additionally, the technologies used in neuroscience may provide lawyers with new forms of evidence that arguably require regulation. Thus, both the technology and applications of neuroscience involve serious questions implicating the fields of ethics, law, science, and policy. Simultaneously, developments in empirical psychology are shedding scientific light on the patterns of human thought and behavior that are implicated in the legal system. The Palgrave Series on Law, Neuroscience, and Human Behavior provides a platform for these emerging areas of scholarship.

More information about this series at
https://link.springer.com/bookseries/15605

Marc Jonathan Blitz · Jan Christoph Bublitz
Editors

The Law and Ethics of Freedom of Thought, Volume 1

Neuroscience, Autonomy, and Individual Rights

palgrave
macmillan

Editors
Marc Jonathan Blitz
Law
Oklahoma City University School of Law
Oklahoma City, OK, USA

Jan Christoph Bublitz
Faculty of Law
University of Hamburg
Hamburg, Germany

Palgrave Studies in Law, Neuroscience, and Human Behavior
ISBN 978-3-030-84493-6 ISBN 978-3-030-84494-3 (eBook)
https://doi.org/10.1007/978-3-030-84494-3

Cover illustration: Sean Gladwell/Getty Images

This Palgrave Macmillan imprint is published by the registered company Springer Nature Switzerland AG
The registered company address is: Gewerbestrasse 11, 6330 Cham, Switzerland

Preface

Freedom of thought is one of the great and venerable notions of Western thought. It remains so in the modern era. Many United States Supreme Court cases mention it and the drafters of the International Bill of Rights enshrined a human right to freedom of thought in Articles 18 of the Universal Declaration of Human Rights and the Covenant on Civil and Political Rights. These articles have inspired similar guarantees in many subsequent human rights instruments. Legal reasoning in the United States, Europe, and elsewhere thus seems to presuppose the freedom of thought.

What it means precisely, however, is anything but clear; surprisingly little writing has been devoted to it. There are no settled definitions about its scope in international human rights law or in the scholarly literature. There are few court cases elaborating it (Bublitz, 2014). This is perhaps because some entertain an overly narrow conception according to which freedom of thought is co-extensive with freedom of conscience or religion. But this overlooks the broader meaning of the concept in the liberal tradition as a right to think as one pleases, free from dictates by authorities, as well as the meaning given to it by drafters of the Universal Declaration, placing it before the rights to conscience and religion. Another reason for the neglect is that other protections, most notably those for freedom of *speech*, secure the *expression* of thought—and philosophers and jurists have assumed that that *thinking itself* needs no legal protection, because it is inevitably beyond government's power to monitor or control. In the words of United States Supreme Court Justice Frank Murphy, "[f]reedom to think is absolute of its own nature" because even "the most tyrannical government is powerless to control the inward workings of the mind" (Jones v. Opelika, 1942). Thus, while freedom of thought was assumed to be essential, it did not need to be shielded by a constitutional or legal *right*. There was no need to turn to courts to protect our freedom of thought because nature already protected it.

In fact, the idea that thought is naturally and comprehensively insulated from external control has a long heritage. It finds expression in the Roman legal maxim *cogitationis poenam nemo patitur*—meaning "no punishment for thought." It is also present in a famous German folk song of the early Enlightenment, adapted by Pete Seeger: *No scholar can map them, no hunter can trap them … If tyrants take me, and throw me into prison, my thoughts will burst free*. The key idea is that thoughts are not accessible. They remain a secret to, and cannot be controlled by, others. The body might be imprisoned—thought is free. Even when our thoughts and feelings *can be* known to others, writers have viewed them as the kind of entity that cannot be commanded in the way that external actions can be commanded: While we might aver (in response to a command or threat) that the earth is flat, that doesn't mean we can easily convince ourselves to believe the truth of that assertion. John Locke thus wrote in 1689 that "such is the nature of the understanding, that it cannot be compelled to the belief of anything by outward force." The same is true of emotions: We cannot simply will ourselves to find a piece of music pleasant or to enjoy the taste of certain foods we have long disliked.

But this supposed insulation of cognition and emotion doesn't spare us the need to understand what "freedom of thought" might cover—and whether and how the law should protect it. Even if Justice Murphy was correct in saying that "government is powerless to control the inward workings of the mind," the workings of the mind are not always "inward." As the Justice himself recognized, thought *can* be punished by government when it is expressed in speech, or revealed in other behavior.

Moreover, history provides us with many examples of rulers and governments changing thoughts and beliefs through force and incentives, from the Inquisition to thought reform camps (Lifton, 1953; Taylor, 2017). Coercion is applied externally, through confinement or social exclusion, but it aims at, and sometimes succeeds in, affecting thought.

In fact, the form taken by such efforts at thought manipulation changes—and some writers have worried that modern societies give government officials (and others) more power to blunt the kind of skepticism and independent thinking that might threaten their interests. A century before the publication of this volume, in 1922, the philosopher Bertrand Russell delivered a lecture to a "freethought community" in London, warning that "new dangers, somewhat different from those of the past, threaten [freedom of thought],"and that "unless a vigorous and vigilant public opinion can be aroused in defense of [freedom of thought], there will be much less of [it] a hundred years hence than there is now." The dangers he identified were one-sided state education, state propaganda, and economic pressures to form socially desirable opinions. As an antidote, he preached the "will to doubt" and reminded his audience that "none of our beliefs are quite true; all have at least a penumbra of vagueness and error" (Russell, 1922, 7).

A century later, neither Russell's worries nor his hopes have fully materialized. Likely, Western countries today respect the freedom to think to a greater degree than a hundred years ago: Thanks in part to the Internet, individuals have immediate access to massive amounts of information. Education, including higher education, is available to far more of the population—providing at least some of the skills and knowledge necessary for individuals to question and assess the information they receive from government or from other citizens. Nonetheless, Russell's worries still are a cause for concern: Epistemic humility has not become a widely cultivated virtue and individual thought is still subject to manipulation.

Moreover, whereas Russell worried in 1922 that transformations in society, politics, and education would give rise to new, more powerful kinds of thought manipulation—and perhaps also new ways to assure mental autonomy—there is also cause in 2021 to ask whether the dizzying technological changes of the current era will also reshape not only the threat freedom of thought faces, but also the ways we can protect it, the extent we can create it (where it is absent), and perhaps the way that courts, legal scholars, philosophers, and others should define "freedom of thought."

It is this latter scientific and technological shift in how we understand our thinking—and what it means to have freedom of thought—that this volume, and forthcoming work in our series, will explore. This shift raises some new challenges to the assumption that our unexpressed thoughts need no legal protection because even the most powerful official is "powerless to control the inward workings of the mind."

First, even when our thoughts remain wholly unexpressed, modern neuroscience and psychology may now give government a way to extract information about them—and perhaps, to coerce them. As Bruce Winick wrote in 1989, "given the emergence of the more intrusive mental health treatment techniques, the 'inward workings of the mind' are now within the reach of government control" (Winick, 1989, 20). When government compels prisoners or inmates of psychiatric institutions to use psychotropic drugs, for example, it alters their thinking patterns. The development of functional Magnetic Resonance Imaging (fMRI) and other brain-scanning technology also gives government a new tool to infer the content of our silent thoughts (Boire, 2004; Stoller and Wolpe, 2005; Farahany, 2012). Numerous scholars have argued that these technological developments require adapting the concept of freedom of thought for the twenty-first century. Most notably, Richard Glen Boire (2000) has not only made this argument but also suggested a new name for it: "cognitive liberty." He, Wrye Sententia, and others at the Center for Cognitive Liberty and Ethics have argued that we need a new conception of freedom of thought to meet the challenges of the twenty-first century. As Sententia puts this point, the right to "cognitive liberty" "updates notions of 'freedom of thought' for the twenty-first century by taking into account the power we now have, and increasingly will have, to monitor and manipulate cognitive function" (Sententia, 2004). They have likewise considered how a right to "freedom of thought" or "cognitive liberty" might function as a liberty independent of more familiar constitutional rights to freedom of speech, privacy, or bodily integrity. The publications of the Center in their Journal of Cognitive Liberties are well worth reading two decades after the Center's major work developing this concept. While differences between cognitive liberty and freedom of thought often appear only semantic, some scholars such as Nita Farahany conceive of them as substantive (Farahany, 2019).

Second, thought may also be external to us—and thus within government's power—even when it is not expressed in speech or behavior, if one follows recent philosophical theory advanced by David Chalmers and Andy Clark, and others who embrace their theory of an "extended mind," According to this theory, the physical correlates of the mind may extend outside the brain. Our thoughts may not just arise from electrochemical interactions between neurons, but also from the way we use certain tools and resources in the world outside of brains and bodies. When we compose notes in a journal, for example, the writing isn't simply a product of our thinking, or impetus for future thought. One may conceive of it as part and parcel of the act of thinking—one that, unlike our purely internal imaginings, can be observed and restricted by government (Clark & Chalmers, 1998).

Third, there is still another reason that we cannot assume that freedom of thought is simply guaranteed by nature—and therefore needs no additional aid from law, norms, or other rules of human conduct to flourish. This becomes apparent when one views "freedom of thought" not first from the perspective of legal or political rights, but rather from the perspective of psychology and philosophy. As a psychological idea, freedom of thought may concern the degree to which thinking is constrained or determined—not by government or other external actors—but rather by psychological or neurobiological processes internal to a person. It may concern the extent to which, or the way in which, persons have active conscious control over their fleeting streams of consciousness. This is a form of freedom from *internal* constraints. Even where our thoughts remain entirely free from government control, our freedom from internal constraints might remain—and may raise significant barriers against a certain kind of freedom of thought (Metzinger, 2015). Conceived narrowly, such freedom of thought may require that we be free of certain types of internal experiences or impulses that we might characterize as foreign to our will, our "authentic" self, or that we have active control over the content of our stream of consciousness.

In other words, freedom of thought may be conceived, to use Isaiah Berlin's terms, as a kind of positive liberty (a freedom *to* exercise certain mental capabilities) rather than merely a "negative liberty" or freedom *from* government or other external internal interference in one's thinking (Berlin, 1969). Or, we might make a similar point using Gerald MacCallum's formulation of a liberty as existing where (1) "some agent or agents," has (2) "freedom from some constraint or restriction on, interference with, or barrier to" (3) "doing, not doing, becoming, or not becoming something" (MacCallum, 1967, 314). For many legal and political thinkers, the second element in this triadic relation consists only in constraints or preventing conditions that might come from government or some other actors (other than the thinker). For others who define freedom of thought in psychological terms, by contrast, the constraining or preventing condition might come from an individual's own psychological tendencies, or lack of mental capacity to control certain internal states or processes. They may also disagree about the third element of this triadic relation: For Justice Murphy, we might have freedom of thought whenever we have *any kind of mental experience* (even experience that consists of reflections we wish to but cannot suppress), so long as it remains free from government restriction. For those who view freedom of thought from a psychological perspective, by contrast, freedom of thought may exist only when we are able to think in a *certain* way—for example, one where we can bring our mental states not only into line with our first-order desires, but our second-order or higher-level desires about how we wish to think and feel. This may have implications for a jurisprudence or politics of freedom of thought, because if freedom of thought requires we think in a certain way (and not simply that we be left free of government interference as we think), then it is possible that this freedom can only be achieved with certain kinds of government support—or that government must refrain not only from interfering with our thinking (where technology allows it to do so), but also with the conduct we engage in to give ourselves certain mental capacities, or overcome certain internal psychological barriers.

Fourth, technologies that owe their existence to neuroscience and psychology are relevant here too: Not only might they help the government monitor or shape our thinking (and potentially violate our freedom of thought in the process) (Boire, 2004). Cognitive enhancement, brain-computer interface devices, or brain scanning technology might also help us to *achieve or increase* our freedom of thought, and we might perhaps claim constitutional protection—or raise ethical objections—against government restriction that stops us from doing so (Boire, 2000; Sententia, 2006; Blitz, 2010). At the same time, some people are concerned about societal pressures to use such devices, appealing to freedom of thought to refuse their use (Bublitz, 2013).

In any event, legal and political thinkers can no longer simply take the position that freedom of thought needs no elaboration. Prominent constitutional law sources in Europe, the United States, and other jurisdictions emphasize that individuals have a right to freedom of thought. The U.S. Supreme Court has often referred to "freedom of thought" or "freedom of mind" (see, e.g., Wooley v. Maynard, 1977). And Justice Benjamin Cardozo referred to it, along with freedom of speech, as the "matrix of every other freedom" in the Constitution (Palko v. Connecticut, 1937). Article 18 of the Universal Declaration solemnly proclaims that "[e]veryone has the right to freedom of thought."

The meaning of this right or freedom, however, has not yet been explicated. There are hardly any court cases explaining and applying it. And in a world where government has the means to infringe this freedom—either by monitoring or altering our brain processes, by controlling the external manifestation of thoughts, or by preventing us from monitoring our own mental processes or altering our own mental capacities—courts will have to say more about what this freedom entails. It is time for the renaissance of this neglected right.

This volume provides an interdisciplinary exploration of how we might draw upon, and better understand, different conceptions of freedom of thought or mental autonomy, in order to elaborate this freedom in an age of neuroscience and emerging technologies for shaping, monitoring, and manipulating the mind.

It begins with history. In Chapter 1, Lucas Swaine provides a broad historical view of how freedom of thought was invoked—or restriction on thought criticized—in Western political and religious thought. He first looks at how thought and the freedom of thought were emphasized in key facets of early Western history—in Socrates' trial and the punishment of Socrates' beliefs and in the religious demands that Christian texts and doctrine placed on thinking, and the freedom of thought concerns raised by religious inquisitorial practices and their secular equivalents. The chapter then turns to how freedom of thought was understood in the early modern period and nineteenth century—by writers such as John Locke, Pierre Bayle, Benjamin Constant, the American Founders, Wilhelm von Humboldt, and John Stuart Mill. In doing so, it closely examines arguments by Locke and Constant, as well as by Thomas Jefferson, that freedom of thought is in some respect guaranteed by nature because government not only lacks justification for restraining thought, but also is powerless to do so, as well as arguments to the contrary (such as those that Jonas Proast offered against Locke's A Letter Concerning Toleration). It ends by considering the challenges raised for freedom of thought by emerging technologies, such as brain-scanning technologies, and legal developments, such as modern use of subpoenas and other legal instruments to gather information about a person's memories or beliefs.

In Chapter 2, Simon McCarthy-Jones looks at more recent invocations of, and debates about, freedom of thought. He draws upon political, legal, and social history to address questions about what he calls the "why," "what," and "who" of this freedom—and stresses that to understand why modern societies value freedom of thought, and how to understand it, it is necessary to undertake what Michel Foucault calls "an archaeology" of the concept. While such an undertaking is beyond the scope of a single book chapter, McCarthy-Jones lays a foundation for it by discussing how the conceptions of freedom of thought, thought control, and brainwashing arose in Western reactions to—and fear of—authoritarian countries' methods for eliminating dissent. Drawing on term frequency data from Google Books, McCarthy-Jones discusses how, in the twentieth century, use of the term initially increased in the 1930s, as "the West became concerned about the show trials held by the Communist Party of the Soviet Union," and then in the 1950s, when the Chinese Communist Party embraced the policy of "reeducation" of political opponents. He also analyzes the frequency and uses of terms such as "thought control" (to refer to the Nazis' systematic use of propaganda) and "brainwashing" (to refer to the Chinese government's treatment of American soldiers during the Korean War). The chapter also briefly examines use of the concept of freedom of thought in discussions (by Supreme Court justices and others) as a right those in Western democracies might invoke against their own government's policies and not merely in describing authoritarian governments' practices.

McCarthy-Jones then turns from twentieth-century history to present-day twenty-first-century concerns about freedom of thought. He focuses in particular on "behavior reading," which involves large-scale collection of data and analysis of citizens' digital records or observable behavior, noting the role this plays in surveillance capitalism and micro-targeting of consumers or voters. He also discusses the freedom of thought concerns raised by "brain reading," wherein "the neural activity of individuals is decoded to reveal the thoughts that it corresponds to."

Having analyzed the concerns that have motivated citizens to invoke freedom of thought in the twentieth century, and lead them to emphasize it in the present (the "why" of freedom of thought), the chapter then draws on philosophical analyses of autonomy and the nature of mental activity to consider the "what" of freedom of thought—that is what kind of internal mental activity it shields from external interference, and what kind of external activity might also count as protected thought (such as recording of memories or ideas in a journal or memo). The chapter ends by asking "who" freedom of thought protects—looking in particular at who has invoked freedom of thought in American cases on it, and the legal analysis justifying (and role of political pressure in) denying its protections to sex offenders in cases such as Doe v. LaFayette.

The subsequent chapters take a closer look at the development and possible future of freedom of thought as a constitutional or other legal right—in European, international, and American law. Chapter 3 by Jan Christoph Bublitz, explores the right to freedom of thought as codified in Article 18 of the Universal Declaration of Human Rights and the International Covenant on Civil and Political Rights. Although its importance is widely affirmed, the right does not play a role in legal practice. Turning it from a dead letter into a living right requires developing a theory of the right. This chapter provides some material for it. The first section disambiguates the political-philosophical from the legal concept and presents the central norms in human rights law. It identifies five explananda that a future theory of the right has to explain and possibly justify: The meaning and scope of the right, the peculiar internal/external structure, its absolute nature, the nature of interferences, and the relation to its sister rights, the freedoms of conscience and religion. The second section disassembles the elements of the right: thought, belief, freedom, and interferences, and explores some contradictions and the sparse case law, especially with respect to potential interferences and coercion. The third section submits some suggestions for the interpretation and construction of the right. The scope should comprise thought and thinking as well as beliefs, which ought to be protected against the imposition of cognitive duties, punishment for thoughts, revelations of thought as well as interferences. However, as many mundane actions change thoughts and beliefs, many of them protected through freedom of expression of intervenors, a taxonomy separating permissible from impermissible interferences is required; a rough test is suggested. Moreover, the chapter explores differences between behavior and thought control as well as tensions between various conceptions of the right, which emerge especially with respect to interventions improving free thought. This motivates a reflection on the absolute level of protection that freedom of thought currently enjoys in human rights law; a suggestion for a clearly defined exception is submitted. After all, the strong protection might be one impediment for the lacking practical relevance of the right. Turning it into a living right with a broad scope of application may require the possibility for context-sensitive and nuanced outcomes, which in-principled rights cannot offer. Several further suggestions for the interpretation of the right are developed along the way.

The next two chapters raise questions about the understanding of freedom of thought in American criminal and constitutional jurisprudence. In Chapter 4, Marc Jonathan Blitz looks at the way freedom of thought has been invoked and discussed (often somewhat vaguely) in existing American constitutional law—and asks how a more developed constitutional jurisprudence of freedom of thought might develop. The Court's jurisprudence on American constitutional rights, Blitz writes, has generated a certain template that applies to different rights—such as the First Amendment right to freedom of speech, the Fourth Amendment right against unreasonable searches and seizures, and the "due process" rights against intrusion into bodily liberty. Drawing upon the work of Frederick Schauer and other constitutional law scholars, the chapter looks at how courts distinguish between the "coverage" of each such right—that is what kind of activity it shields—and the "protection" it offers against government regulation or restriction of such activities. It explains that each of these rights tends to have a "core" where protection is strongest and a "periphery" where protection is somewhat reduced. Blitz then looks both at two ways that this template can elaborate our understanding of how a right to freedom of thought might work in practice. First, rather than operating as an independent right, freedom of thought concerns can affect how much protection a certain type of conduct receives under another, more familiar right: A kind of government surveillance that government would normally get substantial leeway to conduct under the Fourth Amendment might instead require a warrant based upon probable cause when it threatens freedom of thought. Compelled treatment that threatens a person's bodily integrity and therefore implicates their Fifth or Fourteenth Amendment due process rights might face more skepticism from courts when it intrudes not just into a person's body, but also into a person's brain (and thus, her thinking processes). Moreover, apart from moving certain activity into the core of another right, a right to freedom of thought may also function as an independent right, with its *own* core and its own periphery. A right to freedom of thought has often been discussed as an "absolute"—and perhaps, when functioning as a barrier against certain kinds of intrusions, such as psychosurgery or intense "brainwashing" techniques—it is. But in other circumstances, for example, when it limits the extent to which government can regulate our use of cognitive enhancement or other tools we might use to shape our own thought, the right's protection may still be meaningful—but less than complete. The right may protect, to some extent, our ability to reshape our thinking, but still leave government with some room to protect our health and safety.

Gabriel S. Mendlow's chapter, Chapter 5, is a shortened version of an essay recently published in Yale Law Journal. It addresses the puzzle of why American jurisprudence (and English jurisprudence before it) has regarded it as clear that the law cannot punish individuals solely for their thought. As Mendlow observes, the justification or this maxim is actually unclear. The chapter begins by explaining the inadequacies of a number of familiar answers. James Fitzjames Stephens and H.L.A. Hart, for example, each noted that punishing thought would be impossible without an oppressive level of surveillance or other government intrusion. But Mendlow notes that such intrusion would follow not just from punishing thought, but also from excessive punishment of action (which government is not barred from punishing). He similarly rejects arguments rooted in John Stuart Mill's harm principle, from the principle that one can only punish a "culpable wrong," and from the "beyond a reasonable doubt" requirement for imposing criminal liability. Thoughts, he argues, can cause the kind of harm that Millian philosophy allows government to regulate, and can be "culpably wrong" when they give rise to the same risks "inside a person's head" as would be punishable if they arose outside of it, and proving thoughts may be difficult but is not always impossible. Mendlow then offers another explanation of why it is impermissible to punish thought, one which relies on what he calls the "Enforceability Constraint." Citing a number Supreme Court cases, he notes that the Court seems to implicitly recognize a right to mental integrity—that is a right to be free of "unwanted mental interference or manipulation of a direct and forcible sort," such as psychosurgery or compelled administration of psychoactive drugs. Mendlow argues that, because the law makes it impermissible for the state to violate individuals' mental integrity with such direct coercion, it is likewise impermissible for it to achieve same end indirectly—by punishing thought.

The subsequent chapters look more closely at how we might understand freedom of thought or mental autonomy—or aspects of this freedom or autonomy—in light of specific neuroscience findings, emerging technologies for shaping or manipulating our minds, or therapeutic methods for treating mental illnesses or reshaping memories. They consider the ethics of mental manipulation or other alteration of our thought processes and when they are compatible with, or instead threaten, a person's autonomy. In Chapter 6, Faye Niker, Gidon Felson,

Saskia Nagel, and Peter Reiner draw on neuroscience and psychology in exploring two conditions of autonomy and their implications for ethical debates over "nudging," that is the arranging of a person's environment to make her more likely to make certain decisions. First, they argue, the exercise of autonomy often requires not simply that we are free to form, and live in accordance, with a coherent set of higher-order desires, preferences, values, and beliefs—what the authors call "pro-attitudes"—but that we be able to revise such pro-attitudes as we encounter new information. A "robust view of autonomy," in other words, "requires that we have the ability to critically reflect upon and to modify our existing pro-attitudes when our experiences or evidence call them into question." The chapter explores both how certain neurobiological processes—especially the deconsolidation and reconsolidation of memory—can undergird such "evidence-responsive critical reflection," and also how better understanding this process of critical reflection can help us understand what conditions are necessary for it to work effectively. Second, they argue, autonomy requires freedom from "undue external influence." But this does not mean that all external influence is undue. Rather, they argue—drawing on previous psychological research (including that of the authors)—that a certain kind of external influence should be understood as autonomy-supporting. It is, they say, "pre-authorized" by the agent. In determining when nudging—whether by others' actions or by algorithms in computer applications—violates autonomy, then, it is useful to ask when such nudges are autonomy-supporting and do so, in part, in light of empirical understandings of critical self-reflection and pre-authorization.

In Chapter 7, Adam J. Kolber explores the ethical questions—and implications for autonomy—that arise when the manipulation of a person's memory comes not from others' attempts to reshape and influence them but rather from that person's own choice to erase or dampen traumatic or other memories. The President's Council on Bioethics had issued an analysis in 2003 warning that leaving individuals with the freedom to erase their own painful memories would have various ethical and practical problems—leading us to lead less coherent and authentic lives, dull us to others' pain, and erase memories we might have an obligation to keep for social good (e.g., when we are a witness to a crime). Kolber argues—in a chapter adapted from a previous article—that these ethical arguments fail to show that memory dampening is as problematic as the Council claims: Our memories constantly undergo a natural process of editing and revision, and Kolber argues that it is unclear that more conscious editing of memories (carried out with memory dampening drugs) would do any more damage to a coherent sense of self than does natural transformation in memories, nor that our lives are any less "genuine" than they would be without more conscious control of which memories we retain. He also argues that the Council's arguments do not justify significant legal restrictions on when individuals (working with psychiatrists or other medical providers) might choose to dampen memories. More generally, Kolber concludes, use of such technologies might be one component of a larger "freedom of memory," that might include a freedom not only to dampen memories, but to acquire them and keep them private from external observers (armed with brain-scanning technology, for example).

In Chapter 8, Mari Stenlund explores how ethical questions about freedom of thought apply to manipulation of thought in the context of psychiatric treatment. As Stenlund notes, in discussions about mental autonomy, freedom of thought, or "cognitive liberty," "the background assumption" has generally been "that the subjects of cognitive liberty are mentally healthy adults." However, when individuals are subjected to certain kinds of compelled psychiatric treatment for delusions, for example, their autonomy is often limited in ways that, in other contexts, would be viewed as an unacceptable violation of both freedom of thought and other liberties. How then should we analyze what freedom of thought requires in involuntary psychiatric treatment? Stenlund analyzes this question through three lenses—looking at freedom of thought or cognitive liberty as (1) a "negative liberty," which protects all beliefs and thoughts from external interference or manipulation, (2) protection for a kind of "authenticity," which protects only those beliefs genuinely formed by the agent, and (3) as understood from "the perspective of the capabilities approach" in political philosophy, which would protect the agent's capacity to decide which beliefs and values to embrace. Stenlund argues that each of these different lenses emphasizes protections for different aspects of a psychiatric patient's thought processes. She also explains that just as not all psychological treatment would violate freedom of thought under all of these lenses for analyzing freedom of thought or cognitive liberty, so not all psychological illnesses should be understood to undermine cognitive liberty under each of these conceptions. At least some of the experiences considered to be an element of mental illness might be a part of, or even conceivably enhance, an individual's autonomy.

In Chapter 9, Andrea Lavazza provides a more detailed look at how various technologies might force courts to develop a more refined understanding of a right to freedom of thought. He examines brain-based mind-reading fRMI and other brain-scanning technology, use of deep-brain stimulation, Neuralink, and other brain implants to treat depression, addiction, or other diseases (but also, potentially, to manipulate those who receive implants), brain-machine interfaces that allow individuals to control their machines with their thoughts, and possibly merge our thinking with the exchange of information that occurs in "the Cloud." To counter such threats, Lavazza first proposes a definition of a right to mental integrity as involving "the individual's mastery of their mental states and brain data so that, without their content, no one can read, spread or alter such states and data in order to condition the individual in any way." He then argues that the defense of this mental integrity—against the threat raised by emerging neurotechnologies—requires not just legal barriers raised by courts, but *technological* requirements that use "technology as a defence against technology itself." For example, writes Lavazza, where emerging technologies give government officials or others a way to extract data about our thoughts, the law should require that officials can "extract brain data only by means of special access keys managed exclusively by the subjects under treatment or by their legal representatives"—or, where some other entity needs access to this data (even where a subject does not consent), it should be "accessible to professionals in charge of their correct use" and "able to be monitored by a specific authority upon the subject's request." In short, he argues, because the technology that threatens freedom of thought will be "so pervasive ... rules and sanctions are not enough." It will, Lavazza argues, be "very difficult to detect possible abuses from the outside" so it will be "necessary to provide users with countermeasures already incorporated in the devices themselves."

Finally, in Chapter 10, J. Adam Carter looks at how our mental autonomy might be threatened by manipulation that arises not merely when government (or other actors) seek to influence our psychology with speech or "nudging," or by reshaping our biological brain activity, but also by exercising control over our "extended mind." When we extend our minds with technology such as brain-computer interface (BCI) devices, how and when might our mental autonomy be infringed by other actors' control over, or design of, those devices? He first draws on past philosophical work (both his own and that of other scholars) to critique a "Cartesian view" of freedom of thought "according to which a thinker alone has privileged and exclusive access to the content of her own thoughts," and "thought itself is in principle unregulatable (apart from regulating against physical injury to the brain)." He explains why philosophical arguments in the late twentieth and early twenty-first century have undercut two variants of this view "internalist" view of the mind—"content internalism" and "cognitive internalism." Given that our minds are not therefore immune to government regulation or interference, Carter uses hypotheticals concerning Elon Musk's Neuralink brain-computer interface technology to develop and refine what he calls a "thought manipulation (sufficiency)" condition that can allow us to identify when a right against manipulation of thoughts or opinions might be violated by others' use of such technology (or similar technology). More specifically, he argues, our right to freedom of thought is violated when such technology is used to make us "acquire non-autonomous propositional attitudes (acquisition manipulation)" or cause "non-autonomous eradication" of "otherwise autonomous propositional attitudes (eradication manipulation)." The chapter illustrates how either of these conditions might arise in a person using Neuralink.

All of these chapters respond to the key challenge discussed earlier that modern neuroscience—and modern philosophical reflection about our minds and brains—is raising for ethics and for law. In short, it is no longer tenable for ethicists, lawyers, and judges to assume that however one defines freedom of thought, its protection will be guaranteed by the *nature* of thought. Brain scans give government the power to observe the inwards working of the mind. Compelled psychiatrist treatment (with drugs or by other means) gives government power to control it. We have also come to realize that protection of thought cannot be assured by a retreat into an "inner citadel" of purely internal thought (Christman, 1989), because certain crucial components of our thinking (including diary entries and computer calculations) are *not* purely internal. And if freedom of thought or cognitive liberty requires not just that government refrain from reshaping our minds, but that we be permitted to reshape our own minds, then a right to freedom of thought has to be defined in a way that secures us against government or other external interference as we do so. Each of the chapters in this volume thus aims to provide guidance in how to think about and refine our understanding of freedom of thought or mental autonomy in an age when neuroscience and neurotechnologies unsettle prior understandings. Some chapters do so by providing a fuller understanding of what freedom of thought has meant to the philosophers and judges who have insisted upon it in earlier ages, as well as in recent case law. Others do so by considering how ethical principles, or legal maxims or holdings, require us (or might require us) to respond to emerging technologies in brain-based mind-reading, brain-computer interfaces, cognitive enhancement technologies, or the empirically-informed manipulation entailed in "nudging." This interdisciplinary exploration of freedom of thought and its future provide a foundation for future exploration about many of these issues—and, in fact, future publications in this book series shall look more closely at some of the aforementioned technologies, and others, and their implications for mental privacy, cognitive liberty, and our capacity to maintain mental freedom in a world where governments and companies alike can use artificial intelligence and other technologies to monitor us, and shape our information sources.

These volumes, in short, will explore the history of the concept of freedom of thought—and focus on how its future is informed by psychology, neuroscience, and reflections about modern neuroscience-related technologies. This inquiry is a broad one, but is necessarily limited in certain ways. As inchoate as the concept of "freedom of thought" is, it is now invoked in numerous contexts—from the regulation of artificial intelligence to academic freedom. In many of these discussions, the term used broadly and unspecifically, not as an actionable right, but in raising a more general concern about the ways that companies may modify behavior, or about whether academic environments are putting pressure on faculty and students to adhere to certain favored ideologies rather than thinking skeptically about multiple views. These volumes cannot and will not fully discuss all of these important questions—and will touch upon them only to the extent they relate to the volumes' focus on how the mind sciences (and the technologies they make possible) will transform freedom of thought in the coming decades. Still, understanding the lessons that modern discoveries and technologies hold for freedom of thought might help scholars and writers as they think about how to define and protect it in other contexts. In closing, we wish to draw attention to the annual thematic report by the UN Special Rapporteur on Freedom of Religion or Belief, Ahmed Shaheed, to the General Assembly, in the year 2021. It is the first UN document of recent times that explicitly addresses the human right to freedom of thought and was published after completion of the manuscript of this volume.

Oklahoma City, USA — Marc Jonathan Blitz
Hamburg, Germany — Jan Christoph Bublitz

References

Berlin, I. (1969). Two concepts of liberty. In *Four essays on liberty*. Oxford University Press.

Boire, R. G. (2000). On cognitive liberty. *The Journal of Cognitive Liberties, 2*(1), 7–22.

Boire, R. G. (2004). Neurocops: The politics of prohibition and the future of enforcing social policy from inside the body. *JL & Health, 19*, 215.

Blitz, M. J. (2010). Freedom of thought for the extended mind: Cognitive enhancement and the constitution. *Wisconsin Law Review, 2010*(4), 1049.

Bublitz, J. C. (2013). My mind is mine!? Cognitive liberty as a legal concept. In E. Hildt & A. G. Franke (Eds.), *Cognitive enhancement* (pp. 233–264). Springer.

Bublitz, J. C. (2014). Freedom of thought in the age of neuroscience. *Archiv Fuer Rechts- Und Sozialphilosphie*, 1–25.

Christman, J. (1989). Introduction to the inner citadel: Essays on individual autonomy. In J. Christman (Ed.), *The Inner Citadel: Essays on Individual Autonomy.*

Clark, A., & Chalmers, D. (1998). The extended mind. *Analysis, 58*(1), 7–19.

Farahany, N. A. (2012). Incriminating thoughts. *Stanford Law Review, 64*, 352–408.

Farahany, N. A. (2019). The costs of changing our minds. *Emory Law Journal, 69*, 75–110.

Lifton, R. J. (1953). *Thought reform and the psychology of totalism: A study of "brainwashing" in China*. University of North Carolina Press.

MacCallum, G. C. (1967, July). Negative and positive freedom. *Philosophical Review, 76*(3), 312–334.

Metzinger, T. (2015). M-autonomy. *Journal of Consciousness Studies, 22*(11–12), 270–302.

Russell, B. (1922, March 24). *Free Thought and official propaganda—Conway memorial lecture*. B. W. Huebsch.

Sententia, W. (2004). Neuroethical considerations: Cognitive liberty and converging technologies for improving human cognition. *Annals of the New York Academy of Sciences, 1013*(1), 221–228. https://doi.org/10.1196/annals.1305.014.

Sententia, W. (2006), Cognitive enhancement and the neuroethics of memory drugs. In W. S. Bainbridge & M. C. Roco (Eds.), *Managing nano-bio-info-cogno innovations*. Springer.

Taylor, K. E. (2017). *Brainwashing: The science of thought control* (2nd ed.). Oxford University Press.

UN Special Rapporteur on Freedom of Belief or Religion. (2021). Interim Report to the General Assembly. UN Doc A/76/380.

Winick, B. J. (1989). The right to refuse mental health treatment: A first amendment perspective. *University of Miami Law Review, 44*, 1–103.

Contents

Notes on Contributors

Marc Jonathan Blitz is the Alan Joseph Bennett Professor at Oklahoma City University School of Law. His scholarship focuses on the implications—for American free speech and privacy law—of emerging technologies. His current work explores how freedom of thought and other constitutional rights apply to (and are illuminated by) virtual and augmented reality, fMRI and other brain imaging, cognitive enhancement technologies, and brain-computer interfaces. He is a series editor for the Palgrave Macmillan Series on Law, Neuroscience, and Human Behavior.

Jan Christoph Bublitz is a Researcher and Lecturer in Law at the University of Hamburg and a Young Fellow at the Center for Inter disciplinary Research (ZiF) in Bielefeld. His work is situated at the intersections between criminal law, human rights law, legal philosophy, and psychology, especially the legal regulation of the human mind as well as the psychological underpinnings of normative thought. He is a regular contributor to debates in neuroethics and was the principal investigator on projects on law and traumatic memories, legal implications of brain-computer interfaces, and is currently examining ethical and legal aspects arising from the blending of artificial and human intelligence in hybrid minds. He is a series editor for the Palgrave Macmillan Series on Law, Neuroscience, and Human Behavior.

J. Adam Carter is Reader in Epistemology at the University of Glasgow, where he works mainly in epistemology. He has published widely on such issues as knowledge-how, extended knowledge, epistemic relativism, social epistemology, and virtue epistemology. His latest book *Autonomous Knowledge: Radical Enhancement, Autonomy, and the Future of Knowing* is forthcoming with Oxford University Press.

Gidon Felsen is Associate Professor and is a neuroscientist and neuroethicist in the Department of Physiology and Biophysics at the University of Colorado School of Medicine. His lab studies how the nervous system makes and acts upon decisions under normal and pathological conditions, using behavioral, electrophysiological, genetic, and computational approaches. He is particularly interested in identifying the contributions of specific cell types, and the neural circuits they comprise, to the functions of subcortical brain regions. Dr. Felsen also examines ethical, legal, and social issues associated with advances in neuroscience. He has focused this branch of his research on examining how an understanding of the neuroscience of decision making can—and cannot—inform public policies designed to improve the predictably biased decisions that people often make.

Adam J. Kolber is Professor of Law at Brooklyn Law School. He writes and teaches in the areas of neurolaw, criminal law, health law, and bioethics. In 2005, he created the Neuroethics & Law Blog and, in 2006, taught the first law school course devoted to law and neuroscience. He has also taught law and neuroscience topics to federal and state judges as part of a MacArthur Foundation grant. Professor Kolber has been a Visiting Fellow at Princeton University's Center for Human Values and at NYU Law School's Center for Research in Crime and Justice.

Andrea Lavazza is a Senior Research Fellow at Centro Universitario Internazionale, Arezzo, Italy, and Adjunct Professor of Neuroethics at University of Pavia, Italy. His main area of research is neuroethics. In this field, he has written on human enhancement, cognitive privacy, and mental integrity, memory manipulation, and cerebral organoids. His general interests are focused on moral philosophy, free will, and law at the intersection with cognitive sciences. He is working on naturalism and its relations with other kinds of causation and explanation in philosophy of mind and philosophical anthropology. He has published over 100 papers and chapters and 11 books as both author and editor. Lavazza has been involved in several international initiatives concerning ethical regulation of neurotechnologies and biotechnologies.

Simon McCarthy-Jones is an Associate Professor in Clinical Psychology and Neuropsychology in the Department of Psychiatry at Trinity College Dublin. Much of his research revolves around the theme of mental freedom, covering topics such as freedom of thought, auditory verbal hallucinations, and child sexual abuse. His most recent books are *Spite: The Upside of Your Dark Side* (Basic Books, 2021) and *Can't You Hear Them? The Science and Significance of Hearing Voices* (Jessica Kingsley, 2017). He has previously worked in the Department of Cognitive Science at Macquarie University, Sydney, Australia, and in the Department of Psychology at Durham University, UK.

Gabriel S. Mendlow is a Professor of Law and Professor of Philosophy at the University of Michigan. He teaches and writes in the areas of criminal law, criminal procedure, and moral, political, and legal philosophy. His current research focuses on issues at the intersection of criminal law and freedom of thought. He has been awarded fellowships by the National Endowment for the Humanities and the American Council of Learned Societies (Burkhardt Residential Fellowship for Recently Tenured Scholars).

Saskia K. Nagel is Full Professor for Applied Ethics at RWTH Aachen University (Germany). She is a member of the Human Technology Center and of the Neuroethics Collective. Professor Nagel holds a Master degree in Cognitive Science and a doctoral degree in Cognitive Science and Philosophy. She was Assistant and Associate Professor at the University of Twente (The Netherlands) for Philosophy and Ethics of Technology. She explores how new human-technology relations—as, for example, enabled by research in the fields of neuroscience, cognitive science, artificial intelligence, or data science—influence human self-understanding and the understanding of values. Her group connects with colleagues across countries and disciplines to investigate questions of autonomy, responsibility, and trust in light of emerging technologies.

Fay Niker is Lecturer in Philosophy at the University of Stirling in the UK. Prior to this, she was a Postdoctoral Fellow at the Center for Ethics in Society at Stanford University. Her main research interests are in practical ethics and social and political philosophy. Her work focuses primarily on the ethics of influence, broadly understood. Within this, Dr. Niker has been working on: the political morality of nudging; autonomy and social embeddedness; trust and consent in interpersonal ethics; the curation of salience and attention, and the epistemic and ethical harms and values of such curation; and the ethics and politics of "caring technologies." She is a member of the Council of the Royal Institute of Philosophy, the Centre for Ethics, Philosophy and Public Affairs at St Andrews, and the Neuroethics Collective.

Peter B. Reiner is Professor in the Department of Psychiatry at the University of British Columbia, a member of the Centre for Artificial Intelligence Decision-making and Action, and founder of the Neuroethics Collective, a virtual think tank of scholars who share an interest in issues of neuroethical import. A champion of applying rigorous quantitative methods to neuroethical issues, Professor Reiner is the author of over 100 peer-reviewed publications including papers in *Science, Nature,* and *PNAS*. Professor Reiner began his career as a member of the Kinsmen Laboratory of Neurological Research where he was the inaugural holder of the Louise Brown Chair in Neuroscience. He went on to become founder, President, and CEO of Active Pass Pharmaceuticals, and in 2007 co-founded the National Core for Neuroethics.

Mari Stenlund is a Finish Social Ethicist and Doctor of Theology. Her scholarship has focused on mental health, human rights, and religion—and much of her academic analysis focused on the freedom of thought of those with psychotic delusions. She has published multiple articles on the ethical and philosophical questions raised by psychotic delusions, involuntary psychiatric treatments, and human rights theory. She has also analyzed human rights issues in spiritual abuse and alternative therapies. She currently works as a teacher of religion.

Lucas Swaine is Professor of Government at Dartmouth College. He is currently pursuing a book-length treatment of freedom of thought, the working title of which is *Freedom of Thought: First of the Liberties*. Swaine has published scholarly articles in a wide variety of academic journals. He is also the author of *Ethical Autonomy: The Rise of Self-Rule* (Oxford University Press, 2020) and of *The Liberal Conscience: Politics and Principle in a World of Religious Pluralism* (Columbia University Press, 2006).

List of Figures

CHAPTER 1

Freedom of Thought in Political History

Lucas Swaine

Introduction

Despite the fundamental importance of thinking to so many facets of human existence, freedom of thought has not received its due in political and intellectual history. Thinking is highly significant in its own right, a noiseless fixation of burgeoning and developed political orders. But freedom of thought has not adequately been recognized or articulated even though it figures prominently in the development of Western democracies. Freedom of thought proves highly significant when one considers whether political officials should be able to call people to account for their thoughts. It matters also when individuals are pressured to disclose what they are thinking. And freedom of thought is especially salient in situations in which people are punished for what they believe. Freedom of thought poses challenges for present and future practices of liberal democracies, with questions about how to handle people's thinking, in political and legal realms, extending back through Christendom to the ancient courts of Athens.

L. Swaine (✉)
Dartmouth College, NH Hanover, USA
e-mail: Lucas.Swaine@dartmouth.edu

M. J. Blitz et al. (eds.), *The Law and Ethics of Freedom of Thought, Volume 1*, Palgrave Studies in Law, Neuroscience, and Human Behavior,
https://doi.org/10.1007/978-3-030-84494-3_1

This chapter charts the progress of freedom of thought in the history of Western intellectual and political life. I argue that freedom of thought is a special and distinctive liberty, one that has been misunderstood and often infringed. I propose furthermore that freedom of thought warrants a more comprehensive articulation and better protection in democratic societies. I begin with a rudimentary account of the nature and significance of thought, following which I consider fundamental questions about freedom of thought that emerge in received accounts of Socrates' trial and punishment. I move then to discuss the place of thinking and of freedom of thought in Christianity, analyzing the centrality of moralized thoughts in the Bible and in the development of Christendom. I consider subsequently key statements on thought and thinking offered by political theorists in the modern era. Several of those statements help to advance our understanding of freedom of thought, but they leave its nature and its value opaque. I conclude with a brief discussion of theoretical and practical challenges to freedom of thought in modern democracies, describing prospects for better articulation of that value and outlining new ways in which to strengthen and defend it.

The Significance of Thought

Human beings are thinking creatures with complex mental interiors. We use speech and behavior to involve ourselves in the world around us, our thoughts exteriorized in the languages we speak and reflected in the outward acts that we perform. But thought is different from conduct, and much of our mental life we neither express to others nor attempt to put into action. A general conception of "thinking" as "mental activity" facilitates this basic distinction. It allows one to differentiate between thought and conduct, and it accommodates a broad array of mental processes under the rubric of thinking. The idea of thinking as mental activity makes space not only for cogitation and deliberation, but also for feeling, desiring, intending, believing, imagining, and other kinds of activity of mind. This understanding has the benefit of simplicity, it allows for distinctions among different kinds and forms of thinking, and it does not overemphasize the reasonableness or the rationality of the human animal.

One's thoughts are not immediately accessible to others: Thinking is elusive and opaque, its details shadowy in comparison with visible conduct. We gain familiarity with the thoughts of others indirectly and imperfectly. And yet these difficulties do not diminish people's concern to try to fathom others' thoughts. For thinking proves highly meaningful for understanding and explaining human behavior. It is in thinking where people reckon what is right and wrong, where they wonder, imagine, desire, and decide for or against courses of action.

Thinking is also morally important. It can prompt or inhibit morally better or worse behavior. What one thinks can factor into whether someone's conduct counts as a particular form of action. Thinking also matters morally even when thoughts are not joined with any outward action at all. When someone discloses what they believe, feel, or desire, others may consider the person morally better or worse, depending on what and how the person seems to think about things. Speech or conduct may testify to a clear conscience or a guilty mind, for instance, with judgments about someone's thoughts becoming judgments about that person's moral character.

People occasionally join their judgments about others' thoughts with worries about safety or wellbeing. Moral and prudential concerns about thinking often exist in combination, intertwined with people's communication and conduct. One can extrapolate these points to peoples and populations. In form and in content, people's thinking matters to nearby others, it contributes to the success or failure of associations and communities, and it is a fixation of social, educational, and religious institutions. These factors support the conclusion that thinking matters both for political stability and for the very legitimacy of a political order.

The Role of Thinking in the Trial of Socrates

Social and political concern with thinking has a long history. It can be traced through time along a central axis of political philosophy, back to the trial of Socrates. The legend of Socrates is established and entrenched. It is well known that he was charged with impiety and with corrupting the youth of Athens, crimes for which he was found guilty and sentenced to death. Plato's famous account of the ordeal depicts a striking confrontation between a thoughtful individual and social and political authority. But key aspects of the story reveal the centrality of thinking itself in Socrates' trial and punishment, with Socrates' thoughts serving as a significant contributory basis for his accusation, conviction, and execution.

Consider the importance of thinking as a basis for Socrates' indictment and punishment, according to received accounts of his trial. There has been scholarly controversy over whether Socrates was indicted for his beliefs or for his failure to worship Athens' gods in the right ways (Burnet, 1924, 5; Burnyeat, 1997; Giordiano-Zecharya, 2005; Vlastos, 1991). But one can see in surviving reports of Socrates' trial social and political concern for what Socrates thought, not just for what he said or otherwise did in terms of outward worship. First of all, the accusation of impiety represented a distinct charge against him, to which was added the indictment that Socrates had corrupted the youth (Plato, Apology, 24b-c; cf. Xenophon, Memorabilia, 1.1.1). Plato relates that the charge of impiety was leveled first against Socrates, sprung by "old accusers" from earlier days (Plato, Apology, 18b-c). Second, in criticizing the impiety indictment, Socrates pressed his antagonist Meletus to clarify whether he stood accused of worshipping other gods or no gods at all (Plato, Apology, 26b-c; Xenophon, Apology, 24; cf. Burnyeat, 1997). Supposing that Socrates was called to account because he did not worship Athenian gods according to local custom, or because he failed to do so in a sufficiently reverential manner, the question remains as to exactly how, or in what ways, Socrates demonstrated impiety. Even if Socrates were a nonstandard Apollonian, as Myles Burnyeat has suggested (Burnyeat, 1988, 18; Reeve, 2000; Woodruff, 2000), that would leave open which roles or aspects of Apollo Socrates might have reverenced (Hedrick, 1988; Reeve, 1989; Plato, Apology, 35c-d; Xenophon, Apology, 11; Xenophon, Memorabilia, 1.1.2; Diogenes Laertius, Lives, 2.42, 2.44), how he went about worshipping or esteeming Apollo, and whether he questioned others pointedly about orthodox practices and ideas regarding Apollo or other Athenian gods (cf. Diogenes Laertius, Lives, 2.21, 2.31, 2.38, 2.42, 2.45).

Third, whether Socrates suggested or revealed to others various of his thoughts on the gods of Athens, or on customary worship and reverence, the accusation of impiety has clear implications for Socrates' thinking. This is because the language reportedly used to address Socrates' impiety "encompasses all behavior that shows proper acknowledgement of the existence of the gods," as C. D. C. Reeve puts it (Reeve, 2000, 28; Plato, Apology, 26b-d, 29a, 32d, 35c-d; Xenophon, Memorabilia, 1.1.5; Diogenes Laertius, Lives, 2.40). Socrates' thoughts are implicated, here, because he would not have merely followed customary or traditional practices "unthinkingly" (Kraut, 2000, 13–17), and, if all of his speech and behavior with respect to the gods had been orthodox, Socrates would not have been charged with the crime of impiety, so described (Bremmer, 1998; Nussbaum, 1985).

Fourth, it may be noted that each of the two accusations against Socrates is logically and conceptually distinct from what Socrates allegedly said to others, and both accusations are distinguishable also from how Socrates spoke prior to his appearance in court and during his trial. In addition, whether Socrates' alleged impiety was implicit in his practice of questioning others or supposed to have contributed to the debasement of Athens' youth, the charge of corrupting the youth appears to be a concern separable from the impiety accusation. Socrates' denunciation for impiety is distinct, it is described as having been asserted first, and it entails disquietude about what Socrates was believed to have thought—it does not simply concern what he uttered or otherwise did. These factors testify to the social and political significance of Socrates' thinking, as a putative basis for his being put on trial and, subsequently, as a contributing factor in his conviction and execution.

The conclusion that Socrates' thoughts were significant factors in his accusation and trial does not diminish the notion that his speech and conduct were important, too. Socrates was clearly in jeopardy for what he said and for how he acted, given the manner in which he reportedly questioned people and appreciating how he riled important figures and influenced the youth of Athens in unpopular ways (cf. Filonik, 2013, 54–57, 80–81). But these considerations, like the fact that Socrates often used his mind and his voice together, do not diminish the distinct and particular importance of his thinking in his trial and punishment. One can coherently affirm that Socrates' thoughts, speech, and conduct all mattered in their own right.

I have considered the tale of Socrates' trial and punishment as others have conveyed it. However, it should be noted that received understandings are partial and fragmentary, and quite imperfect. Our familiarity with the historical Socrates is transmitted through a small set of recorded statements and historical recollections, a considerable portion of which comes from Plato and Xenophon (Filonik, 2013, 32). There is real question as to the extent to which Plato presented a stylized or embellished Socrates in his early dialogues (Filonik, 2013, 57–58; Ralkowski, 2013, 1–19; Waterfield, 2013). It may be noted that Plato assigned great importance to thoughts and ideas in his political theory, with contemplation and other forms of thought playing major roles in the ideal city that he imagined (see generally Plato, Republic). But even if the historical Socrates did not speak or act exactly as he has been portrayed, the significance of thinking in Socrates' courtroom and jail-cell discussions was not lost on raconteurs or their contemporaries. Otherwise, they would have been very unlikely to have produced or reiterated the stories that they did replete with the nuance and interest related specifically to thinking and to Socrates' utilizations of various kinds of thought. This suggests that Plato and Xenophon, and their respective interlocutors, well appreciated not only the importance of thinking but also basic differences between speech and thought (cf. Plato, Apology, 21b; Euripides, Hippolytus, 612; Avery, 1968; Aristotle, Rhetoric, 1416a).

Not only was thinking a highly significant factor in the tale of Socrates' legal ordeal. The story also prompts one to consider the broader question of whether Socrates should have been called to account for what or how he thought. In addition, the accounts raise important questions regarding the extent to which Socrates' thoughts ought to have weighed in the balance, during his trial and punishment. It stands to reason that many of Socrates' thoughts were linked motivationally to how he spoke or acted, or they were tied in other ways to his outward conduct. But some of his thoughts, presumably, Socrates never expressed to others. And various elements of Socrates' thinking may not have influenced his speech or his behavior, or they might simply have been thoughts that Socrates never attempted to put into action. After all, even supposing that Socrates did not say anything privately that he would not say publicly (Plato, Apology, 33b), that does not logically imply that he disclosed to others the entire contents of his mind. The point is bolstered by Socrates' comments on people who shamefully express their feelings in attempts to sway the jury; he implies that people should keep such expressions to themselves (cf. Plato, Apology, 34c-35b). What is more, it stands to reason that Socrates may not have voiced other thoughts he had until he was drawn into the court of Athens and pressured to do so. His stated reluctance to defend himself in court is plausibly an example of a thought of this kind; so are the thoughts he had on what he took apparently to be the difficulty of defending himself, or of persuading others of his innocence (Plato, Apology, 18c-19a).

The tale of Socrates raises monumental questions. It stirs up concerns regarding freedoms of speech and association, religious liberty, freedom of conscience, procedural justice, and rightful forms of punishment. The story prompts one also to ponder fundamental questions regarding the treatment of people's thoughts in political and legal contexts. Should political or legal authorities be able to call people to account for their thoughts? Ought powerful institutions to be permitted to pressure people to disclose their beliefs, religious, or otherwise, along with associated ideas, feelings, notions, desires, or opinions, in courts of law or in other formal settings? Should people be able formally to be accused, tried, or convicted of crimes based primarily, even solely, on what they may think or not-think? Is it acceptable to hold individuals accountable for their thinking, even if the thoughts in question are not connected to speech or conduct, or if the thoughts may never may be put into action? And what sort of political or legal frameworks ought to be employed to address such questions?

Thought, Thinking, and Christianity

The social and political relevance of thinking has extended well beyond the courts of ancient Greek city-states. Cultures and communities across the globe show concern with the interior life of their membership. Thinking proves salient in central domains of complex societies: It is highly important in education, law, commerce and trade, artistry and innovation, and collective action. Religion is another sphere in which people's thoughts, broadly construed, have mattered to communities and their worldly authorities. One can see why this might be so. Social and political concerns have often been entangled with religiosity, with people's religious beliefs and practices in numerous cases transformed into political and legal issues.

Western political orders took religious routes in their progression toward constitutionalism and liberal democracy. Christianity blazed the path, with thinking proving essential in its course, in some nonobvious ways. Early Christian affirmations laid heavy weight on thinking, setting foundations not just for spiritual concern about people's thoughts, but also for observation, intervention, and correction. I do not propose that Christianity countenances persecution of people for their thoughts, much less that it requires or commands it. However, the Bible clearly indicates that thoughts themselves can be evil. Concerns about wicked thoughts have fuelled wayward and improper investigation of people's thinking, under a false sense of the requirements of Christian doctrine. On many occasions, this has led to persecution of those found to have unapproved beliefs, desires, ideas, and imaginings.

The Bible provides that God is neither indifferent to people's thoughts nor concerned merely with salutary affirmations. It specifies that thoughts can be sources of good or of evil, identifying some thoughts themselves as having a good or an evil nature. Statements to this effect can be found in each Testament and they are hardly of passing importance. For example, the Book of Genesis explains that, prior to the Flood, the imaginings, intentions, and inclinations of human beings were evil. God observed the great wickedness of humanity during this period: He saw "that every imagination of the thoughts of [one's] heart was only evil continually" (Genesis 6:5). The concern has not abated. The Decalogue begins and ends with commandments directly pertinent to a person's thoughts: respectively, not to affirm any other gods than God Himself, and not to be covetous of what others have (Exodus 20:3, 20:17). Solomon's proverbs include the admonishment to "[l]ust not after [a strange or evil woman's] beauty in thine heart; neither let her take thee with her eyelids" (Proverbs 6:25). And the vision of the Prophet Isaiah reproves the wicked to "forsake [their] way, and the unrighteous man his thoughts: and let [them] return unto the LORD, and he [sic] will have mercy upon [them]; and to our God, for he [sic] will abundantly pardon" (Isaiah 55:7). In the New Testament, the Gospel of Matthew elaborates that thoughts can be wicked and that various ways of thinking can defile a person. Jesus is said to have remarked that "out of the heart proceed evil thoughts, murders, adulteries, fornications, thefts, false witness, blasphemies" (Matthew 15:19; see 15:17-20; cf. 9:4). And thoughts are clearly implicated in Jesus' admonishment that "whosoever looketh on a woman to lust after her hath committed adultery with her already in his heart" (Matthew 5:28; see 5:27-30 ff.; cf. Proverbs 23:7).

The Old and New Testaments clarify that not just outward acts are good or evil: One's thoughts can have those qualities, too. This includes what goes on in one's mind or, figuratively, in one's heart (i.e., with respect to one's feelings or emotions). The Bible describes cases in which thoughts themselves are evil, where thinking is iniquitous per se and thoughts alone are wicked. Evil thoughts need not be joined in speech or outward conduct, in order to be wicked, although iniquitous thoughts may be accompanied by speech or outward behavior, or they may be made manifest in evil action. Nor are wicked thoughts identical to the acts that may involve or imply the thoughts in question. Thoughts themselves are their own concern.

The Bible teaches that there are various kinds of iniquitous thoughts and numerous evil ways of thinking, suggesting that these can crop up in one's fields of desires, imaginings, intentions, beliefs, and affirmations. One gathers furthermore that thoughts or feelings may lead to further thoughts that are iniquitous, inasmuch as evil thoughts can come "out of the heart," given that the heart thinks only in a figurative sense. In addition, wicked acts appear to be connected to various thoughts, desires, imaginings, or ideas that are evil in themselves (e.g., adultery and its connection to lust). One is called upon to forsake iniquitous thinking, to abandon evil thoughts and inclinations and to steer clear of wayward desires, and to return to God, to repent and to ask for forgiveness, and to acquire God's mercy. One ought to follow Jesus' guidance and endeavor earnestly to be "perfect, even as your Father which is in heaven is perfect" (Matthew 5:48; Luke 10:27).

The power of these Biblical statements is reinforced by God's omniscience. It is written that God knows all the truths of the world. David's charge to Solomon, in the Old Testament, directs Solomon to serve God "with a perfect heart and with a willing mind: for the LORD searcheth all hearts, and understandeth all the imaginations of the thoughts" (1 Chronicles 28:9). Psalm 44 states that God knows "the secrets of the heart" (Psalms 44:21), and Psalm 139 indicates that God knows one's words before one speaks them (Psalms 139:4; see 139:1-4). The Gospel of Luke maintains that nothing is or can be concealed from God, and that all will be revealed to Him: "nothing is secret, that shall not be made manifest; neither any thing hid, that shall not be known and come abroad" (Luke 8:17). This is reinforced by the proclamation in John's First Epistle that God "knoweth all things" (1 John 3:20; cf. Matthew 9:4, 12:25; Luke 9:47, 11:17). God's omniscience seems clearly to cover knowledge of one's thoughts, broadly construed, including knowledge of whether those thoughts are righteous or sinful, what makes them so, and how one shall be judged for them.

I do not argue that it is incorrect to specify some thoughts as good or evil. It stands to reason that thinking can be wicked, even if the thoughts are never made manifest in speech or in outward action. Certain kinds of thoughts can be morally wrong to entertain, or to mull over, or to have in one's mind, just as it may be sinful to think about particular topics in certain ways (Swaine, 2020). Nor do I contend that the Bible is hostile to freedom of thought: There is good reason to believe that the Bible affirms freedom of thought and liberty of conscience alike. What I do propose is that spiritual and earthly authorities utilized Biblical teachings on evil thoughts to investigate thinking and to sanction people found to have ungodly beliefs and desires. This is especially evident in historical treatments of heresy in Christendom. Biblical statements on evil thoughts contributed to the investigation and sanctioning of heretics, with treatment of heresy raising systematic and widespread concerns for freedom of thought.

Early Christians set themselves earnestly to the task of clarifying true doctrine and determining correct elements of faith. People's thoughts were implicated closely in these developments, particularly with regard to the identification of heresy and in efforts to eliminate it. The term "heresy" (derived from the Hellenistic Greek "hairesis") did not originally have pejorative connotations; it denoted "choice" or "[something] chosen," and it was used to describe someone's decision to join a particular religious order or school of thought (Swaine, 2001, 1045). However, over time heresy came to represent theological error and sin. With Constantine's conversion and the establishment of Christianity as the religion of Roman Empire, the Church became able to work in tandem with secular authorities to extirpate heresies. Emperors convened ecumenical councils that defined Christian doctrine and laid structure for excommunication of nonconformists. This began with the First Council of Nicaea, in A.D. 325. Ecumenical councils subsequently developed orthodoxy through antiquity and across the Middle Ages: Correct beliefs were clarified, incorrect ideas repudiated, canon laws delivered, and heresies distinguished, defined, and attacked. Heretics were both anathematized and excommunicated, allowed neither to meet with nor to talk to fellow Christians (Swaine, 2001, 1045).

The Third Lateran Council of 1179 condemned as heretics the Cathars and the Waldensians, two sizeable groups whose religious beliefs and practices were seen as a threat to both religious and secular order. Authorities thoroughly persecuted the sects, attempting to obliterate the offending beliefs and practices. In 1215, the Fourth Lateran Council laid down requirements of at-least yearly confessions to one's priest, and of penance, empowering priests to absolve their parishioners of their sins. The Fourth Lateran also expanded godly rule and instructed secular princes with a variety of directions. It condemned all heretics, defining penalties and forms of disenfranchisement for heresy, and it advised crusaders that they should prepare themselves for action. Pope Innocent IV subsequently issued his papal bull *Ad extirpanda* to sharpen the orders. It decreed that torture may be used to force confession, setting the stage for execution of the recalcitrant and unrepentant at the hands of secular authorities (Swaine, 2001).

The Inquisition thrived in this framework. One finds numerous cases of freedom-of-thought violations in inquisitors' procedures and practices. To take one example, the medieval inquisitor Bernard Gui strove to pin down the beliefs of alleged heretics, warning of terrible punishment for those who resisted his inquiries (see Lea, 1887b; Walsh, 1987, 50–88). Such interrogations generally proceeded under the rubric of a purpose to expunge heretical views from Christian polities and communities. J. B. Bury suggests that inquisitors' motivations were, in many instances, based in the profound conviction that those who did not believe certain dogmas would be punished eternally. This, in turn, led naturally to persecution, according to Bury, given that the inquisitors imagined they faced "enemies of the Almighty" (Bury, 1913, 52, 53; see 51–71). And there was a kind of public rationale for engaging in such endeavors: Some who privately doubted or disbelieved accepted theological views would "[feign] to acknowledge the truth of the ideas which they were assailing," putting themselves and their communities at grave risk (Bury, 1913, 134, 136–139, 148–149, 162–163; see Walsh, 1987, 61–64; cf. Foucault, 2014, 125–161). The tendency to employ violative investigative techniques to inquire into people's thoughts was hardly limited to the Inquisition, of course. The incorrigible heretic Bartholomew Legate learned as much firsthand in the early 1600s. English religious authorities hauled Legate before the Consistory Court, plying him with questions to determine whether he held "various pestilent opinions" (Bury, 1913; cf. Rawls, 1999a, 182–183). Once the Court reached its determination, it relinquished Legate to secular authorities, who burned him alive.

The Inquisition operated in many regions and over a considerable period of time, causing terrible damage to countless individuals and communities. The form, manner, and extent of inquisitors' investigations were multiply problematic, as were the harsh sanctions meted out to their more unfortunate victims. In the first place, even if one were to grant that inquisitors had proper authority to investigate the thoughts of potential heretics, it is very hard to say that they had adequate reason to be concerned with the mere beliefs of people within the societies they inspected. For instance, one finds no plausible cause for inquisitors to pry into the thoughts of "Conversos" after they no longer even attempted to practice Judaism (see Lea, 1887b, 63–64; cf. Lea, 1887a,

555–556; Lupovitch, 2010, 100; Monter, 1990, 23–26 ff.; Walsh, 1987, 151–154). More generally, it seems evident that there was no sufficient reason for investigating the thoughts of supposed heretics. And if that were not enough to qualify such investigations as significant violations of freedom of thought, the threshold is surely passed when one learns that subjects could meet with severe punishment for being unwilling to accede to inquisitors' requests, for refusing to disclose their thoughts or declining to cooperate with the inquisitorial trials otherwise. Threats of punishment—very credible ones, at that—were levied even against those accused heretics who were simply unreceptive to having corrections made to their beliefs (see Lea, 1887a, 541).

Inquisitorial examples may be ghastly but they help to illuminate why interrogation practices can be deeply morally wrong, and they suggest why latter-day investigative methods, similar in their form, manner, or extent, might violate freedom of thought in several different ways. First of all, such investigations can produce highly adverse psychological and emotional effects in their subjects, proving extremely unsettling for the people who undergo them. Psychological and emotional trauma can result from interrogations in which people are probed and pressed to disclose their thoughts, especially when the subjects understand that penalties await anyone who is not adequately cooperative or forthcoming. Because it is so difficult to demonstrate sincerity regarding what one says one desires or does not desire, or believes or does not believe, even those wishing to satisfy interrogators are susceptible to ordeals. What is more, the extent of questioning, if it is too broad, can reach into private or emotionally sensitive areas for the individual under investigation, causing humiliation and producing painful, lasting effects (cf. Walsh, 1987, 168). Traumatic results may also be exacerbated if the person whose thoughts are investigated is prodded to address topics, or to reveal information, to which he or she has a conscientious objection to discussing or disclosing, although not only thoughts covered by conscience would matter here.

One can identify several key freedom-of-thought concerns in the development of the religious sphere of Christendom; but it should be noted that similar issues have arisen in secular realms, too. For example, England's Treason Act of 1351 made it high treason to "compass or imagine" the death of the king, his wife, or his eldest son and heir (Parliament of England, 25 Edward III St. 5 c. 2 (1351); Barrell, 2000, 32). This meant that even just thinking of regicide could be severely punishable, tantamount to lèse-majesté (cf. Cobbett, 1809, 1456–1457). Such examples complement the historical trajectory that I have charted, in which thinking and freedom-of-thought issues have been salient. The broader set of examples suggests the importance of freedom of thought, and it recommends viewing that freedom as a discrete liberty with its own conceptual contours, one that is not simply covered under freedom of speech or enveloped by other rights and liberties. The cases I have mentioned lead one to hold that freedom of thought has distinctive and special value, that it is possible to violate that freedom, and that it can be seriously wrong to do so (see Swaine, 2018a, 2018b).

Freedom of Thought in the Modern Era

I have noted the significance of thinking and of freedom of thought in early Western political history and proposed that freedom of thought is an important and meaningful liberty. I cannot offer here a detailed account of the significance of freedom of thought in the development of the world's many complex political orders, democratic or otherwise. One can distinguish a slowly growing appreciation of freedom of thought through the modern era and into the present, both in the discourse of social and political theory and in terms of the expansion of that freedom under political and legal institutions. But the progress of freedom of thought has not been linear. Its story is one of qualified movement, of partial advancement in some areas and setbacks in others, not of categorical or unreserved success. Operating in the subterranean regions of social and political life, freedom of thought has often been misconstrued, overlooked, threatened, or violated, with scant articulation as a value unto itself.

Theoretical treatments of freedom of thought have proven fractional and scrappy, across the centuries, with contributions scattered across a miscellany of philosophical works and political declarations. One finds limited concentration on freedom of thought in the works of such figures as John Locke, Pierre Bayle, Benjamin Constant, and the American Founders. Wilhelm von Humboldt and John Stuart Mill offered eloquent paeans in the Nineteenth Century (Humboldt, 1993, 66–69; Mill, 1978, 11–12, 15–52). Freedom of thought was given notable mention after the Second World War, in the United Nations Declaration of Human Rights. According to Article 18 of the UDHR, "[e]veryone has the right to freedom of thought, conscience and religion" (G.A. res. 217A (III), UN Doc A/810 at 71 (1948), Article 18; cf. Convention for the Protection of Human Rights and Fundamental Freedoms, Nov. 4, 1950, Europ.T.S.

No. 5; 213 U.N.T.S. 221). John Rawls expanded briefly upon the notion, arguing that freedom of thought should be included as part of a "fully adequate scheme of basic liberties" (Rawls, 2005, 308, 178–182 ff.). But there remains no systematic articulation of freedom of thought, no clear and thoroughgoing analysis of its nature and importance. Political and legal discourse lacks a proper sense of how freedom of thought can be violated, as well as protected and cultivated; of whether there is a right to that freedom and, if so, what kind of a right that might be. Questions endure also as to exactly why freedom of thought is worth protecting; what the threats to freedom of thought encompass; what practices and values support that freedom (Shiffrin, 2010–2011, 283); and what key liberties freedom of thought supports, in turn.

One might object that existing theoretical treatments of freedom of thought have actually shown that there is little need to protect that so-called freedom. The reason to think so, one might propose, is that freedom of thought is by its nature the kind of liberty that cannot be violated. Consider Locke's influential claims in *A Letter Concerning Toleration*, which have provided groundwork for this conclusion and served as a basis for subsequent understandings. Locke proposes that a person "cannot be compell'd" to believe anything through the use of outward force (Locke, 1689, 7). Only "Light and Evidence" can modify people's opinions, he maintains, and such light "can in no manner proceed from corporal Sufferings, or any other outward Penalties" (Locke, 1689, 8). His points appear to cover the faculty of human understanding in general, offering a sense of freedom of thought that includes religious beliefs as well as people's opinions more broadly. But it is far from obvious that force cannot successfully be used to change people's beliefs, or their opinions, or their thinking or their thought-processes more generally. Locke conveys a specious understanding of the nature of freedom of thought, when it comes to the use of force and the changeability of thoughts, and this, too, proves highly relevant to understandings of freedom of thought itself.

In the first place, it is plausible that "duly proportioned" force (Locke, 1689, 7) can indeed modify people's beliefs, as Jonas Proast noted in his reply to Locke in 1690. Such force might be applied against people otherwise unwilling to go to church, for example, compelling them to sit in pews and to listen to preachers (Proast, 1690, 11, 12–14, 16–19, 23). It is one thing to say that coercive measures should not be used for such purposes and another to claim that such measures cannot be effective (cf. Locke, 1689, 6–9, 13, 27, 45–47, 56–61). Important to consider here are such factors as the coercive techniques employed against subjects, whether thinking is to be modified using direct or indirect means, and whether the plan is to change people's thoughts immediately or over time (Swaine, 2006). Communist and fascist regimes have imparted to humanity the terrible lesson that "corporal sufferings," combined with other heavy-handed courses of action, can indeed produce profound and lasting changes in people's mental lives. Coercive frameworks can be used to effectuate serious changes in people's thinking, to alter their minds with so-called reeducation, and to facilitate such measures with drugs and medical techniques.

The capability of using force to alter people's thoughts is not the same as the ability to coerce people to believe something in particular (cf. Locke, 1959, 322–323). But the former seems to be a proficiency that powerful authorities have developed and deployed. As such, Locke's insistence that force is powerless to change opinion proves inaccurate, at least when taken generally. This conclusion casts a pall on subsequent statements about freedom of thought. Consider, for instance, Constant's remarks on the "absurdity of any attempt by society to control the inner opinions of its members" (Constant, 2003, 103). He declares:

> There is no such possibility. Nature has given man's thought an impregnable shelter. She has created for it a sanctuary no power can penetrate. (Constant, 2003, 103)

Constant's point is correct, so far as it goes, but one must be careful not to overdraw conclusions or to extrapolate beyond what is warranted. It is reasonable to suppose that authorities cannot control all of a person's opinions, much less order and manage each of the opinions of every member of an entire society. The human mind cannot be directed in that way, and modifying someone's personal views is not the same as strictly controlling the formation of their opinions. In addition, Constant argues quite plausibly that opinions and reasoning cannot be changed by the immediate application of force, at least not in the way that authorities might desire. Even so, there are three qualifications to keep in perspective. First, it appears possible for powerful parties to change people's thinking over time. For example, societies may use the power of the law to disallow a cultural or religious practice, thereby leading people ultimately to forswear the proscribed practice. Second, authorities can damage people's mental faculties. That may not be a way of controlling opinions, strictly speaking, but humans can degrade and destroy others' capabilities, adversely affecting or even extinguishing their processes of reasoning, their emotions, their imaginative capabilities, and so on. These considerations lead one to conclude that the human mind is not as impervious to external force as Constant's statements might lead one to believe.

Third, while Constant takes an admirable stand against those who would attempt to control others' opinions and views, it remains possible for authorities to persecute people for their thinking. This point Constant seems ultimately to acknowledge, despite his apparent ambivalence on the subject (Constant, 2003, 104; cf. Constant, 1988, 112, 130–126). He decries government's attempt to make itself seem praiseworthy for "allow[ing] us to think what seem[s] reasonable to us" (Constant, 2003, 451). "But how could they stop us doing so?" he demands (Constant, 2003, 451; cf. Constant, 1988, 20–26). Constant is correct to suggest that clumsy threats of violence do not alter people's views of what is reasonable: that sort of coercion cannot be expected to change one's mind in the way that the threatening party might desire. But more systematic strategies and defter techniques can transform people's understandings of what is reasonable, or their conceptions of reasonableness itself, especially when those techniques are used in combination and over lengthier periods of time. When powerful actors are able to threaten, frighten, torment, defame, injure, jail, traumatize, propagandize, manipulate, or gaslight people, and when they can do so in environments over which they have considerable control, they can effectuate many changes in subjects' views, reworking thoughts and thought-processes in a variety of ways.

Factors such as these prompt one to reconsider prominent statements on freedom of thought in the American tradition. Thomas Jefferson's words in "A Bill for Establishing Religious Freedom" are important touchstones in this respect. Jefferson claims there that "Almighty God hath created the mind free," and that "free it shall remain [by being made] altogether insusceptible of restraint" (Jefferson, 1950). God is Lord of both body and mind, Jefferson writes, and He "chose not to propagate [His plan] by coercions on either" (Jefferson, 1950). Once the Virginia Assembly passed Jefferson's bill into law, James Madison wrote to Jefferson to relay the good news, stating in his letter: "I flatter myself [that we] have in this Country extinguished for ever the ambitious hope of making laws for the human mind" (Madison, 1786).

Jefferson's insistence on freedom of the human mind is admirable and one agrees that there are important ways in which the mind is insusceptible of restraint. Even so, that does not mean that people cannot degrade others' thinking or infringe their freedom of thought. Similarly, to agree that the mind cannot be restrained, in some particular respects, is not necessarily to concede that people are incapable of using coercive measures and other techniques to alter opinions or beliefs, or to change or even to extinguish various kinds of thoughts. I have suggested that parties can violate freedom of thought by interfering with people's thinking and their thought-processes, and I have argued that it is possible to breach freedom of thought by going too far in investigating thought or by punishing people for their thoughts alone (Swaine, 2018a, 2018b). It may be observed that Jefferson implies that it would be wrong for people to violate God's decision to make the human mind free (cf. Swaine, 2020, 208–211). He does not, however, stipulate that disrupting the design of God would be the only thing wrong with trying to tyrannize over the human mind; Jefferson's formulation allows that interference in people's thoughts could be wrong for other reasons, too.

Jefferson might not have drawn precisely these distinctions, of course. But he offers special building-blocks for an expanded understanding of the nature and value of freedom of thought, especially where he contends that "the opinions of men are not the object of civil government, nor under its jurisdiction" (Jefferson, 1950). The statement resonates with the American Founders' affirmation of the existence of a right to freedom of opinion. By the late Eighteenth Century, many Americans held freedom of opinion to be an inalienable natural right: Opinions were seen as "sacrosanct because they were understood to be non-volitional," as Jud Campbell puts it (Campbell, 2017, 280; see generally 280–287). This was to the Founders a freedom-of-thought concern, Campbell maintains, because they understood freedom of opinion to be "at its core a freedom against governmental efforts to punish people for their [non-volitional] thoughts" (Campbell, 2017, 281). The burgeoning view was indebted to the work of Francis Hutcheson, who, in his *Inquiry into the Original of Our Ideas of Beauty and Virtue*, proposed that people have a "Right of private Judgment" that cannot be alienated because "we cannot command ourselves to think what either we our selves, or any other

Person pleases" (see Campbell, 2017, 281, 287 n. 189; Hutcheson, 2004, 186, 187, cf. 38, 87, 118–119, 189, 192–193, 194).

These are keen and important ideas, and there is reason to hold that people should not use force to try to modify others' affirmations, to change their judgments, or to alter their faculties (Swaine, 2006, 62–63 ff.). However, supposing that opinions are not subject to commands, and that one cannot think whatever one (or anyone else) pleases, it does not follow that there is a right to freedom of opinion or a right to freedom of thought. Nor does it mean that either right would be inalienable in the sense described here. It is puzzling to think that one might have a right to something, even to something interior and personal about oneself, on the grounds that it is not subject to anyone's command. If nobody has command or control over anyone's opinion, perhaps nobody has a right to their own private judgment or to their faculty of forming opinions. Alternatively, others might have a claim, perhaps even an equal claim, to one's private judgments or to one's faculties, such as they may be. But it stands to reason that people can be at least partly responsible for the formation of their thoughts and their opinions, and for the ways in which they have modified their capabilities, altered their judgments, and so forth, to arrive at the views that hold (see Swaine, 2020, chap. 3, passim).

A new jurisdictional argument with a sounder justificatory basis could be developed to limit the presumed right of authorities to interfere with people's thoughts or to investigate or to punish thinking. A jurisdictional argument of this kind could serve as part of a broader, integrated case supporting a rights-based claim for freedom of thought, as well (Swaine, 2018b). Such argumentation might also prove consistent with Jefferson's and other Founders' views on providential matters, if not strictly depending upon Jefferson's understandings of the will or the design of God in that respect.

Conclusion

Western liberal democracies have developed and protected an extended range of rights and freedoms. They have engendered pluralism and toleration, religious freedom and respect for liberty of conscience, and a working understanding that citizens should not be punished for their thinking, that so-called thoughtcrime is an abomination (Orwell, 1961, 19, 23, 44, 52, 103, passim). Contemporary democratic citizens seem also to appreciate that it is possible for government to use disproportionate means to investigate thoughts, siding with Constant in rejecting such "inquisitorial nosiness" (Constant, 2003, 104). In the United States, freedom of thought has been able to survive under the armor of the Constitution and its First, Fourth, and Fifth Amendments. These protections, such as they are, have been complemented by a patchwork of related practices and laws, along with key constitutional provisions such as the need for overt acts to establish certain forms of criminal action (U.S. Const. art. III, § 3, cl. 1).

However, freedom of thought faces a variety of pressing concerns in key spheres of democratic life. Problems for freedom of thought emerge in such domains as education, free expression, criminal law, immigration, deliberation, and technology. To take the latter as an example, current technologies used to monitor and surveil citizens are being deployed in ways that suppress thinking and put freedom of thought in jeopardy (Sangiovanni, 2019, 56–61, 82–83; Shaw, 2017). Researchers have also recently innovated special interfaces that "extract and deliver information between brains" and allow "direct brain-to-brain communication" (Alegre, 2017, 231–233; Blitz, 2010; Jiang et al., 2019, 1; Ligthart et al., 2020). This generates various difficulties, including concerns about people gaining new ways to investigate others' thoughts beyond acceptable boundaries. And both government institutions and nongovernment researchers have been working to develop ways of inferring people's thoughts without relying on subjects' speech or outward behavior (Blitz, 2017; Cohen, 2020; Mack, 2018; Swaine, 2018a, 425 n. 68). Related concerns are emerging for patients' freedom of thought in medical research (Lavazza, 2018), and there have been freedom-of-thought controversies in prominent court cases regarding forcible administration of psychotropic medications (Washington v. Harper, 494 U.S. 210, 1990; Sell v. United States, 539 U.S. 166, 2003; Gallagher, 2016; Winick, 1989). In each of these areas, novel technologies promise to provide exciting new abilities for individuals and for government agencies; but they also facilitate freedom-of-thought violations, and they carry with them the dark prospect of degrading this vital liberty.

Other long-standing and largely accepted democratic institutions have contributed to the corrosion and degradation of freedom of thought. They must be buffered and restrained for the sake of protecting both freedom of thought and other cognate rights and liberties. For instance, freedom of thought is threatened by a variety of allowances afforded to political officials in their formal and informal capacities as interrogators (Swaine, 2018b). Compulsory testimony requirements continue to operate without a clear understanding of the nature and importance of freedom of thought, too. Even the very institution of the subpoena deserves fundamental reconsideration—regarding which bodies may issue subpoenas, for what purposes, what officials may ask or require of people called to testify, and when and whether individuals may be punished for noncompliance. Freedom of thought is at risk in each of these areas. These problems are part and parcel of creeping encroachments and of insufficient attention paid to ways in which authorities and institutions can go too far in investigating people's thinking or penalizing those who refuse to disclose their thoughts (Newman, 2019; Swaine, 2018a, 2018b). And continuing government interest in people's thoughts, joined by the ever-present specter of government punishing people for their thoughts alone, contributes further to the degradation of the full value of freedom of thought.

The survival of freedom of thought is crucial for vibrant public and private life, for healthy intellectual culture, and for the advancement of free societies. Effective democracy and rightful governance, and indeed the very legitimacy of a political order, quite plausibly depend on freedom of thought. This special freedom must be elevated and drawn out of the subterranean areas of democracy, emerging to flourish in the discourse of contemporary rights and liberties. With broader articulation, freedom of thought can become a fuller part of the living tapestry of democratic values, intertwined with other rights and freedoms and strengthening the liberal-democratic panoply.

The act of bringing freedom of thought to light, of giving it more complete philosophical and legal expression, can assist in solving primal questions raised at the outset of Western political thought. Should political or legal authorities be able to hold one to account for one's thoughts? Is it right or fair to pressure people to disclose their ideas, feelings, or beliefs, and to penalize them if they refuse to comply? Ought people to be able to be accused or convicted of crimes, based simply on thoughts they have, or which they may lack? We need more than just a Delphic sense of what the answers to such questions may be.

References

Articles, Books, and Online Sources

Alegre, S. (2017). Rethinking freedom of thought for the 21st century. *European Human Rights Law Review* (3), 221–233.

Aristotle, *Art of rhetoric*, trans. J. H. Freese, revised by G. Striker. Harvard University Press (2020/1926).

Avery, H. C. (1968). My tongue swore, but my mind is unsworn. *Transactions and Proceedings of the American Philological Association, 99*, 19–35.

Barrell, J. (2000). *Imagining the king's death: Figurative treason, fantasies of regicide 1793–1796*. Oxford University Press.

Blitz, M. J. (2010). Freedom of thought for the extended mind: Cognitive enhancement and the constitution. *Wisconsin Law Review*, 1049–1117.

Blitz, M. J. (2017). *Searching minds by scanning brains: Neuroscience technology and constitutional privacy protection*. Palgrave MacMillan.

Bremmer, J. (1998). 'Religion', 'ritual' and the opposition 'sacred vs. profane': Notes towards a terminological 'genealogy'. In F. Graf (Ed.), *Ansichten griechischer Rituale: Geburtstags-Symposium für Walter Bukert* (pp. 9–32). Teubner.

Bowden, H. (2015). Impiety. In E. Eidinow & J. Kindt (Eds.), *The Oxford handbook of ancient Greek religion* (pp. 325–338). Oxford University Press.

Burnet, J. (1924). *Plato's* Euthyphro, Apology of Socrates*and* Crito. Clarendon Press.

Burnyeat, M. (1988). Cracking the Socrates case. *New York Review of Books, 35*(5), 12–18.

Burnyeat, M. (1997). The impiety of Socrates. *Ancient Philosophy, 17*(1), 1–12.

Bury, J. B. (1913). *A history of freedom of thought*. Henry Holt and Co.

Campbell, J. (2017). Natural rights and the First Amendment. *Yale Law Journal, 127*(2), 246–321.

Center for Cognitive Liberty & Ethics. https://www.cognitiveliberty.org/.

Cobbett, W. (1809). Evidence against the Queen of Scots. In W. Cobbett & D. Jardine (Eds.), *Cobbett's complete collection of state trials and proceedings for high treason and other crimes and misdemeanors, from the earliest period to the present time, vol. 1 [A.D. 1163–1600]* (21 volumes, pp. 1456–1457). T. C. Hansard.

Cohen, N. (2019, March 7). Zuckerberg wants Facebook to build a mind-reading machine. *Wired*. https://www.wired.com/story/zuckerberg-wants-facebook-to-build-mind-reading-machine. Accessed December 6, 2020.

Constant, B. (2003/1815). *Principles of politics applicable to all governments* (D. O'Keeffe, Trans.). Liberty Fund.

Constant, B. (1988/1814). *The spirit of conquest and usurpation and their relation to European civilization*, in Constant, *Political writings* (B. Fontana, Trans. & Ed.). Cambridge University Press.

Diogenes Laertius. (1972). *Lives of eminent philosophers*, Vol. 1 (2 volumes) (R. D. Hicks, Trans.). Harvard University Press.

Euripides. *Children of Heracles, Hippolytus, Andromache, Hecuba*, ed. and trans. D. Kovacs (Harvard University Press, 2005/1995).

Filonik, J. (2013). Athenian impiety trials: A reappraisal. *Dike, 16*, 11–96.

Foucault, M. (2014). *Wrong-doing, truth-telling: The function of avowal in justice* (S. W. Sawyer, Trans.). University of Chicago Press.

Gallagher, M. (2016). No means no, or does it? A comparative study of the right to refuse treatment in a psychiatric institution. *International Journal of Legal Information, 42*(2), 137–172.

Giordiano-Zecharya, M. (2005). As Socrates shows, the Athenians did not believe in gods. *Numen, 52*(3), 325–355.

Havelock, C. M. (1995). *The aphrodite of Knidos and her successors: A historical review of the female nude in art*. University of Michigan Press.

Hedrick, C. W., Jr. (1988). The temple and cult of Apollo Patroos in Athens. *American Journal of Archaeology, 92*(2), 185–210.

Herodotus. *The Persian wars*, Vol. 3 (4 volumes), trans. A. D. Godley (Harvard University Press, 2005).

Humboldt, W. (1993/1850). *The limits of state action* (J. W. Burrow, Ed.). Liberty Fund.

Hutcheson, F. (2004/1725). *An inquiry into the original of our ideas of beauty and virtue[,] in two treatises* (W. Leidhold, Ed.). Liberty Fund.

Jefferson, T. (1950/1779). A bill for establishing religious freedom. In J. P. Boyd (Ed.), *The papers of Thomas Jefferson, Vol. 2: January 1777 to 18 June 1779*. Princeton University Press.

Jiang, L., Stocco, A., Losey, D. M., Abernathy, J. A., Prat, C. S., & Rao, R. P. N. (2019). BrainNet: A multi-person brain-to-brain interface for direct collaboration between brains. *Scientific Reports, 9*(6115), 1–11.

Kraut, R. (2000). Socrates, politics, and religion. In N. Smith & P. Woodruff (Eds.), *Reason and religion in Socratic philosophy* (pp. 13–23). Oxford University Press.

Lavazza, A. (2018, February 19). Freedom of thought and mental integrity: The moral requirements for any neural prosthesis. *Frontiers in Neuroscience, 12*(82). https://doi.org/10.3389/fnins.2018.00082.

Lea, H. C. (1887a). *A history of the inquisition of the middle ages*, Vol. 1 (3 volumes). Harper & Brothers.

Lea, H. C. (1887b). *A history of the inquisition of the middle ages*, Vol. 2 (3 volumes). Harper & Brothers.

Lefkowitz, M. R. (1989). 'Impiety' and 'Atheism' in Euripides' dramas. *Classical Quarterly, 39*(1), 70–82.

Ligthart, S., Douglas, T., Bublitz, C., Kooijmans, T., & Meynen, G. (2020). Forensic brain-reading and mental privacy in European human rights law: Foundations and challenges. *Neuroethics*. https://doi.org/10.1007/s12152-020-09438-4.

Locke, J. (1689). *A letter concerning toleration*. Awnsham Churchill.

Locke, J. (1959/1690). *An essay concerning human understanding*, Vol. 1. Dover Publications.

Lupovitch, H. N. (2010). *Jews and Judaism in world history*. Routledge.

Lysias. *Collected Works*, trans. W. R. M. Lamb (Harvard University Press, 1930).

Mack, E. (2018). You can talk to MIT's mind-reading headset without ever opening your mouth. *Forbes*, April 6, 2018. https://www.forbes.com/sites/ericmack/2018/04/06/talk-to-mit-alterego-mind-reading-headset-without-ever-opening-your-mouth. Accessed December 6, 2020.

Madison, J. (1786). From James Madison to Thomas Jefferson, 22 January 1786. *Founders Online*, National Archives, https://founders.archives.gov/documents/Madison/01-08-02-0249. Original source: The Papers of James Madison, Vol. 8, 10 March 1784 – 8 March 1786, ed. R. Rutland & W. Rachal (University of Chicago Press, 1973).

Mill, J. S. (1978/1859) *On liberty* (E. Rapaport, Ed.). Hackett Publishing Co.

Monter, W. (1990). *Frontiers of heresy: The Spanish inquisition from the Basque Lands to Sicily*. Cambridge University Press.

Muston, A. (1866). *The Israel of the Alps: A complete history of the Waldenses of Piedmont, and their colonies*, Vol. 1 (2 volumes) (J. Montgomery, Trans.). Blackie and Son.

Newman, D. (2019). Interpreting freedom of thought in the Canadian charter of rights and freedoms. In *Supreme Court Law Review*. Second Series (Vol. 91).

Nussbaum, M. (1985). Commentary on Edmunds. *Proceedings of the Boston Area Colloquium in Ancient Philosophy, 1*(1), 231–240.

Orwell, G. (1961/1949). *1984*. Signet Classics.

Parke, H. W., & Wormell, D. E. W. (1956). *The Delphic Oracle. Vol. 1: The history*. Basil Blackwell.

Plato.*Euthyphro, Apology, Crito, Phaedo*, ed. and trans. Chris Emlyn-Jones and William Preddy (Harvard University Press, 2017).

Plato. *Laches, Protagoras, Meno, Euthydemus*, trans. W. R. M. Lamb (Harvard University Press, 1924).

Plutarch. *Pericles and Fabius Maximus, Nicias and Crassus* (Vol. 3), trans. B. Perrin. Harvard University Press (1916).

Proast, J. (1690). *The argument of the letter concerning toleration, briefly consider'd and anser'd*. George West and Henry Clements.

Ralkowski, M. (2013). The politics of impiety: Why was Socrates prosecuted by the Athenian democracy? In J. Bussanich & N. Smith (Eds.), *The Bloomsbury companion to Socrates* (pp. 301–327). Bloomsbury.

Rawls, J. (1999a/1971). *A theory of justice* (Rev. ed.). Harvard University Press.

Rawls, J. (1999b). *Collected papers* (S. Freeman, Ed.). Harvard University Press.

Rawls, J. (2005/1993). *Political liberalism* (Expanded ed.). Columbia University Press.

Reeve, C. D. C. (1989). *Socrates in the apology: An essay on Plato's apology of Socrates*. Hackett Publishing Co.

Reeve, C. D. C. (2000). Socrates the Apollonian? In N. Smith & P. Woodruff (Eds.), *Reason and religion in Socratic philosophy* (pp. 24–39). Oxford University Press.

Roth, N. (2002/1995). *Conversos, inquisition, and the expulsion of the Jews from Spain*. University of Wisconsin Press.

Sangiovanni, A. (2019). Democratic control of information in an age of surveillance capitalism. *Journal of Applied Philosophy, 36*(2), 212–216.

Shaw, J. (2017). The watchers: Assaults on privacy in America. *Harvard Magazine, 119*(3), 56–61, 82–83.

Shiffrin, S. (2010–2011). A thinker-based approach to freedom of speech. *Constitutional Commentary, 27*, 283–307.

Swaine, L. (2001). Heresy. In D. Jones (Ed.), *Censorship: A world encyclopedia* (pp. 1045–1046). Fitzroy Dearborn.

Swaine, L. (2006). *The liberal conscience: Politics and principle in a world of religious pluralism*. Columbia University Press.

Swaine, L. (2018a). Freedom of thought as a basic liberty. *Political Theory, 46*(3), 405–425.

Swaine, L. (2018b). Legal exemptions for religious feelings. In K. Vallier & M. Weber (Eds.), *Religious exemptions* (pp. 74–96). Oxford University Press.

Swaine, L. (2020). *Ethical autonomy: The rise of self-rule*. Oxford University Press.

Vlastos, G. (1991). *Socrates: Ironist and moral philosopher*. Cornell University Press.

Waterfield, R. (2013). The quest for the historical Socrates. In J. Bussanich & N. Smith (Eds.), *The Bloomsbury companion to Socrates* (pp. 1–19). Bloomsbury.

Walsh, W. T. (1987/1940). *Characters of the inquisition*. Tan Books and Publishers.

Winick, B. J. (1989). The right to refuse mental health treatment: A First Amendment perspective. *University of Miami Law Review, 44*(1), 1–103.

Woodruff, P. (2000). Socrates and the Irrational. In N. Smith & P. Woodruff (Eds.), *Reason and religion in Socratic philosophy* (pp. 130–150). Oxford University Press.

Xenophon. *Memorabilia, Oeconomicus*, trans. E. C. Marchant; *Symposium, Apology*, trans. O. J. Todd, revised by Jeffrey Henderson (Cambridge, MA: Harvard University Press, 2013 [1923]).

Legal Sources

U.S. Const. art. III, § 3, cl. 1.
Parliament of England, 25 Edward III St. 5 c. 2 (1351).
Cline v. State, 204 Tenn. 251 (Tenn. 1958).
Chavez v. United States, 275 F.2d 813 (9th Cir. Cal. 1960).
People v. Olson, 232 Cal. App. 2d 480 (Cal. App. 5th Dist. 1965).
Sell v. United States, 539 U.S. 166 (2003).
State v. D'Ingianni, 217 La. 945 (La. 1950).
United States v. Eucker, 532 F.2d 249 (2d Cir. N.Y. 1976).
Washington v. Harper, 494 U.S. 210 (1990).
Universal Declaration of Human Rights, G.A. res. 217A (III), UN Doc A/810 at 71 (1948).
Convention for the Protection of Human Rights and Fundamental Freedoms, Nov. 4, 1950, Europ.T.S. No. 5; 213 U.N.T.S. 221.

CHAPTER 2

Freedom of Thought: Who, What, and Why?

Simon McCarthy-Jones

> The most dangerous man to any government is the man who is able to think things out for himself, without regard to the prevailing superstitions and taboos. Almost inevitably he comes to the conclusion that the government he lives under is dishonest, insane, and intolerable and so, if he is romantic, he tries to change it. And even if he is not romantic personally he is very apt to spread discontent among those who are.
>
> H. L. Mencken (1982)

Freedom of thought is a secular deity. Among the judiciary of the Supreme Court of the United States, it has been an object of reverence. The "right to think," the Supreme Court has proclaimed, "is the beginning of freedom" (*Ashcroft v. Free Speech Coalition*, 2002). There is no principle, the Court has preached, that "more imperatively calls for attachment" than "the principle of free thought" (*United States v. Schwimmer*, 1929). "Our whole constitutional heritage" the Court has gloried, "rebels at the thought of giving government the power to control men's minds"

S. McCarthy-Jones (✉)
Trinity College Dublin, Dublin, Ireland
e-mail: simon.mccarthy-jones@tcd.ie

M. J. Blitz et al. (eds.), *The Law and Ethics of Freedom of Thought, Volume 1*, Palgrave Studies in Law, Neuroscience, and Human Behavior,
https://doi.org/10.1007/978-3-030-84494-3_2

(*Stanley v. Georgia*, 1969). And as Justice Jackson put it in his adoration, "The priceless heritage of our society is the unrestricted constitutional right of each member to think as he will" (*American Communications Assn. v. Douds*, 1950).

As is the case with deities, worship far exceeds understanding. Legal scholars have put forward conceptions of what a right to freedom of thought should look like (Nowak, 1993; Vermeulen, 2006). Yet, the validity of such efforts is questionable given no clear conception of freedom of thought has been reached by those with expertise in relevant areas such as philosophy of mind and psychology. Creating a concept of freedom of thought is too important to be left for legal scholars to create by proxy. There needs to be a genuinely interdisciplinary enterprise to this end.

The first step of such a project will involve developing important questions to be addressed (Szostak, 2012). This chapter seeks to add a voice into this conversation by suggesting what questions appear to be pressing from the current author's perspective. These are framed as the 'why,' 'what,' and 'who' of freedom of thought.

The Why of Freedom of Thought

The first 'why' question we encounter is why freedom of thought is an important ability to have, and therefore, why the right to freedom of thought is important. As these matters have been inquired into elsewhere (Blitz, 2010; Bublitz & Merkel, 2014; McCarthy-Jones, 2019), I will not consider them here. What I will raise instead is the question as to why we start discussing threats to freedom of thought at a given point in time.

Why Have People Historically Been Concerned About Freedom of Thought?

The events that cause us to think about freedom of thought will inevitably shape our conclusions. We therefore need to consider why we are thinking about freedom of thought and why we are thinking about it the way we are. This requires an archaeology of freedom of thought (cf. Foucault, 2005). Although this undertaking is beyond this current chapter, some initial observations will be made relating to why the issue of freedom of thought has been raised in the past.

The relative frequency with which the term 'freedom of thought' has appeared in books published in the English language has changed over the years. A search using the Google Books Ngram Viewer (Michel et al., 2011) shows a peak in the relative frequency of this term in the mid-1950s (see Fig. 2.1).

The story of this peak begins in the 1930s. As Taylor (2006) has noted, this was when the West became concerned about the show trials held by the Communist Party of the Soviet Union. These had managed to obtain seemingly sincere confessions from former party officials (Taylor, 2006). Yet, it was the ability of the Soviet Union's counterpart, the Chinese Communist Party, to influence minds that came to be of primary concern to the West because the Chinese rulers placed less emphasis than the Soviets on eliminating perceived opponents of the regime. They put more emphasis on attempting to 'cure the disease and save the man' (Schein, 1960). China did this, in the 1950s, through a program of *szu-hsiang kai-tsao* ('thought reform'), which undertook the re-education of millions of people, attempting to remake them into New People (Lifton, 1989). Investigations of these actions, by bodies such as the United States Congress' House Committee on Un-American Activities, bought the issue of freedom of thought to the fore.

A range of terms were used to refer to matters of freedom of thought during the twentieth century. The popularity of such terms has waxed and

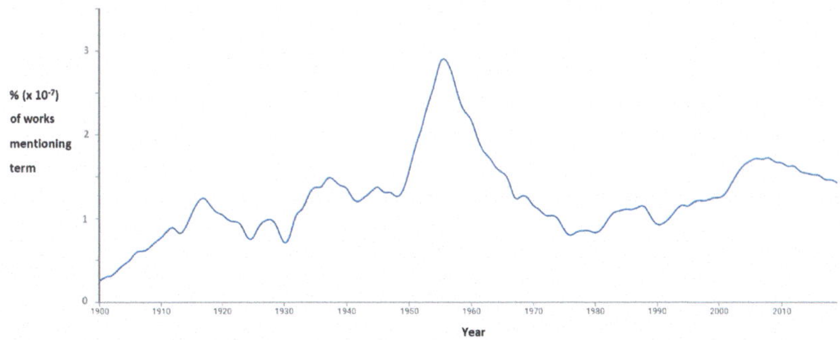

Fig. 2.1 Relative frequency of 'freedom of thought' in English language books post-1900 (*Note* Search conducted using the search term "freedom of thought" in the English [2019] corpus of Google Books Ngram Viewer with a smoothing factor of 3)

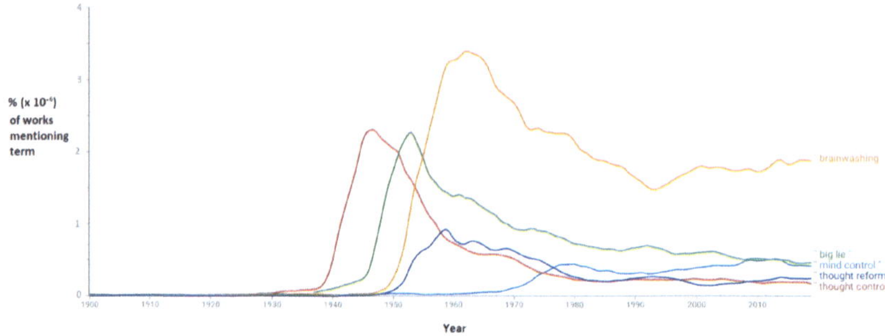

Fig. 2.2 Relative frequency of thought control related terms in English language books post-1900 (*Note* Search conducted using the English [2019] corpus of Google Books Ngram Viewer with a smoothing factor of 3)

waned, driven by political events. As Fig. 2.2 shows, in the mid-1940s the term 'thought control' became popular. This was used to refer to the extensive Nazi propagandizing during the Second World War. The term 'thought control' was widely applied to Joseph Goebbels' use of propaganda to influence the German people. For example, on February 14, 1937, an article on Goebbels in the New York Times appeared with the sub-line "Thought Control Now Achieved in the Reich" (Ross, 1937). Yet, as Fig. 2.2 shows, after the end of the Second World War the relative use of the term decreased. At about this time, another term used by the Nazis, the 'big lie' became popularized, before also dropping away again.

In the place of these terms arose another term. This came from another military and political conflict. Through the 1950s, and peaking in 1960, the term 'brainwashing' became popular. The term was first used in 1950 by an American journalist, Edward Hunter, to translate a term '*his nao*' (literally 'wash brain') which Chinese informants told him was being employed by the Chinese Communist Party (Lifton, 1989). This term came to America as a result of the Korean War. When some American soldiers and civilians, imprisoned by the Chinese during the Korean War, were found to either still subscribe to Communist ideas after repatriation, or even refused repatriation, concerns over 'brainwashing' became prominent in the United States (Schein, 1960). As Fig. 2.2 shows, this term captured the public imagination and has remained popular ever since. It also captured the professional imagination. In 1980 it appeared in the third edition of psychiatry's bible, the Diagnostic and Statistical Manual of Mental Disorders. This recognized that dissociated states could occur in people who had been "subjected to periods of prolonged and intense coercive persuasion (brainwashing, thought reform, and indoctrination while the captive of terrorists or cultists)" (APA, 1980, p. 260).

As Fig. 2.2 shows, although the term 'thought reform' also became popular at the same time as the term brainwashing, due to the activities of the Chinese Communist Party, this less vivid term has not been frequently used since. Figure 2.2 also shows that in the 1970s, peaking in 1980, the term 'mind control' became and remained popular. The reason for this change is of terminology and its impact is worthy of investigation.

Writings on freedom of thought in the twentieth century in the English language were hence very much driven by the fear that competing ideologies could influence people's minds, both inside and outside the boundaries of 'enemy' countries. The cri de coeur of "freedom of thought" could be used to argue that opposing regimes were illegitimately gaining people's consent to their ideologies. This was a powerful rhetorical device. In fact, it was such a powerful tool to use against one's ideological opponents that, after the defeat of Nazi Germany, the disappearance of Soviet Russia, and the development of a more accommodating attitude toward China, we should not expect this tool to have been fully downed. We would expect to see it being wielded domestically.

What might this look like? Consider, for example, the Cambridge Analytica scandal (Cadwalladr, 2019). It is questionable to what extent this represented a non-partisan outcry at voter's thoughts being manipulated. Back in 2012, during President Obama's re-election campaign, there had already been a "dramatic technological shift," which had "greatly advantaged the Obama campaign" (Bimber, 2014). Specifically, more use than ever before was made of microtargeting. Yet this did not lead to widespread protest. Only when, in 2016, were these tactics used to support non-liberal campaigns such as Trump and Brexit did a furore erupt. A case can be made that liberals used terms associated with freedom of thought to claim that the supporters of populist campaigns had been 'brainwashed,' as a mean to delegitimize their perspective.

Claiming that political opponents could only believe what they claim to believe if they have been manipulated is a terrible place to start a political debate. Supreme Court Justice Oliver Wendell Holmes Jr. famously spoke of the need to protect "not free thought for those who agree with us but freedom for the thought that we hate" (*United States v. Schwimmer*, 1929). Similarly, those engaged in political debate need to recognize that free thought does not just lead to positions one agrees with but can also lead to thought one hates. The idea that correct thinking inevitably leads to one's own view, and therefore anyone who disagrees must be either wrong or brainwashed, itself contains the seeds of totalitarianism.

Why Are People Concerned About Freedom of Thought Today?

Stepping aside from political reasons why there may be a contemporary interest in freedom of thought, there are also clear technological reasons why this topic has come to the fore of the minds of many. The relevant technological advances can be referred to as 'brain-reading' and 'behaviour-reading' (McCarthy-Jones, 2019).

In behavior-reading, large quantities of data are collected on citizens. This data is then analyzed by machine learning algorithms to make accurate predictions about their inner worlds. Research shows that people's observable behavior, including their facial expressions, possessions, purchases, musical preferences, internet browsing data, and the words they use and posts they 'like' on social media, can be used to infer what is happening in their hitherto private inner world (Golbeck et al., 2011; Kosinski et al., 2013; Rentfrow & Gosling, 2003; Wang & Kosinski, 2018). For example, one study found that detailed personal information about individuals could be accurately predicted based on a knowledge of what pages on Facebook they had 'liked' (Kosinski et al., 2013).

While studies using individual variables can make predictions about inner states with statistically significant but limited accuracy, prediction accuracy is greatly enhanced when data is available on a wider range of variables. Today, the advent of 'big data' has allowed the accumulation of huge datasets on individuals. Data may be drawn from a variety of sources including individuals' digital footprints and their purchasing habits. Advanced machine learning algorithms can then be used to infer the inner workings of individuals by analyzing this data. Given the history of governmental interest in the thoughts of citizens, the access of the state to the almost unimaginable volumes of data gathered by large scale covert surveillance operations such as XKeyscore, as detailed by the Snowden revelations (Greenwald, 2013), poses potential threats to the freedom of thought of citizens. Whether the potential increased security this yields is a price worth paying needs to be a matter for public debate.

In addition to the national security state, behavior-reading has also been seized on by corporations whose activities take the form of surveillance capitalism (Zuboff, 2019). This approach claims human experience as free raw material that can be used to deduce "thoughts, feelings, intentions, and interests," predict behavior, and then be monetized (Zuboff, 2019, p. 81). These corporations are motivated to "produce tractable, predictable citizen-consumers whose preferred modes of self-determination play out along predictable and profit-generating trajectories" (Cohen, 2013). Quite how much data technology behemoths such as Google holds on their users is unclear. However, data-analytics company i360 boasts that with their "1800 unique data points on all 290 million Americans, we can create the most comprehensive profile of your target making sure you're working with the full picture" (i360, 2019). Such datasets promise to reveal significant data about the inner worlds of individuals and hence pose a potential threat to freedom of thought.

There is public concern over the use of such technologies to facilitate microtargeting; the use of personality data gathered on individuals (through means such as social networking sites, other websites, and offline data brokers) to deliver political adverts to them that should be maximally effective. However, it is unclear how effective microtargeting techniques are at changing political views. Such effects appear likely to be small (Hersh & Schaffner, 2013; Liberini et al., 2018). Indeed, most methods of persuasion used in electoral campaigns seem to have minimal, if any effect (Kalla & Broockman, 2018). We must represent the size of the effect of microtargeting accurately.

What voter targeting does seem to be able to effect is turnout. This is important in the context of increasingly tight elections (the 2000 US Presidential election was effectively won by 537 votes). For example, a 2012 study performed a randomized controlled trial of the effects of showing political mobilization messages to 61 million adult Facebook users during the 2010 US congressional elections (Bond et al., 2012). Facebook users were either shown no voting message at the top of their newsfeed, an informational message about voting, or an informational message plus up to six small profile pictures of their friends who had voted. The effect of showing these profile pictures was estimated to lead to 340,000 extra votes. On-line political mobilization works, the authors concluded. Thus, although it is difficult to separate self-interested promotion from fact, the use of behavior-reading poses a plausible and present danger to individuals' freedom of thought.

Another technological threat to freedom of thought looms on the horizon in the form of brain-reading. Here, the neural activity of individuals is decoded to reveal the thoughts that it corresponds to. There has been significant progress in the ability to infer individual's inner states from their neural activity. People's neural activity can be used to predict what novel, natural images (i.e., not restricted to pre-defined or previously seen objects) they are looking at (Kay et al., 2008; Naselaris et al., 2009). Neuroimaging technologies have advanced from being able to create rudimental reconstructions of what an individual is seeing (Miyawaki et al., 2008) to now being able to create "remarkable" reconstructions (Nishimoto et al., 2011). The potentially invasive uses of this neuro-technology can be seen in a preliminary (unpublished) virtual reality study by Haynes and colleagues (see Smith, 2013). This asked participants in an fMRI scanner to tour several virtual reality houses and then to tour another selection. The research team were then able to work out which of the houses the participants had been to before.

In 2012, Haynes cautioned that the idea of a universal brain-reading machine, which could read off the thoughts of someone in real-time, was still a long way off. One potential barrier Haynes (2012) identified was the necessity to know the neural activation underpinning a thought before it could be detected within someone. This would require creating a dictionary of the neural signatures of a basic vocabulary of tens of thousands of words. A second barrier was that this neural dictionary could be different for everyone. That is, the neural activity underpinning thought A in person 1 could be different to that underpinning thought A in person 2. In recent years, significant cracks in both these barriers have emerged.

There have been advances in the ability to decode novel thoughts a person is having, after having only trained a decoder to recognize the neural activity of a relatively small set of words, sentences, or concepts (Anderson et al., 2016; Oota et al., 2018; Pereira et al., 2018). Such approaches are based on the principle that words are represented in the brain as 'semantic vectors.' In theory, any given word can be represented by the extent to which it is associated with a finite number of specific features such as 'speech,' 'audition,' 'face,' 'body,' 'biomotion,' 'motion,' 'audition,' 'pleasant,' 'unpleasant,' 'human,' 'fast,' 'happy,' 'pattern,' 'arousal,' 'shape,' etc. (Anderson et al., 2016). For example, concepts such as dog, horn, and thunder would all score highly on the 'audition' feature, whereas clouds, flowers, and tomatoes would score low (Anderson et al., 2016). It is possible to represent most words using a unique pattern of vector scores. If you can determine the neural activity associated with each feature, you can decode the words someone is thinking. Imagine a person thinks about a boy, a concept whose neural activity a decoder has not been trained to specifically recognize. The decoder can still infer from high levels of neural activity already known to be associated with the concepts 'male,' 'human,' and 'child,' that the person is thinking about a boy. Importantly, there is also now evidence that the neural activation of a specific 'thought' (i.e., a semantic vector, concept, or mental state/task) in one person is similar to the neural activation associated with this same 'thought' in another person (Anderson et al., 2016; Poldrack et al., 2009; Shinkareva et al., 2008). The door to a universal brain-reading machine is now ajar.

The 'What' of Freedom of Thought

Many important 'what' questions can be asked about relating to freedom of thought. We may ask what freedom of thought is. However, the philosopher Ludwig Wittgenstein (1953) suggested that we should not ask the meaning of a word, but ask its use. This implies we should not ask 'what does freedom of thought mean?' but rather 'what are people trying to do when they invoke the concept of freedom of thought?'. Presumably, as touched on in the previous section, in addition to those using the term freedom of thought to support human rights in a disinterested manner, some are using it to further their interests. There is much to be learnt, using qualitative methodologies such as various forms of discourse analysis, about what people try to use this term in conversation to achieve. This cannot be attempted here, so instead, I will briefly touch on another 'what' of freedom of thought; what is thought and how can it be free?

As Loucaides (2012) has observed, "there is no adequate material in the preparatory works of the drafters of the European Convention regarding the concept of 'thought'" (p. 80). To create a definition of free thought, we first need to understand the nature of thought. Part of this can involve identifying the elements of thought central to the conception of free thought that we are aiming to reach. At the heart of this is autonomy. Metzinger (2013) has defined mental autonomy as "the specific ability to control one's own mental functions, like attention, episodic memory, planning, concept formation, rational deliberation, or decision making, etc."

Metzinger (2013) has argued that mental autonomy is comprised of two abilities: attentional agency and cognitive agency. Attentional agency is the ability to control one's focus of attention. If one cannot control one's attention, one cannot control one's thinking. And, as Metzinger (2015) has claimed "for as long as one cannot control one's own thought one cannot count as a rational individual" (p. 272). Attentional agency is needed for cognitive agency. This is the ability to control goal/task-related, deliberate thought (Metzinger, 2013). To do this, we need to be able to think about our thinking and to be able to perform what are called second-order mental actions.

In "hierarchical" accounts of autonomy (e.g., Frankfurt, 1971), thoughts, desires, and impulses that spontaneously pop up within us are called "first-order" mental actions. To act on these unreflectively is to fail to display autonomy. But, if we think about these first-order mental actions (hence performing a "second-order" mental action), to determine if they are consistent with our own chosen values and goals, then they can become authentic. Second-order mental actions allow us to structure our thoughts, undertake logical trains of thought, and guide our behavior (Metzinger, 2013). Second-order mental actions should hence inform the development of the concept of freedom of thought.

This first-/second-order distinction also looks ahead to the issue of our responsibility for our thoughts (and hence any discussion of their punishment, cf. Mendlow, this volume). We are not responsible for many of our thoughts. Yet, it is important to note that some thoughts, which we are not culpable for, could be deemed to increase the risk of harm to others. For example, consider first-order thoughts that are intrusive and unwanted. An early study of people reporting unwanted intrusive thoughts (UITs) found these included thoughts of acts of violence during sex, of throwing a child out of a bus, and of jumping in front of a train (Rachman & deSilva, 1978). Further research found unwanted intrusive thoughts about violence and sex to be widespread. Sixty percent of people were found to have UITs about running a car off the road, 46% had UITs about hurting family members, and 26% had UITs involving fatally pushing a stranger (Purdon & Clark, 1993). Other studies have echoed such surprising findings. One study found that 6% of people reported having UITs about sex with animals or non-human objects, 1 in 5 men had UITs about a sexual act with a child or minor, and 1 in 3 men had UITs about forcing another adult to have sex with them (Byers et al.,

1998). Such findings show the truth of Nagel's (1998) claim that "civilisation would be impossible if we could all read each other's minds" (p. 4).

First-order thoughts can be understood as involuntary experiences. Indeed, we may even have evolved the tendency to have such abhorrent thoughts. The evolutionary psychologist David Buss has proposed that "all of us house in our large brain specific specialised psychological circuits that lead us to contemplate murder as a solution to specific adaptive problems" (Buss, 2006). People can be seen to be the victims of a range of unwanted, unacceptable first-order thoughts, rather than the perpetrators. By this reasoning, people are not culpable for such thoughts.

Another important matter is to realize that thought is not just 'in the head.' This has been argued by Clark and Chalmers (1998) using their concept of the extended mind. Clark and Chalmers argue that our minds extend into the world and hence our thinking can take place outside of our body. As they put it, if "a part of the world functions as a process which, were it done in the head, we would have no hesitation in recognizing as part of the cognitive process, then that part of the world is (so we claim) part of the cognitive process" (p. 8). In the case of re-arranging Scrabble tiles to make a word, they claim that "[i]n a very real sense, the re-arrangement of tiles on the tray is not part of action; it is part of thought" (p. 9).

To take another example, for some individuals the process of writing is akin to the process of thinking. To take a literary example, in George Orwell's dystopian novel *1984*, the protagonist, Winston Smith, does his thinking by writing a diary. If we accept that diaries and memos are thoughts, then these would receive absolute protection under the right to freedom of thought. This would make them inviolable, which they are currently not when protected by mere privacy laws.

This concept could have powerful legal ramifications. For example, taking a distant historical example first, we can consider the 1683 trial of Algeron Sidney. Accused of treason against Charles II, Sidney could not be convicted by law unless two accusers could be found (Houston, 1991). In the absence of a second accuser, his unpublished private writings were introduced as a second "witness" to his crime, under the justification of the novel principle that *scribere est agere* (to write is to act). If his writings had been viewed as protected thoughts, Sidney may have been safe. That said, Chief Justice George Jeffreys argued that the king should not be cursed, even "in thy thoughts" (Houston, 1991), so Sidney was probably in trouble either way.

To take a more modern example, consider *State v. Dalton* (2003). In this case, Dalton was sentenced to seven years in prison "for creating and possessing a personal diary containing violent sexual fantasies involving children" (Calvert, 2005, p. 136). The verdict in the *State v. Dalton* case was widely criticized, including by Laurence Tribe who called it "as close as you can get to creating a thought crime" (Tribe, as cited in Calvert, 2005, p. 136).

Finally, we may pause to reflect that, in the United States, due to section 215 of the PATRIOT Act which allows officials conducting foreign intelligence investigations to obtain library records of Americans, reading records are not absolutely private (Kaminski & Witnov, 2014). An individual's internet browsing history has been used to help establish a defendant's state of mind (Kaminski & Witnow, 2014). Furthermore, in one case the US Attorney subpoenaed Amazon to find the book purchases of over 24,000 people, as part of an investigation into a crime committed by just one person (Kaminski & Witnow, 2014). Such state actions chill thought. They should not be permitted to be deemed permissible violations of freedom of thought until there has been an explicit public debate on the extent to which security should be balanced against freedom. Such debates are necessary to preserve the autonomy of individuals in a democracy, by allowing them to meaningfully determine the laws that bind them.

The 'Who' of Freedom of Thought

The 'who' of freedom of thought is one of the lesser considered areas. We may ask broad questions, such as whether there has been a cultural shift resulting in fewer people wanting to be free thinkers (McCarthy-Jones, 2019). We may also ask more specific questions, such as what the right to freedom of thought means for people with intellectual disabilities, and how their right to freedom of thought can be exercised on an equal basis with all others (cf. Ward & Stewart, 2008). However, the question I will focus on here is whose freedom of thought is most likely to be defended in court, and the problems this may pose.

The limited case law involving freedom of thought in the United States typically relates to one of two stigmatized groups. The first are cases where freedom of thought is used to defend people who had sexual thoughts about minors (e.g., *Doe v. City of Lafayette, Indiana*, 2003; *United States v. Bredimus*, 2003; *United States v. Kaechele*, 2006; *United States v. Stokes*, 2013; *United States v. Tykarsky*, 2006). The second are cases where freedom of thought is employed to argue against the necessity for a person diagnosed with schizophrenia to take mind-altering antipsychotic medication (e.g., *Rennie v. Klein*, 1981; *United States v. Charters*, 1987). Given the public stigma often attached to these two groups of people, this would make it easy for the public to clamber for such individual's right to freedom of thought to be limited.

Although there is a principle of law that hard cases make bad law, any public debate over the right to freedom of thought is likely to be framed in terms of hard cases. This will encourage limitations being put in place on freedom of thought. For example, it is easy to imagine the issue of the right to freedom of thought being framed as one involving a terrorist who has planted a nuclear bomb in a major metropolitan area (the 'Ticking Time Bomb' scenario). The argument would then run that his/her right to freedom of thought would be outweighed by national security and public safety interests. This would justify the non-consensual monitoring of his/her thoughts for clues as to where the bomb is hidden. A worry here is that once a permissible limitation has been introduced, based on an extreme case which is known to be able to separate people from deeply held ethical values (Opotow, 2007), this opens the floodgates to more limitations being implemented.

Yet, it is more likely that debate will be pushed to focus on the more emotive topic referenced above; child sexual abuse. A 2005 Gallop Poll found that the percentage of Americans who were "very concerned" about sex offenders was nearly double the percentage of Americans who were "very concerned" about terrorism (Human Rights Watch, 2007). As we have already seen, child sexual abuse is also one of the few categories of crime where freedom of thought is often invoked, reinforcing the idea that this issue would be a key frame for a public debate on freedom of thought.

It is highly likely that many people would agree with the sentiment that everything possible should be done to stop the rape of children, including the enforced monitoring of convicted offenders' thoughts. Many would likely be troubled by the use of the right to freedom of thought to allow a convicted sex offender to continue to watch children in public parks while thinking sexual thoughts about them (*Doe v. City of Lafayette, Indiana*, 2003). Similarly, many may query the failure to punish thought in the form of intent in the case of State v. Kemp (2001), in which the defendant was acquitted of attempted child molestation, despite having allegedly agreed to meet a minor at restaurant parking lot, driven there, and bought condoms. For understandable reasons, there is hence likely to be a tranche of the population who will argue that the risk posed by such thoughts to the safety of children outweighs the right to freedom of thought of a potential offender.

The problem with this view is that its emotive force could potentially overcome stronger arguments for absolute protection of freedom of thought. In essence, the strong political pressure on legislators to violate the freedom of thought of pedophiles may be more than reasoned argument could ever overcome. Take, for example, the reaction of politicians to the ruling of *Doe v. City of Layfayette, Indiana* (2004), in which the ban from public parks of the convicted sex offender, Doe, was upheld, effectively punishing him for his thoughts. It was reported in the press that "[t]wo candidates who want to represent Lafayette in the Indiana House praised a federal court ruling barring a convicted child molester from city parks but said a statewide version of the ban is needed" (Calvert, 2005, p. 130). A Democratic candidate, by stating that he "would be supportive of legislation" effectively, stated his willingness to change the US Constitution (Calvert, 2005). Politicians hence not only supported the judiciary rejecting constitutionally protected freedom of thought, but also looked to cement and expand the ruling. As Calvert notes "It is easy to run for office and to support legislation when it is strategically and narrowly framed, such as the concise and visceral frame of "protect children from a pedophile" rather than the more complex and less emotionally appealing frame of "protect a constitutional right from legislative usurpation" (p. 130).

To take another example, after the rape and murder of seven-year-old Megan Kanka in the United States in 2004 by a convicted sex offender, legislation was introduced requiring the police to notify communities of registered sex offenders. This was done on the basis that it would help the public to protect themselves from sexual crime. In reality, the evidence base for the effectiveness of this policy is at best mixed, and its flawed design, which should have been apparent from the start, has led to significant problems (Cohen, 2018; Cull, 2018; Human Rights Watch, 2007; Levenson & Tewksbury, 2009; Zgoba et al., 2018). Yet, as Human Rights Watch (2007) has noted, "when community notification came up for discussion in the US House of Representatives, only one representative voiced opposition and the bill eventually passed 418-0." To be clear, the issue here is not that arguments cannot be made against violating freedom of thought. They can. The issue is that reasoned debate is threatened by political expediency.

More generally, once the door is opened to freedom of thought being re-designated as a non-absolute right, needing to be balanced against other considerations such as national security, a panoply of hard questions arise. One series of questions, as the Editors of this volume have suggested to me, is whether some types of freedom of thought violation are of greater concern than others, hence necessitating differential regulatory/legal responses. For example, is there reason to believe that the active manipulation of thoughts is more problematic than the passive monitoring of thoughts, and if so, why?[1] To take a hypothetical example, imagine a State has valid concerns about the intentions of a citizen who is walking through an infrastructure-critical area. Is it more justifiable to violate the privacy of the citizen's thoughts, by monitoring them to check if they are planning a terrorist act, than it is for the state to actively manipulate the citizen's thoughts by inserting pacifying thoughts into their mind? If autonomy of thought is what the right to freedom of thought is aiming to protect (McCarthy-Jones, 2019), then the latter approach by the State could be considered more objectionable, i.e., monitoring thought allows that people can think what they like, though what they think may have consequences, whereas manipulating thought takes away people's ability to think what they want to. In terms of monitoring people's thoughts, if brain-reading is able to more accurately identify thoughts than a more probabilistic behavior-reading approach, then should the former be more strictly regulated? Similarly, people have some experience in being able to physically act in a way that conceals their thoughts. Yet, who has much experience at trying not to think thoughts? Arguably this makes us more vulnerable to brain-reading than behavior-reading technologies. The expertise required to properly consider such issues again stresses the need for interdisciplinary collaboration in this area.

[1] Admittedly, this is a false dichotomy as monitoring thought is effectively a way of manipulating thought by encouraging people to self-censor (McCarthy-Jones, 2019).

Conclusions

Our biologically evolved ability to access the inner worlds of others is in the process of taking a qualitative leap forward. We now have technologies that can infer our inner world from our behavior. These are already being widely deployed for profit under surveillance capitalism. The Snowden revelations suggest they are also of great interest to the national security organs of states. We will shortly have technologies that can infer our thoughts from our neural activity. These too will inevitably be utilized by surveillance capitalism. Such technologies come with great promise but even greater threats. The traditional assumption that the mind is secure from external intrusions and not in need of legal protection is no longer tenable. The right to freedom of thought, which is supposed to guard citizens' mental autonomy in the face of such threats, is so underdeveloped as to be not fit for purpose. This right urgently needs to be defined to protect thought. This process needs to be guided by the light of what the right is trying to protect: mental autonomy (McCarthy-Jones, 2019).

One recent strand of thought has been that the novelty of the twenty-first century means that we need new rights to protect thought. Candidates offered include rights to "mental self-determination" (Bublitz & Merkel, 2014), "cognitive liberty" (Boire, 2001), "freedom of mind" (*Wooley v. Maynard*, 1977), "mental privacy," "mental integrity," and "psychological continuity" (Ienca & Andorno, 2017). Others have argued that there is no need to design new rights and that we need clearer guidance and development of the meaning of the right to freedom of thought today (Alegre, 2017). My reflections on this stem from the somewhat artificial distinction between speech and thought. While there are notable differences between speech and thought, with thought not simply being silent speech (Fernyhough, 1996; Jones & Fernyhough, 2007), in other ways speech and thought are not mutually exclusive. In Ancient Greece, those who wished to think would seek out Socrates and speak with him. Two minds interacted, dialogically (Fernyhough, 1996; McCarthy-Jones & Fernyhough, 2011), through speech, to think together. This echoes what Jeffrey Rosen has observed was one of Supreme Court Justice Brandeis's favorite sayings; "come let us reason *together*." This highlights that thought is a social process. Is there, we may wonder, a case to combine free speech and free thought rights into a wider right, which one could term, the right to seek truth. Any legal scholar would, of course, be quick to see potentially fatal flaws in such an idea. However, the point I want to make is that the distinction between speech and thought is not as clear as it may first appear. We should not let this distinction rest within human rights law quite so tranquilly as it does currently.

This chapter has raised what I consider to be some of the important questions that should be asked at the start of the modern venture to define and defend freedom of thought. Many other forms of questions could also have been added. We may wish to add in the 'how' of freedom of thought. How do we get government to support a right that in many ways threatens power? How can the law best develop the right to freedom of thought? These are questions that must wait for another occasion. However, thanks to the momentum of scholars who have been writing on freedom of thought, and the contribution of this volume, it is clear there will be more occasions to continue and develop this conversation.

References

Alegre, S. (2017). Rethinking freedom of thought for the 21st century. *European Human Rights Law Review, 3*, 221–233.

American Communications Assn. v. Douds. (1950). 339 U.S. 382, 70 S. Ct. 674, 94 L. Ed. 925.

American Psychiatric Association. (1980). *Diagnostic and statistical manual of mental disorders* (3rd ed.). American Psychiatric Association.

Anderson, A. J., Binder, J. R., Fernandino, L., Humphries, C. J., Conant, L. L., Aguilar, M., et al. (2016). Predicting neural activity patterns associated with sentences using a neurobiologically motivated model of semantic representation. *Cerebral Cortex, 27*, 4379–4395.

Ashcroft v. Free Speech Coalition. (2002). 535 U.S. 234, 122 S. Ct. 1389, 152 L. Ed. 2d 403.

Bimber, B. (2014). Digital media in the Obama campaigns of 2008 and 2012: Adaptation to the personalized political communication environment. *Journal of Information Technology & Politics, 11*(2), 130–150.

Blitz, M. J. (2010). Freedom of thought for the extended mind: Cognitive enhancement and the constitution. *Wisconsin Law Review*, 1049.

Boire, R. G. (2001). On cognitive liberty. *The Journal of Cognitive Liberties, 2*(1), 7–22.

Bond, R. M., Fariss, C. J., Jones, J. J., Kramer, A. D., Marlow, C., Settle, J. E., et al. (2012). A 61-million-person experiment in social influence and political mobilization. *Nature, 489*, 295–298.

Bublitz, C. (2015). Cognitive liberty or the international human right to freedom of thought. In J. Clausen & N. Levy (Eds.), *Handbook of neuroethics* (pp. 1309–1333). Springer.

Bublitz, J. C., & Merkel, R. (2014). Crimes against minds: On mental manipulations, harms and a human right to mental self-determination. *Criminal Law and Philosophy, 8*(1), 51–77.

Buss, D. M. (2006). *The murderer next door: Why the mind is designed to kill.* Penguin.

Byers, E. S., Purdon, C., & Clark, D. A. (1998). Sexual intrusive thoughts of college students. *Journal of Sex Research, 35*(4), 359–369.

Cadwalladr, C. (2019, March 17). Cambridge Analytica a year on: 'A lesson in institutional failure'. *The Guardian*. Retrieved from https://www.theguardian.com/uk-news/2019/mar/17/cambridge-analytica-year-on-lesson-in-institutional-failure-christopher-wylie

Calvert, C. (2005). Freedom of thought, offensive fantasies and the fundamental human right to hold deviant ideas: Why the Seventh Circuit got it wrong in Doe v. City of Lafayette, Indiana. *Pierce Law Review, 3*, 125–160.

Clark, A., & Chalmers, D. (1998). The extended mind. *Analysis,58*(1), 7–19.

Cohen, J. E. (2013). What privacy is for. *Harvard Law Review, 126*(7), 1904–1933.

Cohen, L. (2018). Cruel and unusual: The senseless stigmatization of youth registries. *Criminal Justice, 33*(1), 46–47.

Cull, D. (2018). International Megan's Law and the identifier provision—An efficacy analysis. *Washington University Global Studies Law Review, 17*, 181–200.

Doe v. City of Lafayette Indiana. (2003). 334 F.3d 606 (7th Cir.).

Doe v. City of Lafayette Indiana. (2004). 377 F.3d 757 (7th Cir.).

Fernyhough, C. (1996). The dialogic mind: A dialogic approach to the higher mental functions. *New Ideas in Psychology, 14*(1), 47–62.

Frankfurt, H. G. (1971). Freedom of the will and the concept of a person. *Journal of Philosophy, 68*, 5–20.

Foucault, M. (2005). *The order of things*. Routledge.

Golbeck, J., Robles, C., & Turner, K. (2011). Predicting personality with social media. In *CHI'11 Extended abstracts on human factors in computing systems* (pp. 253–262). https://doi.org/10.1145/1979742.1979614

Greenwald, G. (2013, July 31). XKeyscore: NSA tool collects 'nearly everything a user does on the internet'. *The Guardian*. Retrieved from https://www.theguardian.com/world/2013/jul/31/nsa-top-secret-program-online-data

Haynes, J. D. (2012). Brain reading. In S. Richmond, G. Rees, and S. J. L. Edwards (eds.). *I know what you're thinking: Brain imaging and mental privacy*, (pp. 29–40). Oxford: Oxford University Press.

Haynes, J. D., Sakai, K., Rees, G., Gilbert, S., Frith, C., & Passingham, R. E. (2007). Reading hidden intentions in the human brain. *Current Biology, 17*, 323–328.

Hersh, E. D., & Schaffner, B. F. (2013). Targeted campaign appeals and the value of ambiguity. *Journal of Politics, 75*, 520–534.

Houston, A. C. (1991). *Algernon Sidney and the Republican Heritage in England and America*. Princeton, NJ: Princeton University Press.

Houston, A. C. (2014). *Algernon Sidney and the republican heritage in England and America*. Princeton University Press.

Human Rights Watch. (2007). *No easy answers: sex offender laws in the US*. Retrieved from https://www.hrw.org/report/2007/09/11/no-easy-answers/sex-offender-laws-us

i360. *The database*. Retrieved from https://www.i-360.com/the-database/ [June 13th, 2019].

Ienca, M., & Andorno, R. (2017). Towards new human rights in the age of neuroscience and neurotechnology. *Life Sciences, Society and Policy, 13*(1), 5.

Jones, S. R., & Fernyhough, C. (2007). Neural correlates of inner speech and auditory verbal hallucinations: A critical review and theoretical integration. *Clinical Psychology Review, 27*(2), 140–154.

Kalla, J. L., & Broockman, D. E. (2018). The minimal persuasive effects of campaign contact in general elections: Evidence from 49 field experiments. *American Political Science Review, 112*, 148–166.

Kaminski, M. E., & Witnov, S. (2014). The conforming effect: First amendment implications of surveillance, beyond chilling speech. *U. Rich. L. Rev., 49*, 465.

Kay, K. N., Naselaris, T., Prenger, R. J., & Gallant, J. L. (2008). Identifying natural images from human brain activity. *Nature, 452*, 352–355.

Kosinski, M., Stillwell, D., & Graepel, T. (2013). Private traits and attributes are predictable from digital records of human behavior. *Proceedings of the National Academy of Sciences of the United States of America, 110*, 5802–5805.

Levenson, J., & Tewksbury, R. (2009). Collateral damage: Family members of registered sex offenders. *American Journal of Criminal Justice, 34*(1–2), 54–68.

Liberini, F., Redoano, M., Russo, A., Cuevas, A., & Cuevas, R. (2018). *Politics in the Facebook era. Evidence from the 2016 US presidential elections* (CAGE Online Working Paper Series [389]).

Lifton, R. J. (1989). *Thought reform and the psychology of totalism: A study of 'brainwashing' in China*. University of North Carolina Press.

Loucaides, L. G. (2012). The right to freedom of thought as protected by the European Convention on Human Rights. *Cyprus Human Rights Law Review, 1*, 79–87.

McCarthy-Jones, S. (2019). The autonomous mind: The right to freedom of thought in the twenty-first century. *Frontiers in Artificial Intelligence, 2*, 19.
McCarthy-Jones, S., & Fernyhough, C. (2011). The varieties of inner speech: Links between quality of inner speech and psychopathological variables in a sample of young adults. *Consciousness and Cognition, 20*(4), 1586–1593.
Mencken, H. L. (1982). *A Mencken chrestomathy*. Vintage.
Metzinger, T. (2015). M-autonomy. *Journal of Consciousness Studies, 22*, 270–302.
Metzinger, T. K. (2013). The myth of cognitive agency: Subpersonal thinking as a cyclically recurring loss of mental autonomy. *Frontiers in Psychology, 4*, 931.
Michel, J. B., Shen, Y. K., Aiden, A. P., Veres, A., Gray, M. K., Pickett, J. P., ... & Pinker, S. (2011). Quantitative analysis of culture using millions of digitized books. *Science, 331*(6014), 176–182.
Miyawaki, Y., Uchida, H., Yamashita, O., Sato, M. A., Morito, Y., Tanabe, H. C., et al. (2008). Visual image reconstruction from human brain activity using a combination of multiscale local image decoders. *Neuron, 60*, 915–929.
Nagel, T. (1998). Concealment and exposure. *Philosophy & Public Affairs, 27*(1), 3–30.
Naselaris, T., Prenger, R. J., Kay, K. N., Oliver, M., & Gallant, J. L. (2009). Bayesian reconstruction of natural images from human brain activity. *Neuron, 63*, 902–915.
Nishimoto, S., Vu, A. T., Naselaris, T., Benjamini, Y., Yu, B., & Gallant, J. L. (2011). Reconstructing visual experiences from brain activity evoked by natural movies. *Current Biology, 21*, 1641–1646.
Nowak, M. (1993). *UN Covenant on Civil and Political Rights: CCPR Commentary*. N. P. Engel.
Oota, S. R., Manwani, N., Bapi, R. S. (2018). fMRI semantic category decoding using linguistic encoding of word embeddings. In L. Cheng, A. Leung, & S. Ozawa (Eds.), *Neural Information Processing. ICONIP 2018*. Lecture Notes in Computer Science (Vol. 113030). Springer.
Opotow, S. (2007). Moral exclusion and torture: The ticking bomb scenario and the slippery ethical slope. *Peace & Conflict, 13*(4), 457–461.
Pereira, F., Lou, B., Pritchett, B., Ritter, S., Gershman, S. J., Kanwisher, N., et al. (2018). Toward a universal decoder of linguistic meaning from brain activation. *Nature Communications., 9*, 963.
Poldrack, R. A., Halchenko, Y. O., & Hanson, S. J. (2009). Decoding the large-scale structure of brain function by classifying mental states across individuals. *Psychological Science, 20*(11), 1364–1372.
Purdon, C., & Clark, D. A. (1993). Obsessive intrusive thoughts in nonclinical subjects. Part I. Content and relation with depressive, anxious and obsessional symptoms. *Behaviour Research and Therapy, 31*(8), 713–720.

Rachman, S., & de Silva, P. (1978). Abnormal and normal obsessions. *Behaviour Research and Therapy, 16*(4), 233–248.
Rentfrow, P. J., & Gosling, S. D. (2003). The do re mi's of everyday life: The structure and personality correlates of music preferences. *Journal of Personality and Social Psychology, 84*, 1236–1256.
Ross, A. (1937). https://www.nytimes.com/1937/02/14/archives/goebbels-edits-the-popular-mind-in-germany-thought-control-now.html
Schein, E. H. (1960). *Brainwashing*. Massachusetts Institute of Technology.
Shinkareva, S. V., Mason, R. A., Malave, V. L., Wang, W., Mitchell, T. M., & Just, M. A. (2008). Using fMRI brain activation to identify cognitive states associated with perception of tools and dwellings. *PLoS One, 3*(1), e1394.
Smith, K. (2013). Brain decoding: Reading minds. *Nature News, 502*, 428–430.
Stanley v. Georgia. (1969). 394 U.S. 557, 89 S. Ct. 1243, 22 L. Ed. 2d 542.
State v. Kemp, 753 N.E.2d 47 (Ind. Ct. App. 2001).
Szostak, R. (2012). The interdisciplinary research process. In A. F. Repko, W. H. Newell, & R. Szostak (Eds.), *Case studies in interdisciplinary research* (pp. 3–20)
Taylor, K. (2006). *Brainwashing: The science of thought control*. Oxford University Press.
United States v. Schwimmer. (1929). 279 U.S. 644, 49 S. Ct. 448, 73 L. Ed. 889.
Vermeulen, B. (2006). Freedom of thought, conscience and religion (article 9). In P. van Dijk, F. van Hoof, A. van Rijn, & L. Zwaak (Eds.), *Theory and practice of the European convention on human rights* (4th ed., pp. 751–772). Intersentia Press.
Wang, Y., & Kosinski, M. (2018). Deep neural networks are more accurate than humans at detecting sexual orientation from facial images. *Journal of Personality and Social Psychology, 114*, 246–257.
Ward, T., & Stewart, C. (2008). Putting human rights into practice with people with an intellectual disability. *Journal of Developmental and Physical Disabilities, 20*(3), 297–311.
Wittgenstein, L. (1953). *Philosophical investigation* (G. E. M. Anscombe, Trans.). Blackwell.
Wooley v. Maynard, 430 U.S. 705, 97 S. Ct. 1428, 51 L. Ed. 2d 752 (1977).
Zgoba, K. M., Jennings, W. G., & Salerno, L. M. (2018). Megan's Law 20 years later: An empirical analysis and policy review. *Criminal Justice and Behavior, 45*(7), 1028–1046.
Zuboff, S. (2019). *The age of surveillance capitalism*. Profile Books.

CHAPTER 3

Freedom of Thought as an International Human Right: Elements of a Theory of a Living Right

Jan Christoph Bublitz

Introduction

Only few political and philosophical notions match the grandeur of freedom of thought. With roots reaching at least to Roman times, it is perhaps *the* slogan of the Enlightenment; *sapere aude*, in Kant's famous phrase, having the courage to think for oneself rather than blindly believing authorities. Intimately related to freedom of speech, freedom of thought paves the way for liberal legal orders and the scientific method, for democracy and the disenchantment of the world. It thereby profoundly altered the *conditio humana*. In this sense, freedom of thought lies at the ground of modern societies.

J. C. Bublitz (✉)
University of Hamburg, Hamburg, Germany
e-mail: christoph.bublitz@uni-hamburg.de

M. J. Blitz et al. (eds.), *The Law and Ethics of Freedom of Thought, Volume 1*, Palgrave Studies in Law, Neuroscience, and Human Behavior,
https://doi.org/10.1007/978-3-030-84494-3_3

Freedom of thought is also one of the core human rights. Since its adoption in the Universal Declaration of Human Rights (UDHR, henceforth the "Declaration") in 1948 and the Covenant on Civil Political Rights (CCPR, "Covenant") in 1966, it has been reiterated in most major instruments. However, content and meaning of the right are not well-defined. Neither case law, nor substantive commentary from Committees, Councils, or Rapporteurs elaborate upon its scope and limits; not even legal scholarship devotes much attention to it. In fact, a single case at the international level in which it was the decisive issue is hard to locate—freedom of thought might well be the only human right without real application (Bublitz, 2014). This may surprise as interferences with it are well conceivable. For instance, the second half of the twentieth century saw bitter political ideological struggles over "men's minds." With the dawning of modern psychology in the early twentieth century, thoughts and thinking became the objects of systematic scientific study as well as targets of manifold attempts to modify them. Especially the shaping the public opinion has been at the fore since the days of Lippmann (2007) and Bernays (2005) a century ago. Today, entire subfields of psychology, psychiatry, and neuroscience seek ways to influence and alter how people think and act, and so do multibillion non-medical fields such as marketing. Human rights law acknowledges dangers to freedom of thought posed by severe practices associated with the former—"brainwashing," "reeducation," and "indoctrination"—but remains largely silent about the latter, even though—or perhaps because—many people are exposed to such stimuli on a daily basis, governments may resort to such means in a variety of contexts, and they stand in a latent tension with the idea of democracy (Paulo & Bublitz, 2016). While many of such influences may not rise to the level of seriousness of a violation of Articles 18 UDHR and CCPR, some well do. Identifying them requires a firmer understanding of the rights that oppose such influences. Freedom of thought is one of them. The time is ripe for a renaissance of the right in light of various challenges posed by psychology, psychiatry, and neuroscience today and in the near future.

Turning Articles 18 UDHR and CCPR into living rights requires a theory of freedom of thought. This chapter provides some material and several suggestions. To begin, different conceptions of freedom of thought are disambiguated and an overview of the norms in international covenants and treaties is presented, the focus of the chapter lies in Articles 18 UDHR and CCPR. Five explananda that every theory of the right to freedom of thought must address are suggested. The second section lays out what is known and unknown about the right and points to several underexplored yet foundational problems: What do "freedom" and "thought" mean in the context of Art. 18, how do they relate to "belief," what interferes with the right? It discusses three relevant cases before the UN Human Rights Committee (HR Committee) and the European Court of Human Rights (ECtHR), as well as the meaning of "coercion" in Art. 18.2 CCPR. The third section submits suggestions for the construction of the right. It should protect *thoughts* and *thinking* against the imposition of duties over and punishment for thought, interferences with thought as well as revelations of thought. This should include, freedom of belief, widely understood, as a subform to which special rules apply. Elements for a taxonomy identifying impermissible inferences and a rough test for interferences with the right are suggested. Furthermore, tensions between different conceptions of freedom of thought can arise; some narrow exceptions to the categorical ban of interferences are suggested. The chapter concludes with reflections on the absolute nature of the right.

Meaning of the Right

Many Freedoms of Thought

At the outset, it is worth noting that several conceptions of freedom of thought need to be kept apart. In grand political proclamations and historical writings, freedom of thought often denotes, broadly and loosely, societal conditions conducive to the flourishing of free thinking, e.g., an open climate for discourse and exchange, freedom of speech and press, a marketplace of ideas that includes and tolerates diverse views; *freedom is the freedom of those who think differently*, in the words of the German socialist Rosa Luxemburg.[1] This broad sense of freedom of thought is alluded to when the European Union awards the Sakharov Prize for Freedom of Thought to human rights activists. By committing to freedom of thought, states may incur political and moral obligations to facilitate

[1] Similarly, Justice Holmes writes "if there is any principle of the Constitution that more imperatively calls for attachment than any other it is the principle of free thought—not free thought for those who agree with us but freedom for the thought that we hate" (*United States v. Schwimmer*, at 655, dissenting). For the history of the broad idea see Bury (1947), with respect to freedom of expression Waclawczyk (2019).

such societal and institutional conditions. However, these commitments do not straightforwardly translate into specific and operational legal claims of individuals that courts can apply and governments must observe.[2] Most western states can, by and large, claim to embrace this broader idea of freedom of thought. Nonetheless, they may regularly violate the more specific *legal right* of individuals. Therefore, when speaking about freedom of thought, one has to be precise as to whether one refers to a larger political-societal idea, to a moral or natural right, or to a distinct legal right. In the latter case, is a technical concept with peculiar features, embedded in, and constrained by several legal frameworks. As their parameters prefigure constructions, the right can only be interpreted in the context of a specific legal order. The broader political and philosophical ideas surrounding freedom of thought may become relevant within these confines as material inspiring and influencing interpretations.

Turning Articles 18 UDHR and CCPR into a living right not only requires transforming the political-philosophical idea into a legal right, but also transforming the abstract and general human right to the level of individual cases. This move from the universal to the particular is not a straightforward application of a right to a case since it presupposes intermediate interpretative steps. Courts can render rights more precise when deciding a concrete case, but they seem hesitant doing so, likely because the meaning of the right is too ambiguous or multi-layered. It should also be noted that lawmakers could (and possibly should) render the idea of freedom of thought more precise by domain-specific legislation. Thus, it is often impossible to deduce from first principles how a human or constitutional right applies to a particular case, as many context-specific considerations may come in. The main task for a theory of the right is sketching these transformative steps and putting their inherent normative considerations to discussion. This is the aim of the following.[3]

[2] And possibly other actors; the question of applicability of Art. 18 CCPR in the horizontal relation between citizens, either directly or indirectly via positive obligations of the state to protect freedom of thought against interferences by private actors, is left aside in this chapter.

[3] This may allow a remark on some recent suggestions invoking freedom of thought with respect to worries over influences on thoughts and opinions through online advertisement or breaches of data protection laws. Without doubt, some of those practice may violate the right to freedom of thought; however, many may not rise to the level of seriousness of a human rights violation and are better addressed by norms of ordinary positive law. Again, the broad idea of freedom of thought is implicated, but not necessarily the

The Landscape of Norms

To complicate matters, there are several legal rights to freedom of thought, often enumerated or implied in domestic Constitutions. In the United States, it has been argued that a right to freedom of thought might be implied by the First Amendment to the Constitution (Blitz, this volume). It would then inherit some of its features and hence not be an absolute, unconditionally protected right. The Preamble to the Indian Constitution proclaims ensuring "liberty of thought" as one of the aims of the Constitution. In German Constitutional law, freedom of thought is considered as implied in the right to human dignity pursuant to Art. 1.1 Basic Law (BVerfG 1989, at 40, dissenting opinion). It is thus an absolute right, sometimes not even waivable by rightholders. These examples demonstrate that rights to freedom of thought may differ in relevant nuances; arguments from one jurisdiction might not generalize to others.

The present interest lies in the right in international human rights law. Even there, freedom of thought is codified in several norms, as it is enshrined in the core international and most regional human rights treaties. Its *urform* is Art. 18 UDHR:

> *Everyone has the right to freedom of thought, conscience and religion; this right includes freedom to change his religion or belief, and freedom, either alone or in community with others and in public or private, to manifest his religion or belief in teaching, practice, worship and observance.*

The first aspect to note is that Art. 18 UDHR protects freedom of thought alongside its sisters, freedom of conscience and religion. They are often referred to in the singular, as one right or freedom. Although they are interconnected and overlapping, it is suggested to consider them as distinct freedoms as scopes and possible interferences might vary. Art. 9.1 of the European Convention on Human Rights (ECHR), Art. 10 of the European Charter of Fundamental Rights and Freedoms (ECFR),

human right. Inflating concerns is not conducive to the development of a persuasive and coherent human rights framework that, because of its nature, can only cover substantive and sufficiently clear cases. Nonetheless, given the lack of case-law, such scenarios may have heuristic value as they exemplify interferences.

Art. 22 of the Asean Human Rights Declaration (AHRD) more or less echo the wording of Art. 18 UDHR.[4]

Art. 18 CCPR slightly differs. Of its four paragraphs, the first two are relevant for present purposes:

> Art. 18.1: *Everyone shall have the right to freedom of thought, conscience and religion. This right shall include freedom to have or to adopt a religion or belief of his choice [...]*

The first paragraph is largely repetitive of Art. 18 UDHR. Apart from stylistic matters, the freedom "to have or adopt" replaced the freedom to "change" religion or belief. The main difference is the addition of a second paragraph outlawing coercion:

> Art. 18.2: *No one shall be subject to coercion which would impair his freedom to have or to adopt a religion or belief of his choice. [...]*

This formulation is mirrored, e.g., by Art. 1 of the Declaration on the Elimination of All Forms of Intolerance and Discrimination Based on Religion or Belief (1981). The fact that Art. 18.2 CCPR specifies interferences raises two questions: Does Art. 18 UDHR *not* ban coercion, and is coercion the only type of interference, i.e., does its explicit mentioning rule out other infringements? Both pertain to the larger question whether the scope and protection provided by Art. 18 UDHR and its regional counterparts are identical to Art. 18 CCPR. The difference in wording would allow for a difference in construction.

The American Convention on Human Rights (ACHR, adopted 1969) is a slight exception as it protects the freedoms of conscience and religion without "thought" (Art. 12). Instead, it couples freedom of thought with free speech in the "right to freedom of thought and expression" (Art. 13). Freedom of expression is a separate right in Declaration and Covenant (Articles 19). The ACHR speaks of "thought" where the latter speak of "opinion." Nonetheless, this difference does not seem to be based on substantive considerations.

[4] Two exceptions: The African Charter of Human Rights, adopted in 1981, does not enumerate freedom of thought, only freedom of conscience and religion (Art. 8). The Arab Charter on Human Rights, adopted in 2004, protects "freedom of thought, conscience and religion" but allows for restrictions provided by law (Art. 30.1).

In the interest of coherence, it is suggested to avoid incompatible constructions of these rights and consider freedom of thought as the same right across instruments. Without further specification, the following discussion refers to the original norm, Art. 18 UDHR, but it should equally apply to its regional counterparts such as Art. 9 ECHR and to Art. 18 CCPR.

Scope of the Right

Key Features

The right possesses some salient features. The first is its two-sided structure: It comprises an *internal* side of thought, conscience, and religion (sometimes referred to as the *forum internum*), as well as an *external* side of actions manifesting thoughts, religious, or conscientious beliefs (*forum externum*). The *forum internum* is a metaphorical term for (parts of) the mind or a person's "inner space". Originally developed in the context of freedom of religion and conscience, it denotes the inner connection to, and space of dialogue with God as well as the "inner court" where sins are confessed, as in today's picture of conscience. The text of Art. 18 UDHR refers to the inner side in "change religion or belief," which is understood as the espousing or rejecting faith and, more generally, the forming, holding, and discarding of beliefs and unexpressed thoughts.[5]

The external side comprises actions in the world that manifest religious or conscientious beliefs such as worship. Internal and external sides of the norm are not symmetrical. The three internal elements—thought, conscience, religion—are not mirrored at the external side, which only speaks about religion and belief. The understanding is that thoughts are manifested through expression, which is protected separately in Articles 19 UDHR and CCPR. The crucial aspect of the *forum externum* is that it privileges actions in the external world because of their inner relation to religion and belief with the effect that such behavior—with all social consequences—might be permissible whereas the same behavior might be curbed without an religious or conscientious grounding. In other words,

[5] There are a few excellent works of scholarship on freedom of religion (C. Evans 2001; M. D. Evans 1997; Lindkvist 2017; Taylor 2005) and conscience (Hammer 2002), but they deal with freedom of thought at best peripherally.

rightholders are exempted from some behavioral duties because of religion or belief, e.g., in conscientious or religious objection to military service.[6]

The second key feature of Art. 18 UDHR is that the internal side, the *forum internum*, is considered off-limits for state interventions, the protection is *unconditional* or *absolute*, whereas external manifestations can be restricted for various purposes according to limitation clauses such as Art. 18.3 CCPR.[7] Unconditional guarantees are rare in international human rights law. Even more, the right is also non-derogable pursuant to Art. 4.2 CCPR, which means that it cannot be restricted even in times of public emergencies threatening the life of the nation (affirmed by the HR Committee, 2001, at 7). The inner side enjoys an extraordinary level of protection; it takes priority over virtually all other interests of individuals or society, legitimate and pressing as they might be. This underlines the importance of the right, but also calls for a well-grounded justification. A traditional argument is that the internal side is not of direct relevance to social life, the regulation of which is the main rationale of the law. The line between *internum* and *externum* is thus the line between the private sphere of the individual outside of governmental regulation and the social sphere.

So much for a first impression of the right. It shows five key characteristics that a theory of freedom of thought has to explain and possibly justify: the meaning of freedom of thought and the scope of the right, its peculiar internal and external structure, the absolute protection of its inner side (*forum internum*), coercion and potential interference as well as the relation of thought to conscience, religion, and opinion. Although some of these *explananda* are better understood than others, there are many open questions about all of them.

Case Law and Scholarship

The significance and the exalted status of the right to freedom of thought are widely avowed. During the drafting of the Declaration, the later

[6] One may wonder why such a privilege is justified with respect to religion (Leiter 2013) or conscience (Boucher & Laborde, 2016), a question not further pursued here.

[7] This view was affirmed by the HR Committee in General Comment No. 22 ("does not permit any limitations whatsoever," at 3) and by the Special Rapporteur on Religion or Belief (2010, at 53).

Nobel-Laureate and President of the ECtHR, René Cassin, called it the "origin of all other rights" (Commission on Human Rights, 1948, 13). This stands in contrast to the lack of practical relevance of the right. Litigation on Art. 18 CCPR almost exclusively concerns the external sides of its sister freedoms of religion and conscience, i.e., the manifestation of religion or belief.[8] The few cases about the internal side primarily concern the special case of involuntary *external* actions that may interfere with the *internal* sides of conscience or opinion, e.g., military service of conscientious objectors. But how precisely such external actions affect the internum is controversial (see, e.g., judgment and opinions in *Atasoy and Sarkut v. Turkey; Kim v. Republic of Korea*). Religious conversion, proselytism, and Art. 9 ECHR are addressed below. In general, commentators diagnose—and criticize—the lack of engagement with the *forum internum* by courts.[9]

There is no relevant jurisprudence on the internal side of thought (Alegre, 2017; Bublitz, 2014; Loucaides, 2012; O'Callaghan & Shiner, 2021; Schabas, 2016), with few exceptions discussed in a moment. Apart from them, freedom of thought is largely a dead letter.

The scholarly literature provides rough sketches of the right. In his commentary on the CCPR, Nowak describes it as the right "to develop autonomously thoughts and a conscience free from impermissible external influence" and notes that delineations between permissible and impermissible influences are not easy (Nowak, 2005, 412). With respect to the European Convention, the right is summarized as the guarantee that "the state may never interfere in this most intimate and inner sphere, for instance, by dictating what a person has to believe, by taking coercive steps to make him change his beliefs [...] or by using inquisitorial methods" to discover thoughts (Vermeulen & Roosmalen, 2018, 738). The former judge of the ECtHR, Loucaides, notes that "very little has been written about this freedom and there is not much substantive discussion

[8] For an overview of most relevant matters, see Joseph and Castan (2013) and the summary by the Special Rapporteur on Religion or Belief (2017). Freedom of thought, as a distinct area of protection, has neither been addressed in the Rapporteur's annual reports to the Human Rights Committee or to the General Assembly of the last decade, nor in the Rapporteur's Digest (2011) on the years 1986–2011. The right will be addressed for the first time in the 76th report to the General Assembly in 2021.

[9] E.g., Taylor (2005, 202) "the fundamental nature of the *forum internum* has been undermined by European institutions through persistent avoidance of principles that permit the *forum internum* rights to be asserted".

about it in the case-law of judicial organs including the European Court of Human Rights" (Loucaides, 2012, 80). These remarks largely restate the norm. The intriguing problems emerge when it is rendered more concrete, which requires disassembling and reconstructing its elements.

Elements

Thought and Belief: An Inconsistency

Basic questions about the right are not settled: Already the object of protection, "thought", is ambiguous. Does it refer to a mental faculty (reason), to mental activities (thinking), to the contents of occurrent mental states (thoughts), or to the entirety of subjective experience—and does "thought" comprise all mental states or only some, e.g., rational ones, and what about affective states? For clarity in the following discussion, I wish to suggest already at this stage that the scope should comprise thoughts as mental states and thinking as a mental action. It is helpful to consider both when addressing specific questions.

With respect to protected thoughts, the HR Committee provides some guidance in General Comment No. 22. It writes that Art 18.1 CCPR is "far-reaching and profound; it encompasses freedom of thought on all matters" (at 1). In a similar vein, the European Commission commented with respect to the name parents wish to give to their child that "taking into consideration the comprehensiveness of the concept of thought, this wish can be deemed as a thought in the sense of Article 9" (*Salonen v. Finland*, p. 3). These remarks favor a wide understanding of "thought."

Another central element in Articles 18 UDHR and CCPR is *belief*; the right is often referred to as the freedom of religion or belief. But the relation between "thought" and "belief" is rarely explicated. In ordinary language, believing roughly means taking a proposition to be true. If a person believes X, she thinks that X is the case (or is likely the case). Philosophers view beliefs as favorable attitudes toward a proposition (Schwitzgebel, 2019). Beliefs are also the elements of knowledge, which is often defined as justified true beliefs. Beliefs are *occurrent* when a person is consciously entertaining them or *dispositional* when she could do so. The former beliefs, and possibly the latter, seem to be prime examples of thought (e.g., having the belief "Covid is more than a mere flue" is a belief and a thought).

However, in the context of Art. 18, it is widely assumed that "belief" has a narrower technical meaning akin to conviction, as in the authoritative French version of the Declaration. It comprises only significant personal beliefs such as those experienced as binding dictates of consciousness or those that relate to wider belief systems one adheres to, such as atheism, feminism, or socialism. The rationale behind this narrower view of "belief" is that not every action related to mundane beliefs should be privileged by Art. 18. This privilege, after all, means setbacks to rights of others and public interests as it exempts rightholders from general duties. It is thus only justified with respect to serious and significant beliefs. The ECtHR adopted this narrow view in its jurisprudence; a belief must "attain a certain level of cogency, seriousness, cohesion and importance" (*Eweida v. United Kingdom*, 2013, at 81).

This narrow understanding of belief as conviction is not consistent with the wide understanding of thought "on all matters." All beliefs, cogent or trivial, are thoughts. This creates a *thought-belief inconsistency* in Articles 18 UDHR and CCPR and regional counterparts. There are three solutions to it. The first is considering belief as *lex specialis*, so that the right covers thoughts on all matters, but not belief on all matters. This would create an oddly fragmented scope comprising thoughts on all matters as long as they are not beliefs. But beliefs, in the wide ordinary sense, are surely among the most important thoughts. Another solution is understanding the entire right to freedom of thought narrowly as a right to freedom of conviction, only pertaining to important beliefs. This would curtail the scope of Art. 18 considerably, leaving little scope for "thought" independent from "conscience." The text also speaks against this approach as "belief" is listed as a non-exhaustive example ("includes"). Furthermore, there is no indication that such a narrow understanding was intended by drafters or courts. Moreover, the Declaration aspires to be a document understandable to ordinary people, which suggests interpreting "thought" as what is ordinarily considered as such, a type of mental state and a bundle of mental activities—thinking.

The third and preferred way to solve the inconsistency is by drawing a distinction between the *internum* and the *externum*. The reason motivating the narrow understanding of belief is that it privileges actions in the external world and that this privilege must remain exceptional. But this rationale does not apply to thoughts as internal mental states; privileging them does not cause direct setbacks to others. Accordingly, "thought" should be understood widely with regard to the internal side and encompass beliefs "on all matters," whereas it should be construed narrowly with regard to the external side. This interpretation harmonizes the views of the HR Committee, the ECtHR, and scholarship. It seems to be the best textual and teleological interpretation.

Freedom of Thought

"Freedom" of thought is equally ambiguous. Subtly diverging meanings pull into different directions and may affect the core understanding of the right and potential interferences. The first question is whether "freedom" refers to a normative property—to a liberty in a legal-technical sense—or to a descriptive or factual property of thought—free as opposed to unfree or involuntary. Common sayings such as "thoughts are free, no one can touch or know them" refer to factual properties, the physical untouchability and perceptive inaccessibility of thoughts. The wording of Art. 18.2 CCPR seems to do likewise since "coercion" cannot "impair" the normative, but only the *factual* freedom to have or adopt a belief. In addition, the formulation a "right to freedom of thought" seems to refer to a factual property, since a right *to* a legal liberty appears tautological—a liberty is part of a bundle of positions called a right.[10] These aspects suggest that "freedom" in Art. 18 refers to a descriptive or factual property of thought.[11] But what might this property be—when is thought free?

Several understandings are possible. It could mean, in analogy to free will, *indeterminate* thought, in the sense that an occurring thought was neither fully determined by preceding thoughts and psychological states, nor by the underlying physiological and psychological mechanisms. It could also mean, in analogy to the Principle of Alternate Possibilities (Frankfurt, 1969), that thought is free if a thinker could have thought differently, *ceteris paribus*. These are interesting but also demanding conceptions; it is not evident that freedom of thought in these forms exists at all.

A weaker, but by no means undemanding understanding considers free as *voluntary* thought, in analogy to free actions. This may seem attractive. Free thinking is then the voluntarily controlled performance

[10] According to the standard model based on Hohfeld (1913), *supra*.

[11] The idea of a normative liberty will be taken up *infra*.

of various mental actions that qualify as thinking. However, it is important to note that a large share of thoughts is not under voluntary control, they come and go unbidden in the stream of consciousness. Our minds constantly wander; keeping thoughts focused and ordered is effortful and often short-lived (for a revealing view at the limits of control people have over minds see Metzinger [2015]).

If the scope of the right was limited to free thoughts in these senses, it would be narrow and not provide protection against interferences with other, "non-free" thoughts, e.g., those over which people lack voluntary control. Such a narrow scope runs counter to the rationale of the protection, which seems necessary especially with respect to aspects over which people lack control. Voluntary control over thought should be protected where it exists, but the scope of the right should not be limited to it.

Alternatively, freedom of thought could be understood as free thought or freethinking, summary terms for modes of thought and reasoning that are committed to rational standards and the search for truth, historically associated with the freethinker movement. The Nobel Laureate Bertrand Russell describes the hallmark of free thought as "freedom from the force of tradition and the tyranny of one's own passions; free thought does not mean not absolute freedom, but thought within the intellectual law" (Russell, 1957, p. 45). In other words, free thought is critical thinking, open-minded and open-ended reasoning that neither accepts externally prescribed results, ideologies, dogmatism, nor distortions of thought from cognitive distortions, biases, or emotions. As "free" here primarily means rational, this view is henceforth called the *rationalist conception* of freedom of thought.

Is this conception appropriate in the context of Art. 18? It would narrow the scope to specific classes of thought, thinking, and rational reasoning, and it would exclude non-rationalist forms. Some of the latter, however, appear as prime candidates for protection by Art. 18, e.g., artistic, associative, imaginative, or non-linear forms of thought "out of the box." These modes of thinking should not be excluded from the scope *ab initio*. Moreover, the inclusive spirit of the remarks by the HR Committee may likely not only refer to the content of thoughts "on all matters" but also to the type of thinking "in all forms".

However, the grand political-philosophical concept may have something to contribute to legal interpretation here. Free thought and reason according to the rationalist conception was an idea integral to the Enlightenment, the Age of Reason. It was the inspiration for adopting a right to freedom of thought and may therefore shape its interpretation. To Nowak, the right demonstrates that the Covenant "is based on the philosophical assumption that the individual as a rational being is master of his or her own destiny" (Nowak, 2005, 408). The rationalist conception should thus be considered a central category of the right, even though it may not exhaust its scope. We will return to this with respect to freedom of belief.

Further possible understandings of freedom of thought are sometimes explained with reference to Isaiah Berlin's influential distinction between negative and positive liberties—freedoms *from* and *to* (Berlin, 1969). In legal contexts, this distinction invites misunderstandings because in the law, "liberty" is a technical term and positive dimensions of a right refer to claims of rightholders against others to the performance of an action (correlatively, "positive obligations" denote duties to act), whereas in Berlin's usage, positive freedoms refer to capacities of self-mastery. It is thus helpful to speak of freedom from interferences with thought and of freedom to think in the sense of thinkers having, controlling, and exercising capacities for thought. The former corresponds to rights to non-interference, the standard legal way of understanding freedoms (Nowak, 2005). It suggests a broad scope that protects all kinds of thoughts against external interferences. The positive understanding, by contrast, protects the freedom to think, the performance of diverse mental actions which qualify as thinking, and arguably also the mental capacities and powers underlying and enabling them.

Thus, freedom of thought can be understood differently—as specific forms of reasoning, voluntary control, freedom from interferences, or capacities—and the subtly different conceptions may shape the scope of the right, interferences, and (intuitive) evaluation of cases. These conceptions of freedom allow for graduations: A person can have more or less cognitive abilities, an interference can be more or less invasive or effective. (This has the odd consequence that thought might be free to different degrees.)

Freedom of Belief

Articles 18 UDHR and CCPR are often also referred to as freedom of belief. As suggested earlier, the meaning of belief is ambiguous. With respect to the external side, it has to be construed narrowly in the sense of conviction. But with respect to the internal side, it should be construed widely, as a special kind of thought. Freedom of belief is thus not a homogeneous concept. Furthermore, beliefs possess some peculiar features that require additional remarks. As said, beliefs are affirmative attitudes toward a proposition; believing X means taking X to be true or correct. Moreover, beliefs can refer to different matters, in the context of Art. 18 to three: religious beliefs, conscientious beliefs, and beliefs about facts of the world. These beliefs differ in some respects, e.g., whether they can be true or false, correct or incorrect, and the respective standards for assessing this. Religious beliefs, for instance, are matters of faith precisely because they can be neither proven nor disproven. Conscientious beliefs have a peculiar standard of correctness, correspondence to an inner experience or a "voice of conscience." These aspects do not apply to ordinary beliefs about the world, which are truth-apt, i.e., they can be true or false. These are of interest in the following.

Importantly, forming, holding, or discarding beliefs is, to a large extent, a *non-voluntary* exercise. This is evident with respect to religious or conscientious beliefs which are sometimes defined as binding dictates of conscience—*here I stand, I can no other*. But notably, the same is true, *mutatis mutandis*, for ordinary beliefs. Usually people cannot choose at will what they take to be true, believing requires supporting reasons, evidence, and consistency with other beliefs. It is psychologically impossible to consider a random proposition to be true in the absence of or even against evidence. Whenever one tries to form a belief, one searches for evidence supporting or refuting it. In this sense, people lack voluntary control over belief formation (so-called *doxastic involuntarism*). Rather, the cognitive system seems to form and revise beliefs largely automatically, non-consciously, and without voluntary control in response to experiences in the world. Therefore, people have all sorts of belief without having consciously formed them.

The rules by which beliefs are formed are not transparent to believers. By contrast, there are rules by which beliefs *should* be formed, rules of rational belief formation or *epistemic rationality*. Its standard is the truth or correctness of beliefs. Controversial in detail (e.g., Bondy, 2018), rules of epistemic rationality demand, among others, that beliefs are adjusted to the strength of available evidence and are revised if necessary. Psychology and life-experience shows that belief-forming mechanisms are susceptible to a range of factors that do not observe epistemic rationality, such as one-sided reasoning, biases, and rationalizations.

However, thinkers have some indirect forms of control, such as selectively attending to pieces of evidence, encouraging or stifling doubts. In particular, they can call their beliefs into question and scrutinize them from various perspectives. Such powers exist, but they are limited. They not only require cognitive resources, but they are also constrained by features of the belief-forming mechanisms; they do not confer thinkers control over the belief, but trigger an internal belief revision program. Thereby, they provide some indirect influence over one's belief formation.

What does this mean for Art. 18? Well, it raises the question what *freedom* of belief refers to. Strictly speaking, adopting a belief of one's choice—as guaranteed by Art. 18 CCPR—is often *impossible*. People neither freely choose their convictions, nor their beliefs about the world. Thoughts can be commanded, but beliefs cannot. This insight should motivate a wider and less literal understanding of the provision, and it underlines why special consideration of freedom of belief, in addition to freedom of thought, may often be necessary. Moreover, freedom of belief may mean the absence of interference with the belief-forming system, or the capacity to rational belief formation (i.e., the rationalist conception applied to beliefs). Both may lead to different scopes (a point we will return to).

Interferences

Another relevant element are interferences with the right. The literature refers to a few drastic examples: brainwashing (whatever it means precisely), indoctrination, reeducation camps (Nowak, 2005, 413). This confers the impressions that interferences necessitate severe and powerful measures, less severe means appear insufficient. Such a restrictive view, however, is not self-evident as the converse is at least equally plausible: A great many actions seek to change other peoples' thoughts and beliefs, and often succeed doing so, from persuasion in written communication over psychological pressure to coercive administration of thought-altering drugs. Such actions (henceforth "interventions") are ubiquitous, but that does not place them beyond concern. Accordingly, a different perspective is suggested: Rather than conceiving of thought and thinking as largely invincible, only intrudable by powerful means, the malleability and vulnerability of human thought as well as its in-principle openness to external influence should be acknowledged. People change each other's minds all the time on a myriad of ways. The challenge lies in separating permissible from impermissible interventions. This requires developing normative criteria which should be put to discussion, to be refined and defended. This normative groundwork is still largely outstanding (see discussions in Bielefeldt et al. [2016] and Bublitz [2020a]).

A crucial aspect is that some interventions are themselves protected by rights of intervenors, e.g., as exercises of freedom of speech and expression (Articles 19 UDHR and CCPR). This creates a tension between the right to send potentially mind-altering stimuli to others—free expression—and the right to remain free from such stimuli—freedom of thought. This tension is underappreciated in the scholarly literature, but it is important as it sets limits to freedom of expression (e.g., as rights of others pursuant to Art. 19.3 CCPR). Conflicts of rights are common features of legal orders that are usually resolved by methods of balancing or reconciliation. The peculiar problem in the present case is that the absolute nature of Art. 18 does not allow them since interferences cannot be justified; every action that interferes with freedom of thought *eo ipso* violates the right. The balancing stage in which adequate and context-specific solutions can be found is unavailable. This has the unintended and methodologically questionable, but practically inevitable consequence that such considerations affect the definition of interferences. The alternative, not accommodating potential rights of intervenors, would lead to absurd outcomes.

Art. 18.2 CCPR speaks of "coercion"—unlike Art. 18 UDHR and regional counterparts such as Art. 9 ECHR. This might be read as a specification of potential interferences, which raises the question what coercion means in this context, whether it is the only possible type of interference, and whether the scopes of the rights vary across documents. Coercion is a complex concept that roughly means to get a person to perform an action against her will through the use of force or unlawful threats. As the HR Committee explains in General Comment 22, coercion includes "the use of threat of physical force or penal sanctions to compel believers" to maintain or recant their beliefs (1993, at 5). So much is settled. The problem is that coercion of belief, in this strict sense, is often not possible, given that people are frequently impotent to change their belief at will (*supra*). Even at gunpoint, one cannot get oneself to believe that the Earth is flat. If coercion were the only modality to interfere with Art. 18 CCPR, it has a narrow scope of application. However, this narrow interpretation seems to miss the point of the guarantee of Art. 18.2 CCPR. It is primarily not a norm against coercion, but for the protection of beliefs. This suggests that coercion might not be the only form of interference. With this in mind, let us look at three leading cases regarding Art. 18 CCPR and Art. 9 ECHR.

***Kang v. Korea*—Coercion**

One of the few cases explicitly addressing the right to freedom of thought under the CCPR is *Kang v. Korea*. The complainant was held in solitary confinement for 13 years for terrorist charges and on the allegation (which he rejected) of being a communist. He was detained in a prison which ran an "ideology conversion system." Benefits, including parole, were offered if he renounced his beliefs and took a "law-abiding" oath. The Human Rights Committee recognized the "coercive nature of such a system [...] applied in discriminatory fashion with a view to alter the political opinion of an inmate by offering inducements of preferential treatment within prison and improved possibilities of parole" (at 7.2.). Consequently, it found a violation of Art. 18.1. and Art. 19.1. CCPR, in conjunction with Art. 26 CCPR (non-discrimination on political grounds).

Presumably, 13 years of solitary confinement violate human rights *per se*. But how does this treatment interfere with freedom of thought or belief more precisely, and does it amount to coercion? The facts of the case are not entirely clear as to whether Kang was punished for holding a belief—a clear violation of freedom of thought. The communication by the HR Committee rather speaks about "offering preferential treatment" and withholding of a benefit (release). Whether offering benefits or preferential treatment can constitute an unlawful threat is controversial ("coercive offers"). But let us suppose that it is in the context of Art. 18.2 CCPR. How then does the offer affect freedom of thought?

It might seem that the offer does not undermine the freedoms set out above because it weakens neither thoughts nor thinking. The complainant may be motivated to profess a belief he does not hold (renouncing communism). This interferes with the *forum externum*, it coerces an (unwanted) expression, but it does not hinder the complainant to continue to believe in communism. Nonetheless, coercing someone to profess a belief is sometimes said to interfere with the *forum internum* (as an instance of an "indirect interferences").[12] Why could this be the case?

[12] Indirect interferences with the *forum internum* are not further analyzed here as the category is vague and tailored to religious and conscientious beliefs. The most salient case is mandatory military service for conscientious objectors. Does it interfere with the *external* manifestation of conscience—and hence be justifiable under specific conditions,

One way in which forceful expressions are problematic is that they lead to wrongful confessions. Given the history of the Inquisition and attempts to elicit false confessions, all attempts to obtain them should be banned in-principle. But the extraction of a confession is not at stake here. A different argument may point to the psychological harm such professions may cause (Bielefeldt et al., 2016, 80).

Another line holds that coerced expressions indirectly harm the thinker (Shiffrin, 2011; but also see Mawhinney, 2016). Drawing on the foregoing remarks about freedom of belief, here is a variation of this thought: The main point of concern about coercion in light of Articles 18 UDHR and CCPR is that it creates an inner conflict, the temptation to not only profess a belief, but to truly change beliefs, without evidence to do so. Although changing beliefs may not be possible at will (*supra*), there are indirect routes and psychological mechanisms that may cause belief changes. These mechanisms can be triggered by the psychologically burdening situation that creates pressures to alter beliefs in exchange for the satisfaction of other psychological needs—unmet, in this case, because of the long solitary confinement. In other words, the offer exploits a vulnerability to change beliefs for inadequate reasons, which means, roughly, against rational and personal standards. Psychological needs are no good reasons for changing a belief (provided they are unrelated to its content). Of course, renouncing communism may well be practically rational for a person in such a situation as it advances her overall interests. But it is not from the perspective of epistemic rationality. The offer strives to have the person abandon her own judgment and accept authority instead, without adducing reasons for the correctness of the belief—the opposite of freedom of thought.

Accordingly, the interference with freedom of thought lies in the creation and exploitation of psychological weaknesses which may move persons to (non-consciously) form beliefs on inadequate (non-rational) ways. This is worth noting as it is not a case of coercion in the classic sense, but rather a form of *psychological manipulation*.

Art. 18.3 CCPR—or does it interfere with the conscientious beliefs themselves? Under some conditions, it might be the latter as contributing to killing may cause grave inner turmoil and pangs of conscience. For the latter, see *Kim v. Korea*; and the concurring opinion of Kälin (fearing that a wide understanding of indirect interferences dilutes and jeopardizes "the very core meaning of conscience, namely that the *forum internum* must be protected absolutely"). For the former (*forum externum*), *Bayatyan v. Armenia* (ECtHR).

Kokkinakis and *Larissis v. Greece*—Proselytism

A second example concerns two leading cases on proselytism before the ECtHR. Although they address interferences with freedom religion, they are material to the present inquiry. The applicant in *Kokkinakis v. Greece*, a Jehovah Witness, was repeatedly convicted for proselytism. To protect freedom of belief, Greek law penalized proselytism, defined as "any direct or indirect attempt to intrude on the religious beliefs of a person of a different religious persuasion with the aim of undermining those beliefs, either by any kind of inducement or promise of an inducement or moral support or material assistance, or by fraudulent means or by taking advantage of his inexperience, trust, need, low intellect or naïvety" (at 16).

In the concrete case, the applicant and his partner called at the door and "engaged in a discussion" with a resident, the wife of an Orthodox cantor. They told "her that they brought good news; by insisting in a pressing manner, they gained admittance to the house and began to read from a book on the Scripture [...], encouraging her by means of their judicious, skillful explanations" to change her beliefs (at 9). The attempt remained unsuccessful; the woman testified that "the discussion did not influence my beliefs" (at 10). Nonetheless, the applicant was convicted to several months in prison.

The ECtHR had to solve the conflict between different elements of freedom of religion, the freedom to proselytize and propagate one's religion versus the freedom of the *forum internum*. To this end, it drew a distinction between proper ("bearing witness") and improper forms of proselytism. The latter include "exerting improper pressure on people in distress or in need," as well as "the use of violence or brainwashing." By contrast, merely discussing beliefs and teachings with others is not improper. As Greek authorities failed to establish additional aggravating elements, the Court found that the conviction violated applicant's freedom of religion.

The *Kokkinakis* judgment was not unanimous. To some judges, governments may curb even such basic conversion attempts, whereas to others, the state should not intervene in such conflicts at all.[13] The decision attracted many scholarly criticisms (e.g., Evans, 2017; Taylor, 2005).

[13] The partly dissenting opinion of Judge Martens suggests that the state should not intervene in conflicts between different religions because, among others, improper spiritual conversion is difficult to establish (at 18). But that would forgo the protection of the

By and large, however, the judgment points in the right direction. The tension between protection of the *forum internum* and the right to religious practice is only solvable by separating proper and improper means of influence, and the criteria proposed by the Court, vague as they are, appear adequate. The gray areas need to be rendered more precise, but this is a context-specific task that defies simple abstract definitions (Judge Pettit, concurring; Taylor, 2005, 67; Bielefeldt et al., 2016).

A few years later, the Court upheld convictions based on the same anti-proselytism law in *Larissis v. Greece*. The applicants, superiors in the army, read the bible to subordinates and encouraged them to visit church services, so that the latter felt obliged to do so. The Court held that their special role may suffice to turn an otherwise proper conversion attempt into undue influence: "the hierarchical structures which are a feature of life in the armed forces may colour every aspect of the relations between military personnel, making it difficult for a subordinate to rebuff the approaches of an individual of superior rank or to withdraw from a conversation initiated by him. Thus, what would in the civilian world be seen as an innocuous exchange of ideas which the recipient is free to accept or reject, may, within the confines of military life, be viewed as a form of harassment or the application of undue pressure in abuse of power" (at 51).[14] While the Court's worry about undue pressure is understandable, it is worth remarking that reading the bible or encouraging church visits is hardly describable as a form of coercion, at least in the absence of threats. The Court's judgment appears nonetheless reasonable in light of the powers of social psychology and the psychological pressure such encouragements may generate.

The jurisprudence on proselytism allows for some lessons: Firstly, the two cases show that fine and context-specific lines of undue influence need to be drawn. Secondly, the incriminated measures are no forms of coercion *sensu stricto*, but rather forms of manipulation or exploitation of psychological weaknesses that may interfere with Art. 9 ECHR. This

forum internum as long as no other offenses are committed. States would fail to discharge their duty of protection. Gray areas are hardly an argument against drawing boundaries.

[14] See also Judge Valticos, partly dissenting, "any attempt going beyond a mere exchange of views and deliberately calculated to change an individual's religious opinions constitutes a deliberate and, by definition, improper act of proselytism, contrary to" Art. 9. "Attempts at 'brainwashing' may be made by flooding or drop by drop, but they are nevertheless, whatever one calls them, attempts to violate individual consciences and must be regarded as incompatible with freedom of opinion."

means, thirdly, that potential interferences are not restricted to coercion in the classic sense. Of course, the jurisprudence of the ECtHR concerns Art. 9 ECHR which does not contain a clause equivalent to Art. 18.2 CCPR specifying "coercion." But the findings are nonetheless transferable.[15] One reason is that the formulation of Art. 18.2 CCPR pertains to the often impossible adoption of a belief of one's choice (*supra*). As *Kang* demonstrates, there are equally problematic measures that should trigger Art. 18 CCPR protection.[16] Another reason is the following: The introduction of "coercion" and Art. 18.2 CCPR was a political compromise to appease worries of some (mainly Muslim) countries about religious proselytism; a worry that motivated their abstention from the Declaration (see for the Declaration, Morsink, 1999, 24; for the CCPR Nowak, 2005, 416; Taylor, 2005, 75). Art. 18.2 CCPR was meant as a clarification, making explicit what Art. 18 UDHR implicitly contained.[17] It was not meant to change the scope of, or possible interferences with, the right. On the contrary, it was supposed to strengthen and reinforce the protection of beliefs pursuant to Art. 18 UDHR and Art. 18.1 CCPR precisely against undue conversion attempt. It should thus not be read as restricting potential interferences to coercion. If Art. 18 UDHR or Art. 9 ECHR can be interfered with by non-coercive means, so should Art. 18 CCPR. Accordingly, non-coercive means such as "improper proselytism" may interfere with Art. 18 CCPR.[18]

Fourthly, one may wonder how the jurisprudence on proselytism relates to interferences with freedom of thought and non-religious beliefs. Interferences with religious beliefs are presumably not identical to those with other beliefs. What is permissible in proselytism may not be so

[15] Cf. the debates about the meaning of "coercion" during drafting in the report of the General Secretary, A/2929 at 110.

[16] The HR Committee also hints at a non-strict understanding of coercion when it writes that Art. 18 bars coercion and "[p]olicies and practices having the same intention or effect" (at 5). Furthermore, it is sometimes wondered why the HR Committee has not found a violation of Art 18.2 in *Kang* (Nowak, p. 417). The reason according to the present suggestion is that Art. 18.2 is not a separate right, it just illustrates a key part of the protection of Art. 18.1 CCPR.

[17] See the records of the meeting UN Doc. E/CN.4/SR.319; the retrospective report A/2929 at 108 et seq.; Hammer (2002, 42).

[18] Taylor (2005, 2020) might support a different view insisting on "coercion", as his criticism of the ECtHR case-law on proselytism draws on the point that actions were not coercive.

in other domains; convincing someone to vote for a political party or to buy a product by talking to them about death or existential dread is presumably impermissible. However, with the exception of context-specific considerations, interferences with these freedoms share common ground. The rough criteria established by the ECtHR for improper proselytism—violence, psychological pressure, exploiting weaknesses, influence in institutional hierarchies—also provide guidance about interferences with freedom of thought and conscience.

***Mockutė v. Lithuania*—Coercive Psychiatry**

Finally, attention is drawn to a recent case before the ECtHR, *Mockutė v. Lithuania*, which concerns the use of psycho-corrective methods to promote critical attitudes and self-reflection. The applicant was involuntarily placed in a psychiatric hospital due to an acute psychosis for a little less than two months. During hospitalization, she was forcibly administered antipsychotic medication and physically restrained, but in conformity with medical standards. The doctors suspected that her involvement in a spiritual meditation group was among the causes of her mental health problems. By contrast, the applicant experienced it—especially the meditation—as a source of inner peace. At the beginning of therapy, she showed uncritical and "categorical" attitudes toward her psychotic behavior and her situation, i.e., she did not understand her condition, a typical symptom of psychosis. The treatment aimed at moving her to develop a critical attitude toward her condition, including her spiritual group. To this end, doctors discouraged her from meditating (whether it was prohibited remains unclear) and applied "psycho-corrective methods" which are unfortunately not described in more detail. The treatment was successful insofar as the applicant developed understanding for her condition so that she agreed to further voluntary treatment post-release; but she did not change her categorical views about the meditation group. The case also concerns breaches of privacy through the dissemination of medical information and, more broadly, the restrictive stance Eastern European countries take against new religious movements. These aspects are left aside here.

With respect to psycho-corrective methods, the Court notes twice that a "State cannot dictate what a person believes or take coercive steps to make him change his beliefs" (at 119, 129). Given the circumstances of the involuntary hospitalization, it was satisfied that "pressure was exerted on her to change her religious beliefs and prevent her from manifesting them," which interferes with Art. 9 ECHR (at 123).

However, after concluding that the "interference contravened Article 9 of the Convention" the Court continues examining whether interferences were justified. It draws on a provision of the Lithuanian Constitution according to which persons possess an inviolable sphere of private life that may not be limited in any way (at 129). The Court writes that it is "prepared to accept that the needs of psychiatric treatment might necessitate discussing various matters, including religion, with a patient, when he or she is being treated by a psychiatrist. That being so, it does not transpire from Lithuanian law that such discussions might also take the form of psychiatrists prying into the patients' beliefs in order to 'correct' them when there is no clear and imminent risk that such beliefs will manifest in actions dangerous to the patient or others" (at 129). The Court therefore assumes that the treatment was not in accordance with Lithuanian law, so that the interference cannot be justified for lack of a basis in domestic law.

This reasoning is remarkable. First and foremost, the Court examines justifications although interferences with the *forum internum* are not open to them. Unfortunately, it does not explain its approach. The Court might not have considered the measures as interfering with the *forum internum*, the term is not mentioned in the judgment. However, "psycho-corrective measures" that pressure a person to change her beliefs seem, by all standards, to impinge upon the *forum internum*. After all, in the words of Art. 18.2 CCPR, they impair the freedom to have a belief of one's choice. The reasoning is also surprising because the Court dismisses the measures by invoking an inviolable sphere guaranteed by the Lithuanian Constitution—under the idea of privacy—in lieu of the inviolable sphere guaranteed by Art. 9 ECHR.

A possible explanation for this unconventional reasoning emerges in a broader perspective. As the dissenting opinion by three judges remarks, the case might be primarily seen as "a complaint about the alleged improper treatment at a psychiatric hospital, whereas the religious aspect represents only one part thereof" (at 5). Coercive psychiatric medication is notoriously controversial, and the absence of jurisprudence on it by the ECtHR and other human rights courts is suspicious. Patient movements ("anti-psychiatry") have called for the abolition of coercive practices in psychiatry for years, often invoking freedom of thought. It was also a dominant theme with respect to the Convention on Rights of Persons with Disabilities. The substantive dilemma is that some psychiatric interventions aim at changing thoughts, thought-patterns or beliefs and thus contravene the letter of the law. On the other hand, such interventions do not appear unjustifiable from the perspective of medical ethics. The Court seems to share this affirmative view when it writes: "it is for the medical authorities to decide on the therapeutic methods to be used, if necessary by force, to preserve the physical and mental health of patients who are entirely incapable of deciding for themselves" (at 124). It thus adopts a deferential attitude regarding medically necessary coercive treatments. Here, the dilemma emerges: If the Court had found a violation of the *forum internum* of Art. 9 in the present case, the legal grounds for coercive psychiatry in its entirety would have been seriously undermined. A rational court seeks to avoid precedents with supposedly undesirable and also somewhat unforeseeable consequences. Against this backdrop, the straying reasoning of the Court appears as a doctrinal sleigh-of-hand: The weight of the case is placed on a domestic provision, which is different from freedom of religion and does not have an exact counterpart in the ECHR.[19] It thereby avoids setting precedents.

Moreover, the case touches upon the intriguing question whether encouraging someone to develop a critical attitude may interfere with freedom of thought or religion. The dissenting opinion observes: "The psychiatrist obviously wanted the applicant to reflect on her own mind and behaviour, and such reflection naturally forms part of psychiatric treatment" (at 13). According to this view, undermining of belief does not per se qualify as an interference; gaining understanding of oneself or one's situation; improving abilities for self-reflection may increase freedom of thought in the rationalist conception.

This line of reasoning is not without merits. Historically, the idea of freedom of thought is deeply linked to improving reason and overcoming, in Kant's words, mental immaturity (1784). Promoting critical reflection, overcoming "categorical views" and fixed ideas not open to evidence or counterargument is not necessarily worrying in light of freedom of

[19] The decision corresponds to what one may see as a general strategy of the Court to evade decisions which would provide contours to the *forum internum*, a feature of the jurisprudence criticized by others (Evans, 2017; Taylor, 2005). A related case in-point is *Riera Blume v. Spain*, in which applicants were detained in a hotel to "deprogram" their beliefs about a sect through psychological and psychiatric methods. The Court did not rule on the alleged violation of Art. 9 ECHR, but found a violation of Art. 5.1 ECHR (detention).

thought—on the contrary. Insofar as the psychiatric treatment made the applicant gain understanding of her condition, it may have promoted freedom of thought. However, the means to achieve this may have contravened freedom of thought at the same time. Although the psycho-corrective methods are not described more fully, the setting in which they were applied—involuntarily hospitalized, physically restrained, forcibly administered drugs—must be seen as coercive. The dilemmatic question is thus: Can it be legitimate to interfere with freedom of thought, in the negative dimension, in order to promote the freedom to think in the rationalist conception? A question we will return to.

In the present case, matters are even more complex because the targeted belief was of spiritual nature, leading to a tension between freedom of thought and freedom of religion. Perhaps, such is the nature of religious beliefs that a critical attitude erodes them as it undermines emotional identification and promotes doubts. Interferences with the religious *forum internum* and freedom of thought may differ in precisely this point. This is another reason for developing separate taxonomies of interferences for the sister freedoms of Art. 18.

Finally, the case raises the question about the classification of bodily actions with substantive mental effects such as meditation. One might see them as external manifestations of belief to which limitation clauses apply. However, their strong mental effects—the experience of inner peace and stability,—concern the *forum internum*. Banning such practices may thus amount to an (indirect) interference with the inner side.

To summarize: Interferences with the *forum internum* can take various forms. No attempts are made in the sparse jurisprudence to render "coercion" pursuant to Art. 18.2 CCPR more precise; it seems to be understood loosely, also encompassing undue interference or psychological pressure. Because of the problems of coercing beliefs *sensu stricto*, this approach deserves support. Coercion is thus just one among several potential types of interference with Art. 18 CCPR. Furthermore, *Kokkinakis* and *Mockutė* show that the strict confines of the absolute protection of Art. 18 UDHR and CCPR require creative interpretations, as rights of others or paternalistic considerations may need to be accommodated. Interferences with freedoms of religion and thought may need to be evaluated by different standards, as the promotion of a critical attitude toward spiritual beliefs in *Mockutė* shows. Apart from *Kang*, none of the reported decisions was unanimous. This indicates the high degree of uncertainty in this area, which results from the lack of principled or systematic approaches. Some suggestions are forwarded in the next section.

Suggestions for the Right

Scope: Thought, Thinking, Belief

Drawing on the foregoing, the following develops the contours of a right to freedom of thought. Let us start with "thought." As suggested, it should be understood in two ways, as having specific mental states—thoughts—and as performing various mental activities—thinking. Both concepts are clear at the core but vague at the margins. *Thoughts* might be understood roughly (and aware of the controversies in philosophy of mind) as mental representations. These representations often include semantic content such as propositions, but need not do so, e.g., mental imagery. A key question is whether "thought" also includes affective (emotional) states. Psychology has debunked the traditional dichotomy between emotion and rationality; emotions are important contributors to rational decision-making (Lerner et al., 2015). In spite of this, however, emotions and thought are distinct items of the mental furniture. Including emotions would create a freedom of emotion in Art. 18, a conception significantly different to freedom of thought. Emotions should thus not be included as objects of protection. They may become relevant indirectly, however, insofar as tampering with them affects thoughts and thinking.

Thinking comprises—and requires—a range of cognitive capacities and mental actions, from comprehending language and logic to rules of rationality, from associative over artistic thought to mental stimulation. These capacities and the psychological and neuronal mechanisms that enable and realize thinking should also enjoy protection against negative interferences.

The right further protects *freedom of belief*. Beliefs are understood in the wide ordinary sense (not only as convictions) as attitudes toward propositions about the world which can be true or false (*supra*). It comprises occurring and dispositional (or implicit) beliefs, which form the knowledge base of a person. Religious and conscientious beliefs are special cases of freedom of belief. In addition, Articles 19 UDHR and 19.1 CCPR protect opinions, which should be understood to include value judgments and desires, which stand in close relation to beliefs pursuant to Art. 18. But these aspects must be left aside here. Freedom of belief is not an additional freedom, it derives from freedom of thought and thinking, but in view of the salient role Articles 18 UDHR and CCPR accord to beliefs and their psychological and philosophical peculiarities, it merits explicit mentioning and sometimes special consideration.

Scope: Freedom

It is suggested that "freedom" in Art. 18 refers to both a normative and a factual freedom. In the most abstract formulation, it guarantees a liberty in the technical normative sense, i.e., the absence of claims of others. The quintessence of the right is the following (Bublitz, 2014, 2015):

> *No one else, including the state, has legal claims over the content of a person's thoughts, or the type of her thinking.*

More precisely, a liberty of a person to think means that she is not under a duty not to think, and a liberty not to think means that she is not under a duty to think.[20] Art. 18 encompasses both variations. The correlative of the liberty of the rightholder is a no-claim of the duty-bearer. Accordingly, no one has claims about what another person thinks or, beliefs.

No Cognitive Duties

This interpretation also entails that the state cannot impose on rightholders any duty over thought or thinking. In this sense, it cannot prescribe what to think or not to think, or dictate what a person believes, as the ECtHR writes in *Mockutė* (also see *Ivanova v. Bulgaria*, 2007, at 79). Freedom of thought thus bans any norm of the type "it is prohibited to think T." Therefore, governments cannot argue that a "citizen was under a duty to think T" to justify governmental actions. This is the *no cognitive duties*-principle of Art. 18. It might appear evident at first glance but it is not without questions and counterexamples (in a moment). One may further ask whether changing or influencing thoughts and beliefs could ever be a legitimate governmental aim. This is sometimes denied by claims that thoughts or beliefs are outside of the purview of governments

[20] The concept of a liberty is not uncontroversial after Hohfeld, who spoke of "privileges." But the disputed matters are immaterial for present purposes, see Curran (2010) and Williams (1956).

(see Tussmann, 1977). But *the no cognitive duties*-principle does not entail the impermissibility of that aim, an additional argument to that end would be required. Rather, it is recognized that governments may pursue legitimate purposes through, e.g., information campaigns which influence or motivate thought-change in citizens, as long as limits of interferences are observed (*infra*).

However, the law in fact imposes some cognitive duties and *prima facie* justifiably so. For instance, citizens are expected to consider foreseeable consequences of their actions in almost every situation; the law does not promote thoughtless or careless behavior, it may even punish people for it. The law imposes a multitude of behavioral duties, and their performance may presuppose thought and thinking. A vivid example are duties of witnesses to testify accurately, which entails remembering past events truthfully (Kolber, this volume). How is this duty compatible with the *no cognitive duties*-principle? A distinction between behavior and thought needs to be drawn. Part of the *raison d'etre* of the state is controlling behavior; it imposes and enforces behavioral duties to this end. Complying with these duties is all that is required from persons. Compliance may factually require thinking, e.g., about the situation, but this does not transform the behavioral duty into a cognitive one. That the duty primarily pertains to behavior is also demonstrated by their enforcement at the behavioral level, e.g., through physical restraints, not via interventions into thought. Thinking necessarily related to behavior does not fall under above principle. This requires finer distinctions between cognitive and behavioral duties which cannot be drawn here, but which are well conceivable. However, some duties, such as the one of witnesses, seem to constitute cognitive duties—and thus contravene the *no duties*-principle. However, in ordinary cases, the interference with their freedom to think seems trivial whereas the public interest in fact-finding and law-enforcement seems compelling. The duty of witnesses to remember may thus amount to an exception to the absolute protection of freedom of thought.

In general, the absolute nature of the right to freedom of thought demands that states enforce behavioral duties through means not interfering with thought; persons can be motivated to perform actions by incentivizing or deterring them, or they can be physically constrained, including incapacitation. But the state has to resort to forces working externally on the person, rather than exerting control over them from within.

No Punishment for Thought (no Thought Crimes)
From the no cognitive duties-principle, the old Roman maxim *cogitationis poenam nemo patitur*—no one shall be punished for thoughts—follows.[21] Punishing someone for performing or omitting an action logically requires a prior duty not to perform or omit the action that an offender failed to discharge. Without such a duty, no punishment for the failure to comply. The illegitimacy of thought crimes originates in the lacking legitimacy of cognitive duties.

Yet again, as the state may impose behavioral duties and punish for non-compliance, the borders of the *cogitationis* maxim need to be rendered more precise (*infra*). Also, the question whether Art. 18 bans non-punitive sanctions for thoughts, e.g., loss of employment, requires further examination by future research.

No Interferences with Thought
A liberty allows rightholders to do as they please with respect to the object of the liberty. But it does not entail or ensure that they factually possess relevant capacities or skills, nor the absence of impediments or actions of others that may affect the domain of the liberty. For instance, interferences with thought of rightholders for reasons not presupposing a cognitive duty are possible. The liberty of thought and thinking is, by itself, naked or unprotected. To protect against factual interferences, it must be buttressed by claims against others to non-interference. This is the factual understanding of "freedom" in Art. 18 (interferences are analyzed *infra*).

No Revelation—Privacy of Thought
In regard to freedom of religion, it is widely accepted that it covers the privacy of belief; no one has to reveal one's belief (Loucaides, 2012; Schabas, 2016). This is an ancillary claim that protects the freedom to adopt and discard beliefs against negative sanctions (Evans, 2017). It should analogously apply to thought and thinking: no one has to reveal one's thoughts or the type of thinking one performs.

In ordinary life, people often observe the behavior of others and draw inferences about their thoughts all the time. This cannot be prohibited. Nor can manifested thoughts, in writing or behavior, give rise

[21] It is recorded in the Digests of Justiniam (48.19.18), cf. Gablow, this volume.

to privacy of thought claims which only concern *unexpressed thoughts* (for an exception below). For manifested thoughts, ordinary privacy and data protection laws are the adequate place of regulation. Nonetheless, freedom of unexpressed thought is not without application. For instance, some neuroimaging techniques read out brain states that afford inferences about unexpressed thought or thinking, which have found the attention of law-enforcement agencies.[22] Art. 18 bans their use without permission by rightholders.

Power of Waiver
The foregoing four principles are negative liberties. An important further element of a right is the *power* of rightholders to waive its protection, enabling them to consent to interferences (e.g., to enroll in thought altering cognitive therapy). Rightholders may also enter contractual obligations pertaining to thinking, many jobs in the mental economy in fact require performance of cognitive tasks. However, failures to meet these obligations are not enforceable via interferences with freedom of thought (but rather ground damages for non-performance).

Promoting Preconditions
In addition to these negative liberties, the right to freedom of thought may impose on states positive obligations. The extent of such obligations is controversial and differs across instruments. Under the ECHR and the Covenant on Economic, Social, and Cultural Rights, positive obligations are well-established, this is less so under the CCPR and the Declaration. This general issue is not further pursed here.[23] In substance, it should be noted that freedom of thought, especially the freedom to think and rationalist conceptions, have many preconditions. They require mental capacities and skills that must be acquired and matured through training and experience. In fact, this is an open-ended task, everyone can always become a better, more rational, less-biased thinker. An important aspect

[22] One example is a method called brain fingerprinting, see Farwell (2012) and Rosenfeld (2005).

[23] See Nowak, assuming "horizontal effects" for freedom of opinion, Art. 19.1 CCPR (2004, 441); the HR Committee assumes positive obligations in General Comment No. 31 ("fully discharged if individuals are protected by the State, not just against violations of Covenant rights by its agents but also against acts committed by private persons or entities", at 8). See also Joseph and Castan (2013, 39); for the ECHR Mowbray (2004).

is the possession of knowledge. Beliefs are formed against the background of existing beliefs. The more and the better (true, correct) those are, the more and better beliefs a person forms. As Loucaides remarks: A "person who is ill-informed cannot think freely because, being deprived of all the necessary information, his intellectual process of thinking is barred from developing freely its optimum extent. Therefore, it cannot be emphasised enough, that a prerequisite to the exercise of freedom of thought, is the effective exercise of the right to freedom of information" (Loucaides, 2012, 87). Cutting a long story short: States should, and perhaps must, promote such preconditions of freedom of thought.

Protection Against Interferences by Third-Parties
Furthermore, states have the obligation to protect rightholders from interferences by third-parties. Notwithstanding the extent of such duties, they may do so through various measures, e.g., by passing new legislation that prohibits or even penalizes interferences with thought (Bublitz & Merkel, 2014). The interesting point is that this requires rendering the content of freedom of thought more precise with respect to specific contexts. An example might be regulations of digital services or social media platforms with respect to targeted advertisement. At many places, legal systems already provide protection against undue influence, manipulation, fraud, etc. But it seems that this is done unsystematically and without deeper recourse to freedom of thought. Therefore, some aspects such as the freedom from non-coercive manipulation are likely systematically underappreciated in domestic legal orders (regulations of advertisement are one example). The right to freedom of thought then calls for more recognition by legislators and stricter regulations. In this context, it is important to recall that human rights law only draws outer boundaries of permissible governmental action. Many intriguing questions, however, are not situated at these boundaries, but in the regulatory spaces before them. Shaping them is the prime task of legislators. Art. 18 and its counterparts may have the most impact by influencing regulations in these spaces.

These are the seven main dimensions of protection provided by Articles 18 UDHR and CCPR. In the remainder, only some of them can be examined a bit closer. The most challenging aspect in need of further elaboration are factual interferences with freedom of thought.

Scope: Interferences

Because of the absent balancing stage, definitions of interferences are crucial. A plausible construal of the right has to offer resources to define interferences more concretely; this presumably requires a taxonomy of interferences that accommodates various criteria. On the one hand, the widest possible construction considers every action altering thoughts or beliefs ("intervention") of another ("recipient") as a potential interference. Without qualifications, this leads to the absurd consequence that talking to someone on the street without prior consent could violate Art. 18. On the other hand, if only brainwashing, reeducation camps, and interventions of similar, almost torture-like intensity qualify, the norm would leave much—presumably too much—room for various dubious and worrisome interferences. The previous discussion of the case law has shown that "coercion" pursuant to Art. 18.2 CCPR does not capture the range of possible manipulative interferences. To separate permissible from impermissible ones, a multi-layered taxonomy needs to be developed.[24] Here is a sketch:

Negative Effects on Thought and Thinking

To qualify as an interference, the intervention must have a substantially negative effect on thought and thinking, such as detrimental effects on cognitive abilities, e.g., a drug that weakens attention or causes thought disorders. Effects must pass a *de minimis* threshold; the myriad of stimuli that enter people's minds each day do not qualify for lack of a substantive effect. It is worth noting that the introspective feeling of whether stimuli are strong or effective might not be the best indicator as humans are not very good at introspectively identifying what influences them (and to which extent it does so). What are negative effects on thoughts? Scenarios are conceivable in which particular thoughts are induced or eliminated, e.g., through brain stimulation. But in general, thoughts are fleeting states that may easily vanish simply because the thinker is distracted or shifts attention. These are the limits of working memory. But ordinary and mundane effects on thoughts cannot qualify as interferences.

[24] For a related argument for a theory of freedom of religion to avoid "intuitive" but inconsistent decisions see Evans (2001, 33).

Undermining or Bypassing Control Over Thoughts and Thinking
Furthermore, at the conceptual level, speaking of an interference requires that the effect has been brought about by the intervenor, not the affected person herself. This relates to control of the person over the intervention, e.g., incoming stimuli, and its effect. Control over interventions varies in kind and degrees. People may exert control in many ways, e.g., they have to attend to stimuli or can turn away from them, some effects are easily resistible (elaborated more fully in Bublitz, 2020a). People have much control over a book they read; but less control over the advertisements on billboards which they peripherally perceive in the upper corner of their field of vision when driving at highways; and virtually no control over the effects of a drug that their drink is spiked with. Roughly, when rightholders retain sufficient control over an intervention, it is not an interference. Put conversely: interventions have to *undermine* or *bypass control* of affected person to qualify. This is a necessary, but not a sufficient condition, and it forms part of a test of interference: *Does an intervention respect the other as a free and self-controlled thinker; or does it undermine or bypass control?* The latter interferes with freedom of thought, the former may not.

Interferences with Freedom of Belief
Special considerations apply to beliefs. What does it mean to interfere with freedom of belief? Although forming and changing beliefs is often not under voluntary control (*supra*), this does not mean that all interventions causing changes in beliefs are in-principle dubious. Consider a compelling argument. It is compelling precisely because it does not leave any choice about its evaluation; it is compelling because it must be accepted. Although people lack voluntary control over the changes induced by it, one may say it was still them, not intervenors, who brought them about. After all, their belief-forming system was in control. This shows that a finer understanding of freedom of belief is necessary. It surely commands the absence of interferences impeding the working of the belief-forming system, e.g., via a drug. This would be a negative effect on a cognitive capacity as captured by above definition. More interesting are other manipulative interferences. Any interpretation of freedom of belief has to accommodate the fact that humans influence and potentially change each other's beliefs all the time. In virtually every conversation, the mechanisms forming and revising beliefs are operative and leave thinkers only limited, indirect voluntary control. However, this

does not call for a stop of (unwanted) conversations. Freedom of belief cannot imply the absence of any input into the belief-forming system.

Rather, it is suggested that freedom of belief opposes actions that weaken or undermine the ability of rightholders to form rational beliefs (as in the rationalist conception). Only then, they potentially interfere with freedom of belief. The reason behind this is that freedom of thought can only protect against grave negative effects; bringing someone to rationally form a belief against their will might be a nuisance or have detrimental psychological effects, but cannot trigger freedom of thought protection.[25] After all, one of the justifications for the absolute protection, so I can only suggest here without further argument, is the search for truth.

Another perspective supports this suggestion. The right regulates interpersonal relations. The question is thus: How should people treat each other, given the fact that beliefs are constantly formed and revised without much direct control of believers? The answer must be this: As the default mode, people should respect each other as rational believers, i.e., as people who want to form their beliefs according to evidence and rational standards. This allows them to form correct beliefs, to understand the world and find truth. As long as people respect each other as rational believers, freedom of thought is not implicated.[26]

This understanding neither implies nor presupposes that people are usually rational believers, only that they can be such. Rather, it concerns the ways in which people *should* engage with each other. What does respecting others as rational believers mean? In abstract, it means to refrain from exploiting rational weaknesses and susceptibilities of another person's belief-forming system. This is what happened in *Kang*. While one may not have a duty to counteract those weaknesses, one should

[25] The casebook example of such detrimental effects are parents who are deceiving themselves about the bad character of their children. Another could be coping strategies to alleviate inner conflicts. No one is under a legal duty to be a rational believer (though there might be such ethical duties), but freedom of thought may not, and possibly cannot, protect against mental distress or similar effects.

[26] Further support for this interpretation can be derived from philosophy. Forming beliefs according to rational standards has often been equated with freedom of thought, not only by Russell, but, e.g., also by Pettit and Smith (1996).

neither exacerbate nor exploit them.[27] This allows for a second part of the test of interferences with regard to freedom of belief:

> *Does the intervention respect the recipient (the rightholder) as a rational believer, i.e. as a person who forms beliefs in light of evidence, other beliefs and rational standards; or does it seek to exploit rational weaknesses or move her to form beliefs for other, non-related grounds such as psychological needs?*

The latter interferes with freedom of belief, the former may not. This test reflects default norms for interactions and may need further context-specific adaption to the particulars of a case. Works of art do not have to treat recipients as rational believers, nor do chefs or lovers. Religious and conscientious beliefs may require distinct standards; and so do the forming of desires and emotions. With these caveats, the test provides rough guidance about interferences with freedom of belief.

Countervailing Rights of Intervenors: Free Expression

If the test indicates an interference, a further step has to accommodate the fact that some interventions are themselves exercises of rights of intervenors, primarily freedom of expression (for more on this conflict see Bublitz, 2020a). Freedom of expression is the right to send stimuli that potentially affect thought of recipients, and it is not restricted to stimuli preserving their control or rational belief formation. The scopes of freedom of expression and freedom of thought are thus not neatly separated but partly overlap. None of the two rights can claim lexical priority over the other. Although freedom of thought is unconditional whereas freedom of expression is conditional, the latter deserves a robust scope of application. Striking balances between both is thus unavoidable. A first distinction can be drawn between *actions* and *effects*. Freedom of expression entitles rightholders to actions such as speaking but does not confer any claims about the effects of the speech in recipients (corresponding to the *no claims over others' thoughts*-principle). Speakers may speak but no one has to listen. But if expressions happen to have effects, e.g., because recipients are exposed to them in public, freedom of expression can justify

[27] One may wonder what this implies for providing false information. As such, it does not exploit a weakness in the belief-forming system which checks information but against other beliefs. Systematic disinformation may qualify, as this erodes the ability to check against other, true beliefs.

these effects. The deeper reason is this: Some actions protected by rights imply a *pro tanto* permission to affect others' thoughts. A right to build a house entails that others might see it or be psychologically affected by the architecture; a right to open a shop entails the presentations of goods; expression and communication inherently affect others. In such situations, freedom of expression and freedom of thought need to be reconciled in light of various criteria such as intensity and strength of the effect, the importance of the expression, the degree of control it leaves, whether there are less intense means for expressions. Balances between freedoms may suggest that expressions are permissible provided they observe, as far as possible, freedom of thought of recipients, e.g., by avoiding unwanted communications or captioned audiences, not deploying control-bypassing elements, explicitly informing recipients about stimuli and their effects, etc.

Other methods of intervention, by contrast, are not protected by rights of intervenors (or only weakly so). This is especially true for direct brain interventions such as administering drugs or neuro-interventions in rightholders. These actions usually do not pursue any aim of intervenors other than altering thought and thinking of recipients. To this, intervenors have no claim (*no claims over others' thoughts*). Unlike expression, the intervening action as such is trivial (e.g., injecting a substance, setting up a magnetic field), and the freedom of perform this action does not entail a *pro tanto* permission to affect others. People may play around with electric or magnetic stimulators, but must stop when others are affected by them. As a consequence of this normative difference between interventions, some means to change others' thoughts might be permissible (expression), whereas the same effect brought about by another might not. This adds a last criterion to the test which now reads in full:

> *Does an intervention respect the rightholder as a free and self-controlled thinker or a rational believer who forms beliefs in light of evidence and rational standards– or does it undermine or bypass her control, exploit rational weaknesses or move her to form beliefs for other, non-related grounds such as psychological needs? If so, is the intervention an exercise of an important rights of intervenors which entails a* pro tanto *permission to affect thoughts?*

Consequences

From these considerations, a rough distinction between direct and indirect interventions arises. Indirect interventions are those that reach the mind/brain of recipients via the outward senses, they are often informational inputs into the cognitive machinery of rightholders. Direct interventions are those that reach the mind/brain on other, primarily neurobiological ways, such as brain stimulation or drugs (for a further elaboration see Bublitz, 2020a, and the criticism by Levy, 2020). They differ in virtue of their normative protection and the amount of control recipients can exert over them. People have most control over consciously perceived indirect interventions, e.g., perceptual stimuli, less over no-consciously perceived stimuli (e.g., subliminal stimuli), and almost no control over direct interventions.[28] This leads to the following taxonomy:

1. Direct brain interventions
2. Indirect interventions: non-consciously processed stimuli (subliminal)
3. Indirect interventions: consciously processed stimuli
4. Indirect interventions: Communication fully respecting rationality

The first and—depending on circumstances, the second—class of interventions regularly interfere with freedom of thought, whereas the fourth and—depending on circumstances, the third—may not. Many interventions fall on a spectrum in-between and require evaluation in light of the suggested test and further context-sensitive considerations as those formulated by the ECtHR in *Kokkinakis* and *Larissis*.[29] This taxonomy

[28] The distinction between direct and indirect interventions is not based on crude mind-brain dualism, but on different causal pathways of interventions. That this is a suitable criterion to distinguish between interventions for normative purposes has been disputed (Levy 2007, 2020). However, normative as well as factual differences between interventions are key criteria. The alternative is an assessment solely based on effects. It would neglect normatively different protections of interventions and the privileged status of expressions.

[29] A third category that led to discussions in neuroethics are environmental alterations. They may change thoughts or beliefs, but might not be conceptualized as an intervention. Architecture, for instance, may affect how people feel and think in a place (open space vs. narrow confines). Intervenors may avail themselves of such effects (the prison as a Panopticon), see Bublitz, 2018. Recently, choice architecture through nudges has received much attention (Thaler & Sunstein, 2009). The question in these cases is whether alterations to

corresponds and reconstructs various views on the matter. For instance, the Special Rapporteur on Freedom of Opinion considers "forced neurological interventions" a violation of Art. 19(1) (2018, at 23). Correctly so—these are direct brain interventions that bypass control, do not respect recipients as rational believers and are likely no expression of rights. Subliminal stimuli, banned by laws on marketing and broadcasting, fall into the second category and are prohibited as they bypass control capacities. Moreover, interventions that are by themselves innocuous might be assessed in combination with others: While talking to a therapist for an hour might be beyond concern, participating in several cognitive therapy sessions .changing thought dispositions and involving a range of subtle psychological mechanisms may amount to an interference (and therefore requires informed consent).

With respect to dubious indirect interventions such as advertisement, much depends on the strength of their effects and the mechanisms which produce them. The important general lesson is that such interventions are worrisome even if they fall short of constituting coercion or inducing uncontrollable buying urges, it may suffice that they change people's beliefs about a product on control-bypassing ways not respecting recipients as rational believers. Many mechanisms deployed in marketing raise such worries. A simple example: According to the mere exposure effect, the repeated exposure to a stimulus, say a message, leads people to evaluate it more positively. Simply repeating a message makes it psychologically more believable (Bornstein & D'Agostino, 1992). A so caused increase in the degree of belief is not warranted by standards of rational belief formation. Employing this mechanism by repeatedly exposing people to messages without providing further reasons to change beliefs fails to respect them as rational believers, it exploits rational weaknesses in their belief-forming system. If effects are severe enough, this constitutes an interference. (It seems unlikely that the pursued aim is

the environment affect freedom of thought; but a range of further considerations come into play. For instance, buildings have to be designed in some way, and proprietors have a right to design them, just as store-owners have a pro tanto claim to design their store as they please, including the placement of products. Nonetheless, if such environmental alterations have substantive effects on thinkers, the right to freedom of thought has to be taken into account. This requires context-specific assessments. Rules for prisons are different than for supermarkets. A simple solution would be informing customers about the choice architecture, improving her ability to form rational decision as they become aware of arational influences.

significant enough to justify it.) Another example is stimuli so designed that recipients process them only superficially by so-called peripheral routes of processing (Petty & Cacioppo, 1984). This bypasses control of recipients, fails to respect them as rational believers, and thus raises freedom of thought concerns. These examples show how the developed criteria allow concrete assessments.

Tools and the Freedom to Think

A different form of interference merits attentions as it animated the Cognitive Liberty movement (Boire, 2001; Sententia, 2006, also see Bublitz, 2013). Previously, thinking has been understood as a natural ability. But it is constrained by parameters of the cognitive system and can be greatly enhanced by tools for thought, cognitive artifacts (Clark, 2008). A good example is calculating. Numbers and operations easily become too complex to be carried out in the head; writing them down and calculating with symbols vastly increases powers for calculating. The same is true for ordering thoughts or writing a longer piece of text. Many technological innovations might be viewed as ways to augment cognitive capacities. This cannot be without relevance for the right to freedom of thought, especially the freedom to think.

An intriguing philosophical theory proposes a radical perspective: Thinking is not only taking place inside brain and skull, but in the external world—the mind extends into the world. Accordingly, a piece of paper, a calculator or an iPad can become part of the mind (Clark & Chalmers, 1998; Menary, 2007). If applied to the law, the Extended Mind Thesis would have far-ranging consequences. Material objects, chattel, would become parts of the mind, and thereby, of the person (Blitz, 2010). This view is hardly reconcilable with foundational legal distinctions between persons and objects, nor with the internal/external distinction of Articles 18 UDHR and CCPR, and hence cannot guide delineations of their scopes.

But even though legally, they are not part of the mind or person, cognitive artifacts such as pen and paper or iPads could be enabling conditions of thinking, and therefore fall under the protection of Art. 18. This seems plausible if they enable basic forms of thinking and ordinary cognitive functioning. Depriving a thinker of such tools might amount to an interference with freedom of thought, and states might be obliged to provide such basic tools to rightholders in specific conditions such as prisons.

However, as such external actions concern the social sphere, an absolute right can not be warranted—it can still be a strong right. Moreover, the writing might then fall under the privacy provision of Art. 18, even though it is expressed thought.

This freedom to think also covers bodily practices with strong effects on thought such as meditation. Conceiving of it merely as bodily movement might miss its point. Provided effects are substantive, a prohibition of meditation (as alleged in *Mockutė*), might interfere with freedom of thought. In addition, this dimension comprises medical tools of thought, e.g., medications against cognitive impairments from ADHD to Alzheimer disease. It might also cover tools as those advocated by the Cognitive Liberty movement at least insofar as they enable modes of thoughts or thinking otherwise not attainable (Walsh, 2010). Because of the social dimension, this cannot be an absolute right; but it may be strong and affect drug policy nonetheless (Bublitz, 2016).

Tensions Between Different Conceptions & Exceptions
So much for interferences. A problem that has colored some previous examples and that gives rise to thorny questions arises from inner tensions of the idea of freedom of thought. Different conceptions may pull into different directions. Structurally, the problem arises when an intervention contravenes the negative freedom from interferences but aims at promoting other aspects such as the freedom to think by improving mental capacities or rational belief formation.

The psychiatrists in *Mockutė*, for instance, succeeded in moving the applicant to adopt a critical view on "categorical" thoughts. This supposedly increased her freedom to think different thoughts, overcame internal impediments due to the mental disorders and promoted rational belief formation. But it nonetheless encroached upon the freedom *from* control-bypassing interferences. In a philosophical view, one might say that the applicant's freedom of thought was not violated because her thinking was not free whereas in a legal sense, there was an interference. The question is thus whether the ends can justify the means.

A similar conflict arises with respect to educational institutions such as the picture of schools as places of "thought control." Mandatory schooling fulfills the criteria of an interference: It is conducted in a coercive (mandatory) setting, involves a relationship of unequal power between an authority and vulnerable persons whose abilities of thought are not fully developed, and whose future life courses depend on grades.

This situation, offers significant incentives for adopting one's thinking to the demanded norms. It does not respect schoolchildren as rational believers, they are none yet. It interferes with freedom of thought. However, the larger aim behind schooling is promoting preconditions of freedom of thought: Various cognitive abilities and skills, rational thinking and belief formation, the absence of impediments, a large knowledge base, self-trust and intellectual curiosity. These conditions have to be created and fostered—the prime aim of education, properly conceived. Freedom of thought and especially rational believing demand such interventions in cognitively not fully developed children. This calls for institutions such as schools, even mandatory ones.

Furthermore, the tension can also arise with respect to competent adults insofar as interventions do not seek to exploit but to alleviate weaknesses in belief formation. As an example, people discount bad information and overestimate good information in the updating of belief, creating biases. A study showed that a few pulses of transcranial magnetic stimulation (TMS) applied to the inferior frontal gyrus eliminates this effect (Sharot et al., 2012). If such pulses are applied without consent (and without other side-effects), does it interfere with Art. 18?

A categorical insistence on the absolute protection denying any interference is not persuasive. Plausible constructions of the right have to accommodate the fact that freedom of thought has conditions that may need to be created and promoted through interferences with freedom of thought. There is no way around this insight. As a consequence, strict exceptions might need to be developed and clearly defined. They might be justified because the need for them arises from within the concept of freedom of thought. They promote the value of freedom of thought, not other values or public interests. They might be construed, as Loucaides mentions in passing with respect to the ECHR, as "inherent limitations" of the right (2012, 86). Assuming the general justifiability of paternalistic measures, here is a suggestion:

An intervention contravening the freedom from interferences might be permissible if (a) the person is not competent to make a decision about such interventions herself, (b) the intervention aims at improving the freedom to think by alleviating substantial deficits in thinking abilities or rational belief formation, (c) it is in the best (medical) interest of the person, (d) there are no less invasive means, and (e) the benefits of the intervention outweigh the setbacks, all things considered. These criteria may need refinement for different purposes, from schools to psychiatry (and alignment with domestic mental health laws). In addition, given the dangers of misuse of such exceptions, it should be ensured that (f) interventions do not primarily pursue other governmental goals and (g) that they do not seek to imprint moral, political, or other values, they must strive to be content neutral. The primary and dominant characteristic of the intervention must be the promotion of freedom of thought. Under these conditions, and only then, interferences with freedom of thought may not lead to a violation of the right because it advances freedom of thought in affected persons.

An exception along those lines provides the resources to explain why and under which conditions mandatory education does not contravene freedom of thought. Moreover, it shows why psychiatric interventions, e.g., in acute psychotic states, with the aim to restore freedom to think, might be warranted. The psychiatric measures in *Mockutė* may fall under the exception. However, their attempts to address religious beliefs may not since rules for this particular type of belief may be different (*infra*). The use of the TMS device without consent in competent adults violates Art. 18.

Several further practices likely not falling under the exception merit mentioning: Forcibly administering thought-altering drugs to render persons competent to stand trial (*Sell v. USA*) pursue aims not in the best interest of the person. This is true *a fortiori* for interventions establishing the (cynical) competency to be executed. Furthermore, people, e.g., in institutions such as prisons or care homes are sometimes sedated so that dealing with them is easier—this is not promoting freedom of thought and hence violates it. A particularly thorny issue is criminal rehabilitation of offenders. Special considerations may apply because of the permissibility of punishment, which allows states to treat citizens in ways otherwise prohibited. But in general, human rights including the right to freedom of thought have to be observed by penal institutions, even though this may limit available means to reform offenders. The Ludovico Technique of the novel Clockwork Orange (Burgess, 1962) would flout freedom of thought (Bublitz, 2018).

Reflections on the Absolute Nature

This brings us to concluding remarks on the absolute nature of the right. Without calling its supreme importance into question, several examples have shown that an unconditional protection is hard to maintain in light of real cases: interferences through expressions; paternalistic improvements of free thinking, the duty to remember, and attempts of courts to bypass the perimeters of the *forum internum*. More examples can likely be found. This places justificatory pressure on the right.

Future examinations have to investigate the reasons for the absolute protection with a view on the *travaux préparatoires*, the history of ideas and current debates in philosophy. Some philosophers recently argued that some interferences with freedom of thought might be permissible as they are the most effective way to prevent crimes, reform offenders, or promote other pressing societal goals (Douglas, 2014; Persson and Savulescu, 2012). In their often hypothetical scenarios, these interferences with thought are the only available means to pursue an exceptionally significant goal, e.g., saving the world from climate change by altering how people think about long-term costs of their actions. Under such conditions, the absolute protection of the right becomes indeed arguable. However, such thought-experiments are not necessary guides for good interpretations of a right or for practical policy. There will always be hard-cases and good arguments for exceptions. Every deontological right can be countered by an consequentialist thought-experiment pointing to better overall outcomes. That, as such, is not surprising. What ultimately matters are reasonable regulations for real life. The absolute protection of freedom of thought is predicated on the assumption that the state can mobilize all its physical forces and that this usually suffices to control people's behavior and achieve societal goals.[30] And this assumption seems to be largely true.[31]

However, these thought-experiments have merit as they underscore a somewhat neglected topic in writings on the right: Interferences with freedom of thought can be genuinely benign. In classic treatments, the roles of the good and the bad are clearly and stereotypically distributed: Dictators versus the oppressed, the church versus science, the monarchy versus the Enlightenment—constellations in which one cannot but champion freedom of thought. The challenging cases of today, however, are different: What about an effective but manipulative control-bypassing

[30] According to the HR Committee in General Comment No. 29, Art. 18 CCPR is non-derogable because "derogation can never become necessary" (2001, at 11). This is, ultimately, an empirical claim.

[31] See the discussion in Bublitz (2019) and the reply by Persson and Savulescu (2019).

intervention that alleviates racial or gender biases—should it be mandatory? Many people might consider this a price worth paying for a less discriminatory society. An absolute construction of the right to freedom of thought needs to provide good reasons to show why they are wrong. This requires sustained debate about the foundations of the right and ultimate grounds of the legal order.

Finally, the absolute protection seems to be among the causes for the lacking practical relevance of the right. This is the *tragedy of absolute rights* (Bublitz, 2014). They are so strong that courts will attempt to keep clear of their ambit because options to find reasonable decisions for individual cases are severely limited; they seek to avoid precedents without room for maneuvers in the future. The case law of the ECtHR might well be read in this manner. Turning freedom of thought into a living and practically effective right requires a broader scope and has to accommodate the fact that people change each other's thought on potentially worrisome ways all the time. Finding reasonable solutions for those cases requires more fine-grained and context-sensitive considerations than an absolute right can provide. Perhaps, the absolute protection must be softened to create some "discretionary edges" (Evans, 2017, 88). Perhaps, *forum internum* and *externum* should be seen less as mutually exclusive categories but as an overlapping continuum, as the former Special Rapporteur on the right suggests (Bielefeldt et al., 2016). Perhaps, another non-absolute right such as the right to mental integrity (Bublitz, 2020b; Ienca & Andorno, 2017) should complement freedom of thought and absorb minor cases. In any case, the grounding of the absolute protection need to be revisited for the right to become an effective legal guarantee.

Summary

Freedom of thought is not a homogeneous concept. It comprises several conceptions at multiple layers. The grand political and philosophical idea is not identical with a legal conception, and both should be kept apart in discussions. There are several rights to freedom of thought at the domestic and international level, prefigured and constrained by the legal orders in which they are embedded. International human rights to freedom of thought are modeled after Articles 18 UDHR, with slightly diverging wordings. It is suggested to consider the right as identical across documents, as far as possible, to allow a coherent international understanding. This, of course, will ultimately depend on the courts applying and interpreting the right. The hallmark of Articles 18 UDHR and CCPR as well as regional counterparts such as Art. 9 ECHR is their absolute nature, interferences are not open to justification. This influences the interpretation of the right and is likely one of the reasons for its practical irrelevance.

More concretely, theories of the right to freedom of thought must explain five explananda: its content and meaning, interferences, the internal/external structure, its absolute character as well as its relation to other rights. The foregoing discussion yields some suggestions:

First, the scope of the right should comprise thought and thinking. Second, "belief" plays a salient role in the norm. Its restrictive interpretation as conviction is correct with respect to the privileging of external actions but runs into inconsistencies with respect to mental states since all beliefs are thoughts. Therefore, with respect to the internal side, the right should protect the freedom of *all* beliefs. And as beliefs—affirmative attitudes toward propositions about the world which can be true or false—possess several peculiarities, freedom of belief deserves and requires special consideration. Accordingly, without unduly enlarging the scope, Art. 18 covers freedom of thought, thinking, and belief.

Third, freedom of thought, conscience, and religion should be seen as three distinct freedoms as their scopes and possible interferences may vary. For instance, conscientious and religious beliefs are not truth-apt; attempts to persuade others in spiritual matters, proselytism, might legitimately take forms different to persuasion with respect to scientific or political beliefs. Nonetheless, the three freedoms share common ground, so that doctrines and jurisprudence on one often apply to the others.

Fourth, Art. 18 provides protection in seven dimensions: Its essence lies in a normative liberty, according to which rightholders are not under any thought-related duty. Correlatively, no one has claims over thoughts and thinking of another person. From this, the venerable *cogitations* maxim—no punishment for thoughts—emerges. The right also bars factual interferences negatively affecting thought, thinking, or rational belief formation. It also guarantees the privacy of thoughts. Insofar as states have positive obligations, it calls for the provision of preconditions of free thinking, from education to tools, as well as protection against interference by third-parties.

Fifth, the peculiar inner and outer structure of Art. 18 stems from the idea of a *forum internum* of religion and conscience. The extent to which this metaphor applies to freedom of thought needs further examination.

It roughly denotes the inner psychological space in which persons think and reflect about themselves. Whether this metaphor usefully adds something to defining the scope of the right remains to be shown; the present proposal does not draw on it in any detail. Furthermore, the inner side of thought has to be delineated in several respects: One border concerns the step from mere thought to action and the difference between cognitive and behavioral duties. Intentions might be the dividing line. Another line concerns the external side of thought. The philosophical Extended Mind Thesis that cannot guide the interpretation of the scope of Art. 18 because of its internal/external structure. It is important to note that the law is not bound by supposedly ontological distinctions, as it may cut the world in way pursuant to normative considerations. Because of the internal/external structure of Art. 18 as well as the foundational legal distinction between objects and persons, the Extended Mind Thesis is inapplicable. However, the Extended Mind Thesis demonstrates the extent to which cognition is integrated with artifacts and the environment. This insight calls at least for the provision of simple tools to enable basic cognitive functioning such as pen and paper for prisoners.

Sixth, one of the key challenges for the right is defining permissible and impermissible interferences. Instead of assuming that only powerful interventions may affect thought, interpretations should accommodate the fact that changing others thought and thinking, also in negative ways, is a common occurrence. Art. 18.2 CCPR speaks about coercion. But the analysis of the rare case law, as well as considerations about the nature of beliefs and coercion, show that this is neither a precise nor an exhaustive definition of possible interferences. Many dubious ones are better described as manipulative interferences. Moreover, some interventions are protected by rights of intervenors which entail a *pro tanto* permission to affect other's minds, especially freedom of expression. As the scopes of freedom of thought and expression cannot be interpreted in a way that both do not overlap, balances need to be struck. The following test for interferences is suggested (it may need context-specific modifications):

> *Does an intervention respect the rightholder as a free and self-controlled thinker or a rational believer who forms beliefs in light of evidence and rational standards – or does it undermine or bypass her control, exploit rational weaknesses or move her to form beliefs for other, non-related grounds such as psychological needs? In the former two cases, the intervention may not interfere with freedom of thought, whereas it does so in the latter. Then, one*

has to further ask : Is the intervention an exercise of an important right of intervenors which entails a pro tanto permission to affect thoughts?

Seventh, the absolute nature of the right needs reconsideration. The reasons for it are not entirely clear, and interestingly, nowhere stated more precisely. The discussion of cases as well as legal practice seems to indicate that a living right to freedom of thought with a relevant scope may need to allow more nuanced decision, taking into account competing rights, different social situations, as well as practical considerations. Without firmer and finer explanations of its grounds and limits, courts will likely remain reluctant to apply Art. 18, if only for the fear of unforeseeable precedents.

Finally, the grand political idea as well as the general human right might be too abstract and lofty to provide answers to concrete cases. It is not only the task of courts, assisted by legal scholarship, to render the right more precise, but also of lawmakers in regulating of specific domains, such as advertisement. Human rights can only provide the outer limits of what governments, and by extension, third-parties, might do. But many of the intriguing questions are not situated at these borders. Lawmakers should regulate these gray areas and thereby render the right more precise. In this regard, one may presumably speak of a systematic neglect of freedom of thought in several domains. Although this may need closer examination in detail, freedom of thought may not have received the attention it is accorded to by international human rights law. Novel technologies provide opportunities to remedy these shortcomings. As such issues are complex and easily surpass the horizons of courts in daily business or individual lawmakers, this is a moment for effective scholarship. By offering persuasive operationalizable theories of the right as well as concrete policy suggestions, it can illuminate the path of its further construction and decisively shape the future of freedom of thought.

Cases

Atasoy and Sarkut v. Turkey, Human Rights Committee, Communications 1853/2008 and 1854/2008, Views adopted 29 March 2012.

Bayatyan v. Armenia, European Court of Human Rights, Application 23459/03, Judgment GC, 7 July 2011.

Bundesverfassungsgericht, BVerfG, Beschluss 2 BvR 1062/87 v. 14 September 1989 („Tagebuch").

Eweida and others v. *United Kingdom*, European Court of Human Rights, Applications 48420/10, 59842/10, 51671/10 and 36516/10, Judgment 15 January 2013.

Ivanova v. Bulgaria, European Court of Human Rights, Application 36207/03, Judgment 12 April 2007.

Kang v. Republic of Korea, Human Rights Committee, Communication No. 878/1999, Views adopted 15 July 2003.

Kim v. Republic of Korea, Human Rights Committee, Communication 1786/2008, Views adopted 25 October 2012.

Kokkinakis v. Greece, European Court of Human Rights, Application 14307/8, Judgment 25 May 1993.

Larissis and others v. Greece, European Court of Human Rights, Applications 140/1996/759/958–960, Judgment 24 February 1998.

Mockutė v. Lithuania, European Court of Human Rights, Application 66490/09, Judgment 27 February 2018.

Riera Blume v. Spain, European Court of Human Rights, Application 37680/97, Judgment 14 October 1999.

Salonen v. Finland, European Commission of Human Rights, Application 27868/95, Decision 2 July 1997.

Sell v. United States, United States Supreme Court, 39 U.S. 166 (2003).

United States v. Schwimmer, United States Supreme Court, 279 U.S. 644 (1929).

References

Alegre, S. (2017). Rethinking freedom of thought for the 21st century. *European Human Rights Law Review, 3*, 221–233.

Berlin, I. (1969). Two concepts of liberty. In I. Berlin, *Four essays on liberty* (pp. 118–172). Oxford University Press.

Bernays, E. L. (2005). *Propaganda*. Ig Publishing.

Bielefeldt, H., Ghanea, N., & Wiener, M. (2016). *Freedom of religion or belief: An international law commentary*. Oxford University Press.

Blitz, M. J. (2010). Freedom of thought for the extended mind: Cognitive enhancement and the constitution. *Wisconsin Law Review, 4*, 1049.

Boire, R. G. (2001). On cognitive liberty. *The Journal of Cognitive Liberties, 2*(1), 7–22.

Bondy, P. (2018). *Epistemic Rationality and Epistemic Normativity*. Routledge.

Bornstein, R. F., & D'Agostino, P. R. (1992). Stimulus recognition and the mere exposure effect. *Journal of Personality and Social Psychology, 63*(4), 545.

Boucher, F., & Laborde, C. (2016). Why tolerate conscience? *Criminal Law and Philosophy, 10*(3), 493–514.

Bublitz, J. C. (2013). My mind is mine!? Cognitive liberty as a legal concept. In E. Hildt & A. Francke (Eds.), *Cognitive enhancement* (pp. 233–264). Springer.

Bublitz, J. C. (2014). Freedom of thought in the age of neuroscience. *Archiv Fuer Rechts- und Sozialphilosophie, 100*, 1–25.

Bublitz, J. C. (2015). Cognitive Liberty or the International Human Right to Freedom of Thought. In J. Clausen & N. Levy (Eds.), *Springer Handbook of Neuroethics* (pp. 1309-1334). Springer.

Bublitz, J. C. (2016). Drugs, enhancements, and rights. In F. Jotterand & V. Dubljevic (Eds.), *Cognitive enhancement: Ethical and policy implications in international perspectives* (pp. 309–328). Oxford University Press.

Bublitz, J. C. (2018). The Soul is the Prison of the Body – Mandatory Moral Enhancement, Punishment & Rights Against Neuro-Rehabilitation. In D. Birks & T. Douglas (Eds). *Treatment for Crime: Philosophical Essays on Neurointerventions in Criminal Justice* (pp. 289-320). Oxford University Press.

Bublitz, J. C. (2019). Saving the world through sacrificing liberties? A critique of some normative arguments in "Unfit for the Future". *Neuroethics, 12*(1), 23–34.

Bublitz, J. C. (2020a). Means matter: On the legal relevance of the distinction between direct and indirect mind-interventions. In N. Vincent. T. Nadelhoffer & A. McCay (Eds.), *Neurointerventions and the law: Regulating human mental capacity*. Oxford University Press.

Bublitz, J. C. (2020b). The nascent right to psychological integrity and mental self-determination. In A. von Arnauld, K. von der Decken & M. Susi (Eds.), *The Cambridge handbook of new human rights: Recognition, novelty, rhetoric* (1st ed., pp. 387–403). Cambridge University Press.

Bublitz, J. C., & Merkel, R. (2014). Crimes against minds: On mental manipulations, harms and a human right to mental self-determination. *Criminal Law and Philosophy, 8*(1), 51–77.

Burgess, A. (1962). *A Clockwork Orange*. Heinemann.

Bury, J. B. (1947). *A history of freedom of thought*. Oxford University Press.

Clark, A. (2008). *Supersizing the mind: Embodiment, action, and cognitive extension*. Oxford University Press.

Clark, A., & Chalmers, D. (1998). The extended mind. *Analysis, 58*(1), 7–19.

Commission on Human Rights. (1948). Summary Records of the 60th meeting, 4 June 1948, UN Doc. E/CN.4/SR.60.

Curran, E. (2010). Blinded by the light of Hohfeld: Hobbes's notion of liberty. *Jurisprudence, 1*(1), 85–104.

Douglas, T. (2014). Criminal rehabilitation through medical intervention: Moral liability and the right to bodily integrity. *The Journal of Ethics, 18*(2), 101–122.

Evans, C. (2001). *Freedom of religion under the European Convention on Human Rights*. Oxford University Press.

Evans, M. (1997). *Religious liberty and international law in Europe*. Cambridge University Press.

Evans, M. (2017). The freedom of religion or belief in the ECHR since Kokkinakis, or "Quoting Kokkinakis." *Religion and Human Rights, 12*(2–3), 83–98.

Farwell, L. A. (2012). Brain fingerprinting: A comprehensive tutorial review of detection of concealed information with event-related brain potentials. *Cognitive Neurodynamics, 6*(2), 115–154.

Frankfurt, H. G. (1969). Alternate possibilities and moral responsibility. *Journal of Philosophy, 66*, 829–839.

Hammer, L. M. (2002). *The international human right to freedom of conscience: Some suggestions for its development and application*. Dartmouth Publishing.

Human Rights Committee. (1993). General Comment No. 22, Art. 18, CCPR/C/21/Rev.1/Add.4.

Human Rights Committee. (2001). General Comment No. 29, Art. 4, CCPR/C/21/Rev.1/Add.11.

Human Rights Committee. (2010). General Comment No. 34, Art. 19, CCPR/C/GC/34/CRP.4.

Hohfeld, W. N. (1913). Some fundamental legal conceptions as applied in judicial reasoning. *Yale Law Journal 23*(1), 16–59.

Ienca, M., & Andorno, R. (2017). Towards new human rights in the age of neuroscience and neurotechnology. *Life Sciences, Society and Policy, 13*(1), 5.

Joseph, S., & Castan, M. (2013). *The international covenant on civil and political rights: Cases, materials, and commentary* (3rd ed.). Oxford University Press.

Leiter, B. (2013). *Why tolerate religion?* Princeton University Press.

Lerner, J. S., Li, Y., Valdesolo, P., & Kassam, K. S. (2015). Emotion and decision making. *Annual Review of Psychology, 66*(1), 799–823.

Levy, N. (2007). *Neuroethics: Challenges for the 21st century*. Cambridge University Press.

Levy, N. (2020). Cognitive enhancement: Defending the parity principle. In N. Vincent. T. Nadelhoffer & A. McCay (Eds.), *Neurointerventions and the law: Regulating human mental capacity*. Oxford University Press.

Lindkvist, L. (2017). *Religious freedom and the universal declaration of human rights*. Cambridge University Press.

Lippmann, W. (2007). *Public opinion*. BN Publishing.

Loucaides, L. (2012). The right to freedom of thought as protected by the European Convention on Human Rights. *Cyprus Human Rights Law Review, 1*, 79–87.

Mawhinney, A. (2016). Coercion, oaths and conscience: Conceptual confusion in the right to freedom of religion or belief. In F. Cranmer, M. Hill, C. Kenny,

& R. Sandberg (Eds.), *The confluence of law and religion* (pp. 205–217). Cambridge University Press.
Menary, R. (2007). *Cognitive integration: Mind and cognition unbounded*. Palgrave Macmillan.
Metzinger, T. (2015). M-autonomy. *Journal of Consciousness Studies, 22*(11–12), 270–302.
Morsink, J. (1999). *The Universal Declaration of Human Rights: Origins, drafting, and intent*. Pennsylvania studies in human rights. University of Pennsylvania Press.
Mowbray, A. (2004). *The development of positive obligations under the European Convention on Human Rights by the European Court of Human Rights*. Bloomsbury Publishing.
Nowak, M. (2005). *U.N. covenant on civil and political rights: CCPR commentary* (2nd, rev. ed.). Engel.
O'Callaghan, P., & Shiner, B. (2021). The right to freedom of thought in the European Convention of Human Rights. *European Journal of Comparative Law and Governance*, 1–34.
Paulo, N., & Bublitz, J. C. (2016). Pow(d)er to the people? Voter manipulation, legitimacy, and the relevance of moral psychology for democratic theory. *Neuroethics, 12*, 55–71.
Persson, I., & Savulescu, J. (2012). *Unfit for the future: The need for moral enhancement*. Oxford University Press.
Persson, I., & Savulescu, J. (2019). The irrelevance of a moral right to privacy for biomedical moral enhancement. *Neuroethics, 12*(1), 35–37.
Pettit, P., & Smith, M. (1996). Freedom in belief and desire. *Journal of Philosophy, 93*(9), 429–449.
Petty, R. E., & Cacioppo, J. T. (1984). The effects of involvement on responses to argument quantity and quality: Central and peripheral routes to persuasion. *Journal of Personality and Social Psychology, 46*(1), 69.
Rosenfeld, J. P. (2005). Brain fingerprinting: A critical analysis. *The Scientific Review of Mental Health Practice, 4*(1), 20–37.
Russell, B. (1957). The value of free thought. In B. Russell, *Understanding History* (pp. 44-83). Philosophical Library.
Schabas, W. (2016). *The European Convention on Human Rights: A commentary*. Oxford University Press.
Schwitzgebel, E. (2019). Belief. In E. N. Zalta (Ed.), *The Stanford encyclopedia of philosophy* (Fall 2019 edition). https://plato.stanford.edu/archives/fall2019/entries/belief/
Sententia, W. (2006). Neuroethical considerations: Cognitive liberty and converging technologies for improving human cognition. *Annals of the New York Academy of Sciences, 1013*(1), 221–228.

Sharot, T., Kanai, R., Marston, D., Korn, C. W., Rees, G., & Dolan, R. J. (2012). Selectively altering belief formation in the human brain. *Proceedings of the National Academy of Sciences, 109*(42), 17058–17062.

Shiffrin, S. V. (2011). A thinker-based approach to freedom of speech. *Constitutional Commentary, 27*, 283.

Special Rapporteur on Freedom of Opinion and Expression. (2018). *Report to the General Assembly.* UN Doc. A/73/348.

Special Rapporteur on Freedom of Religion or Belief. (2010). *Report to the Human Rights Council.* UN Doc. A/HRC/16/53.

Special Rapporteur on Freedom of Religion or Belief. (2011). *Rapporteur's digest, 1986–2011.* https://www.ohchr.org/documents/issues/religion/rapporteursdigestfreedomreligionbelief.pdf. January 25, 2021.

Special Rapporteur on Freedom of Religion or Belief. (2017). *Report to the Human Rights Council.* Un Doc. A/HRC/34/50.

Taylor, P. M. (2005). *Freedom of religion: UN and European human rights law and practice.* Cambridge University Press.

Taylor, P. M. (2020). *A commentary on the international covenant on civil and political rights.* Cambridge University Press.

Thaler, R. H., & Sunstein, C. R. (2009). *Nudge: Improving decisions about health, wealth and happiness* (2nd ed). Penguin Books.

Tussman, J. (1977). *Government and the Mind.* Oxford University Press.

Vermeulen, B., & van Roosmalen, M. (2018). Freedom of thought, conscience, and religion. In P. van Dijk, G. J. H. van Hoof, A. B. van Rijn, & L. Zwaak (Eds.), *Theory and practice of the European Convention on Human Rights* (5th ed., pp. 735–763). Intersentia.

Waclawczyk, W. (2019). *Freedom of thought and its relation to freedom of expression.* Wydawnictwo Naukowe.

Walsh, C. (2010). Drugs and human rights: Private palliatives, sacramental freedoms and cognitive liberty. *The International Journal of Human Rights, 14*(3), 425–441.

Walsh, C. (2014). Beyond religious freedom: Psychedelics and cognitive liberty. In B. C. Labate & C. Cavnar (Eds.), *Prohibition, religious freedom, and human rights: Regulating traditional drug use* (pp. 211–233). Springer.

Williams, G. (1956). The concept of legal liberty. *Columbia Law Review, 56*(8), 1129.

CHAPTER 4

Freedom of Thought and the Structure of American Constitutional Rights

Marc Jonathan Blitz

Freedom of thought has long been a core value in American jurisprudence and that of other legal systems. (See Swain, 2021). Thanks to modern neuroscience—and the technologies it makes possible—it may soon also become a legal *right*. That is, the freedom to think might not only be something that we value and celebrate, but something that the judicial system needs to protect—and, in order to do so, more clearly define.

Consider two roles that rights play in American constitutional jurisprudence—and why it is that technological advances may require a right to play these roles in protecting our thought. First, rights generate a barrier of sorts—a judicially administered force field—that keeps state power from entering and exercising control (or monitoring what we do) in spheres where the state is not meant to be, often because such spheres have to be reserved for individual autonomy or privacy. The Supreme

M. J. Blitz (✉)
Law, Oklahoma City University School of Law, Oklahoma City, OK, USA
e-mail: mblitz@okcu.edu

M. J. Blitz et al. (eds.), *The Law and Ethics of Freedom of Thought, Volume 1*, Palgrave Studies in Law, Neuroscience, and Human Behavior,
https://doi.org/10.1007/978-3-030-84494-3_4

Court described this function of constitutional rights in the 2003 case of Lawrence v. Texas: There are "spheres of our lives and existence," it said, "where the State should not be a dominant presence"—spheres that are the realm of an "autonomy of self" and that encompass "freedom of thought, belief, expression, and certain intimate conduct." (Lawrence v. Texas, 2003, 562).

Second, there is another function of constitutional rights which is not to keep the state entirely out of a realm in which it has no place, but to protect certain interests (often in safeguarding individual autonomy or privacy) even when they become intertwined with activities *a state does have a need to regulate* (e.g., in protecting public health, safety, or some other interest of the public). Speech, for example, is constitutionally protected in the United States not only when it is in a book, e-mail, or other text, but also when it comes from protestors or pamphleteers whose actions can impact traffic or other aspects of the shared environment. Personal privacy is constitutionally protected not only when we are in our homes, but also (at least to some degree) when we travel on roadways or surf the World Wide Web and interact with other Web users.

Freedom of thought has had little need for either of these types of rights-based protections. But neuroscience-related technologies are changing this state of affairs. Consider first how brain scanning, brain-computer interface devices, and other technology may create the need for a strong barrier to keep government from manipulating or surveilling our thinking. Until now, this was largely unnecessary. As Justice Frank Murphy wrote in a 1942 Supreme Court opinion, "[f]reedom to think is absolute *of its own nature*" since even "the most tyrannical government is powerless to control the inward workings of the mind." (Jones v. Opelika, 1942, 618). As Frederick Schauer has written, "thought is intrinsically free. The internal nature of the thought process erects a barrier between thought and the power of government sanction." (Schauer, 1982, 93).

Of course, internal thought is often bound up with external action. The words we express are the product of, and also embody and convey, our thoughts—and non-speech conduct is also often preceded and guided by mental deliberation. Speech and non-speech conduct not only embody and result from thought, but they are crucial inputs to it. Most of the raw material for our perceptions and beliefs comes from the outside world—from books we read or speeches we hear, and from actions we take to generate those beliefs. So government can control thought indirectly by controlling the speech and other external conduct that embodies and supports it. But while these outer manifestations of, or inputs to, thought have been shielded by free speech or constitutional protections for liberty of action, "the *inward* workings of the mind" have not required such rights-based liberty protection.

In an age of neuroscience and neurotechnology, however, this is unlikely to remain the case. Functional Magnetic Resonance Imaging (fMRI) and other brain imaging technology may allow officials to monitor and punish even thoughts that remain unexpressed. Emerging technologies, such as new kinds of mind-altering drugs and brain-computer interface devices, may let them manipulate our minds without censoring our words. As Jan Christoph Bublitz points out, freedom of thought is thus more vulnerable in an "age of neuroscience" where the state (or another entity) is "able to peer into the brains and minds of citizens and posses[s] the tools to alter thoughts, beliefs and convictions" (Bublitz, 2014). As Richard Glen Boire has stressed freedom of thought "can no longer exist in a Cartesian quarantine, blind to the connection between our thoughts and our brains." (Boire, 2004; See also Fox & Stein, 2015). Other scholarship has also analyzed the challenges raised by these technologies (See, e.g., Boire, 2001a, b; Sententia, 2004; Kolber, 2006, 2008; Stoller & Wolpe, 2007; Fox, 2008, 2009; Blitz, 2010b; Farahany, 2012a, b; Ienca & Andorno, 2017; Lavazza, 2018; McCarthy-Jones, 2019).

The most straightforward response to these technological changes is one that restores in law the *absolute protection* for thought privacy and integrity that was once provided by nature. As I have written in earlier work, "[t]o the extent fMRI or other brain-based mind reading technologies widen a crack in the wall nature erects around our thought processes, one might argue that the law should seal it up again." (Blitz, 2017). This would call for a fairly simple constitutional right: one which protects thought with an impermeable legal barrier against state interference or monitoring.

But this is too simple a model for a right to freedom of thought. It covers what we might think of as the "core" of such a right, but not the "periphery" or area outside of that core. As noted above, rights don't always simply keep the state entirely out of a certain activity. They also provide for more nuanced protection where the state *cannot* be entirely kept out of certain sphere—but is still prevented from doing any more harm, to expressive liberty or another liberty, than is necessary to further its legitimate interests (and perhaps allowed to act only when the state's interests are unusually strong).

This is more likely to be the case when the "inward workings of the mind" become accessible to the state not only because fMRI or other technology allows the state to *enter into* these mental processes in some sense, but also because our mental processes expand "outward" (or we come to understand ways in which they always have been external to our body in certain respects). Andy Clark and David Chalmers argue, in setting out their theory of the "extended mind," that the physical action that underlies our thinking may occur not just in our brain, but in the environment around it: When a person relies automatically on a smartphone or other computer, and not simply her natural biomemory, to store and retrieve important biographical or factual information, then that computer-stored memory may be just as integral to her mental life as the memories encoded in and retrieved from her neurons (Clark & Chalmers, 1998).

Some of this externalized thinking (or support for thinking), perhaps, merits absolute protection of the kind many assume must cover our internal thoughts. Private mental processes may otherwise become vulnerable to hackers or to the manipulation of companies that manage the links between computer-stored memory and the "cloud." (Carter, 2021; Carter & Palermos, 2016). Certain scholars have thus called for extending absolute or extremely strong protection of thought to cover threats presented by digital technologies. Susie Alegre, for example, has done so in arguing that it is not only the rise of such neurotechnologies, but also other aspects of the "digital age" that create the need for "a thorough assessment of what an interference" into thought "could look like." (See Alegre, 2017). Simon McCarthy-Jones likewise offers, in his chapter in this volume, arguments for robust protection of thought not only in the face of "brain reading" with fMRI and other scanning technology, but also "behavior reading" based on surveillance of our Internet activity (See McCarthy-Jones, 2021). But at least some of thought that is externalized in brain-computer interfaces or other mediums outside of the body may become intertwined with activity the state likely has to be able to monitor or regulate: If, for example, planners of a crime decide to move their planning from electronic or verbal communications (which can be monitored by officials with a warrant) to exchanges that occur through brain-computer interface devices, we should pause before assuming that freedom of thought will give them an impermeable privacy shield there.

This chapter therefore provides a sketch of certain features a jurisprudence of freedom of thought may have if that freedom is protected by a more complex, multifaceted right. For reasons explained more fully below, the first step in providing such a sketch is to explain what it is that the right protects—its "coverage," "scope," or "domain." In *Stanley v. Georgia*, the Supreme Court of the United States said that the American Constitution makes it impermissible for government to "premise legislation on the desirability of controlling a person's private thoughts" (Stanley v. Georgia, 1969, 566). "Our whole constitutional heritage," it stressed, "rebels at the thought of giving government the power to control men's minds" (Id., 565). But what exactly constitutes "giving government power to control men's minds"? If, as the court has suggested, any exercise of such power would violate a constitutional right to "freedom of thought" or "freedom of mind" (Wooley v. Maynard, 1977, 714), what exactly does that freedom protect—or protect against?.[1]

What, as Bublitz asks, "is the content of the right – what falls under its ambit, which measures interfere with it?" (Bublitz, 2014). Is it really a single right? Or is it rather a set of separate related rights, each of which covers different interests and applies in its own way against specific types of interferences? In the American context, for example, a right against direct interference, or manipulation of, individuals' brain functioning may have one constitutional source, and set of legal implications. It is closely related to, and perhaps an extension of, the bodily autonomy protected by the due process clauses of the Fifth and Fourteenth Amendments. Government officials, one might argue, violate a different kind of mental liberty if they manipulate our thinking with external stimuli (such as with "subliminal" stimuli, to the extent that is possible, or with "dark patterns"on Web sites or other means of manipulating thought by manipulating our environment). The right to mental privacy is arguably different. Under current constitutional law, government generally has

[1] Frederick Schauer has likewise observed that "fMRI scans and other techniques of modern neuroscience" may "make inquiring into the topic of freedom of thought (as opposed to the external manifestations of that thought) more important now than would have been the case a generation ago" (Schauer, 2015, 444 n.78). In more recent work, Schauer has expressed skepticism about an "independent principle of freedom of thought," but, in doing so, "defer[s] for the time being the growing possibility that psychotropic, surgical, electronic, and other technological advances might increase the possibility of literally changing an agent's thoughts, and put[s] aside as well the possible technological techniques by which external forces might now or in the future actually know what I am thinking without my exhibiting any external manifestations of my thought" (Schauer, 2020).

more power to monitor what individuals say in Internet communications than to control the content of those communications, Constitutional privacy rights impose some constraint on government surveillance, but not an absolute barrier. Might there be a similar difference between the way (and extent) our constitution protects our mental integrity and the way it protects mental privacy? Moreover, even if we have a right against *others'* shaping or observing our minds, does this *also* mean we have a right to use cognitive enhancement tools, brain-computer interface devices, or other technologies to *reshape our own* minds?

In short, the law may offer very different freedom of thought protections for (1) our "mental integrity" or what some scholars call our right against mental "manipulations" or "interventions," – and may protect it differently when the threat directly targets the biology underlying thinking that when it shapes it with external stimuli (2) our "mental privacy," and (3) our right to voluntary mind modification.[2] Within these categories, courts might draw other distinctions: Law may respond one way, for example, when the threat of thought manipulation comes from government, and another when it comes from a corporation or other private actor. It may leave others—and perhaps even government officials—more freedom to shape others' thinking with education or through other communications of that kind the First Amendment generally protects than through more "direct" interventions into brain operations (See Bublitz, 2014; Bublitz & Merkel, 2014, 69–74). And even where surveillance or shaping of thought that is impermissible in most settings, some have explored the question of whether it should be permissible in unusual situations, where, for example, extraordinary national security interests are at stake (See Lavazza, 2021). Finally, there are numerous other distinctions one might draw within the different categories sketched above: Individuals might claim a right to technologically modify their own minds in a variety of ways—to reinforce their belief in a particular proposition, to change their emotions about another person or

[2] Other writers on freedom of thought have proposed different taxonomies for classifying its components. Ienca and Andorno, for example, describe it as consisting of a "right to mental privacy, the right to mental integrity and the right to psychological continuity" (Ienca & Andorno, 2017). Alegre describes it as including, at a minimum, the right not to reveal one's thoughts or opinions, "the right not to have one's thoughts or opinions manipulated; and the right not to be penalised for one's thoughts" (Alegre, 2017).

task, or to reshape their cognitive capacities or features of their personality, for example. That a right protects one of these modifications doesn't mean it protects others to the same extent and in the same way.

It is also conceivable that the legal protection freedom of thought receives in one jurisdiction will differ from that it receives in another. If and when American courts elaborate upon the First Amendment right to freedom of thought described in Stanley, or perhaps find it in other constitutional provisions, the right they elaborate may be different in its scope and application from that which European courts find in Articles 9 of the European Convention on Human Rights and Article 10 of the Charter of Fundamental Rights of the European Union.

These questions have been insightfully analyzed and explored in the emerging scholarship of freedom of thought—largely through the lens of philosophical thinking about what autonomy interests are at stake, and what justifications government might have to shape or monitor thought, or put restrictions on how we shape it ourselves. In this chapter, I look more closely at these questions about freedom of thought through another lens—namely that of American constitutional doctrines developed for more familiar constitutional rights: The First Amendment right to freedom of speech, the Fourth Amendment right to privacy from government surveillance, and the Fifth and Fourteenth Amendment right to bodily autonomy and other liberties.

In the first part of the chapter, I will describe a template American courts use to understand these constitutional rights that might also shape a future right to freedom of thought—at least as it develops in American constitutional law. I will describe key features of how courts have defined what Frederick Schauer has called the "coverage" of, and "protection" offered by, more familiar constitutional rights (Schauer, 1982, 89)-and explain how it is helpful to understand these two concepts together as often establishing a "core" and "periphery" within the coverage of each right. In short, the First Amendment doesn't provide a shield of equal strength (it doesn't provide the same level of protection) against all government regulation of expression. Rather, it protects what it has called "core political speech," for example, more strongly than it protects commercial speech (See, e.g., Meyer v. Grant, 1988, 420). It firmly blocks government officials from restricting, controlling, or punishing public debate in the streets, or on the Internet, but gives them far more room (albeit not unlimited power) to control the discourse that occurs in public schools or government workplaces (See, e.g., Connick v. Myers,

1983; Hazelwood v. Kuhlmeier, 1988). The Fourth Amendment likewise staunchly protects us in our homes and other private places (See Kyllo v. United States, 2001; Silverman v. United States, 1961), but protects it more weakly in many other settings—such as school hallways, in public employment, and as we travel on public roadways (See Delaware v. Prouse, 1979, 654; Skinner v. Ry. Labor Executives Ass'n, 1989, 625; Vernonia School District v. Acton, 1995, 668).

Second, I will consider how the above-described template might apply to freedom of thought and help us begin to make sense of its complexity—with two different possible approaches. First, an interest in freedom of thought might be important not because it gives rise to an independent right to exercise such freedom, but because of the way it affects courts' analysis of other more familiar constitutional freedoms: Free speech and privacy protections may be more robust when it is not just expression or freedom from surveillance that is at stake, but also our mental autonomy and privacy. Constitutional protections against state control over, or restraints on, our bodies may be stronger if the state is trying not only to control our bodies, but alter the workings of our brains (to control our mental processes).

Alternatively, an interest in freedom of thought can provide the groundwork for an *independent* right to that freedom. When courts elaborate this right, they may do what they have done in the jurisprudence of free speech, privacy, and other constitutional rights: distinguish between a "core," where the right is a strongest, and a "periphery," where it has power but is more likely to give way to other interests. This is at odds with how the right to freedom of thought is sometimes described- as a right that is *always* of absolute strength. But I will argue that where our claims to freedom of thought come with potentially significant costs to others (or make claims to resources that have importance for the achievement of other public purposes), this claim to absolute protection is unlikely to be plausible for courts—although certain components of our freedom of thought may offer protection that is close to absolute.

This might help to explain some of the examples, considered earlier, of ways that government might shape our thinking, or place limits on how we shape it ourselves. When officials limit our access to, or use of cognitive enhancement technology, for example, it may be that they are generally acting in what I am calling the "periphery" of the coverage of a right to freedom of thought. That is, such limits might apply in circumstances where government has significant interests—in protecting health and safety—that may compete with, and sometimes override, the interests in mental autonomy that we pursue using thought-enhancement technology (See Blitz, 2016b, 2021).

That doesn't mean that regulation of such technology will *always* be in this periphery. Matters may be different when government engages in such regulation not with the aim of protecting our health and safety (and with a plausible account of how its regulation does so)—but rather with the aim that Stanley v. Georgia declared constitutionally impermissible: namely, the aim "of controlling a person's private thoughts" (Stanley v. Georgia, 1969, 566). In short, while certain means of mental manipulation—such as compelled psychosurgery—will likely *always* be constitutionally impermissible, other measures, such as depriving individuals of cognitive enhancement technology, may likely be constitutionally impermissible only when the government carries them out with an impermissible motive, or does so in a way that causes far more harm than necessary to autonomy interests.[3]

Third, I will also briefly discuss some additional considerations courts use in deciding what is at the core of a particular right and what is at the periphery. This is in part a matter of where the interests protected by the right (such the autonomy that underlies freedom of speech, or privacy at stake in Fourth Amendment interest) are at their strongest, and the countervailing interests, such as the safeguarding of public safety, are weakest. But it is also in part a matter of social convention: Interventions into our brain, or technologies that shape or observe our unexpressed thoughts, often intuitively seem like the gravest violations of any principle of freedom of thought not just because they cause more harm to autonomy than, say government surveillance of our Internet activities, but rather because they are invading an arena for our thinking where we have long been *used to* being insulated against government surveillance and control. Rights, in other words, often have a status quo bias: One important guide to how they will protect a certain interest is to understand how that interest has been protected in the past, legally or in other ways.

[3] In fact, this kind of "motive analysis" has already been suggested by Jane Bambauer as a way to test the constitutionality of government measures that interfere with freedom of thought by restricting individuals' acquisition of knowledge) (See Bambauer, 2014, 69, 87–89).

To be sure, any sketch of a jurisprudence that hasn't yet emerged is necessarily tentative. It would have been impossible for a writer in the mid-twentieth century to predict the contours of our current twenty-first-century Fourth Amendment law without knowing certain details about how smartphones store information or track movements in public space. Likewise, it is impossible to know how courts will analyze threats to mental liberty without knowing more about how these threats will develop. As Dov Fox points out, for example, brain scanning technology is likely to merit one constitutional analysis when it reveals "a subject's cognitive thoughts and propositional attitudes, such as normative judgments, religious convictions, and hopes or fears for the future" and a very different constitutional analysis when it reveals "the less privileged sphere of sensory recall and perceptual recognition" (Fox, 2008, 2). But to the extent that the large-scale structure of rights jurisprudence in American law remains stable—to the extent that courts continue to use certain techniques for defining the coverage and protection of a right, and tend to divide the coverage of each right into core areas that receive greater protection and other areas where protection is lower, it is at least illuminating to imagine how an emerging right to freedom of thought or "cognitive liberty" may fit this larger structure.

The Structure of Constitutional Rights

Coverage and Protection

In his work on First Amendment doctrine, Frederick Schauer proposes that scholars distinguish between what he calls the "coverage" of a constitutional right and the degree of "protection" that right offers against a type of government intrusion. Schauer illustrates this distinction by analogizing a right to a knight's "suit of armour" (Schauer, 1982, 89). "A suit of armour," he notes, will cover a person and in doing so, provide at least *some* protection to all parts of the body it shields. But that protection may not be absolute: It will protect "against rocks, but not against artillery fire." Still, Schauer points out, the lack of absolute protection doesn't mean that the armor is useless: "The armour does not protect against everything; but it serves a purpose because with it only a greater force will injure me" (Id.). Similarly, he writes, even when a right provides less than absolute protection, it still provides a barrier against government restriction: It requires the government to provide a sufficient justification for regulating whatever is covered by the right (Id.).

Courts, therefore, often have to engage in a two-step inquiry when a challenger invokes a constitutional right—for example, by claiming that their right to freedom of speech has been violated. First, they have to analyze whether the conduct that the government is regulating falls within the coverage of the right. In First Amendment free speech law, for example, courts might ask whether the activity that government is restricting counts as "speech" within the meaning of the First Amendment. In many cases, the answer is clear. If government were to arrest a person for posting an anti-government message in a blog post, a social media message, or a newspaper editorial, there is no question it would be restricting speech. It would, thus, face a First Amendment barrier against such an arrest. Other scenarios present harder cases. Courts have struggled with the questions of whether (and, if so, when) the First Amendment's "freedom of speech" protects individuals who record public events with a cell phone camera (American Civil Liberties Union v. Alvarez, 2012, 595), disseminate computer decryption code that can allow others to duplicate copyright-protected movies (University City Studios v. Corley, 2001), design and sell cakes for customers buying them for weddings, birthdays, or other occasions (Masterpiece Cakeshop v. Colorado Civil Rights Commission, 2018), or provide psychological counseling that most psychologists view as ineffective and harmful (such as therapy aimed at changing individuals' sexual orientation) (King v. Governor of New Jersey, 2014; Pickup v. Brown, 2014).

In these cases, the coverage question is framed as being about whether a certain form of conduct constitutes "speech," protected by the First Amendment. The question is whether a certain activity, like cake design or dissemination of decryption code, is covered by the First Amendment's free speech "armour." But the Court's coverage question might instead be focused not on what government is restricting, but on aspects of the restriction other than its target (such as the government's motive or justification). Free speech law might protect a kind of conduct not as a general matter, but only against certain types of threats from government. A bomb threat, for example, is arguably not covered by the First Amendment. "True threats"—that is threats to commit unlawful violence—are not, as a general matter, constitutionally shielded from government restriction and punishment. But they *are shielded* against government restrictions driven or defined by ideological rather than safety concerns: If, for example, government *only* prosecutes threats made by critics of the government and not its supporters, courts will generally find such ideologically-motivated speech restriction to be unconstitutional (R.A.V. v. St. Paul, 1992, 387–390).

Second, once a court is convinced that a certain type of conduct counts as "speech"—or that the restriction of counts as an infringement of freedom of speech—and is therefore within the coverage of the Constitution's free speech clause, it will then ask how much protection the First Amendment rights-holder receives. In modern American constitutional law, this level of protection is generally defined in terms of what courts refer to as a level or tier of "scrutiny." The strongest level of scrutiny government generally faces is "strict" or "exacting" scrutiny (See, e.g., United States v. Playboy Entertainment Group, 2000, 813; United States v. Alvarez, 2012). This is the kind of scrutiny lawmakers or other officials usually face when they seek to stop individuals from expressing certain ideas—or, in other words, try to suppress speech that has a certain meaning or content. It is almost impossible for government to overcome: Officials can suppress speech on the basis of its content only when they have a government interest of the most extraordinary weight—what the court calls a "compelling government" interest—and, even then, when they cause no more damage to speech than they need to in order to achieve that interest (See id.). By contrast, the level of protection is lower when government regulates speech in a way that is "content-neutral." For example, a city ordinance might bar anyone from entering a public park after 10:00 p.m. Such an ordinance places a limit on protestors or other speakers: It prevents them from holding a protest at a certain place and time. But its restriction isn't targeting particular speech content. It applies to *all* speakers (and other potential park visitors) in the *same way* regardless of what they might wish to say in the park—or whether they wish to say anything at all (See Clark v. Community for Creative Non-Violence, 1989, 295–296; Occupy Fresno v. County of Fresno, 2011, 863). Such speech restriction receives only "intermediate scrutiny." Officials only need a government interest of "substantial" or "significant" weight (not a "compelling government") interest—and their speech restriction needs to be substantially related to that substantial or significant interest, but the fit need not be perfect: They can overshoot a little, and restrict more speech than necessary to achieve the interest, so long as they do not restrict "substantially more" speech than necessary (See Ward v. Rock Against Racism, 1989, 799).

The level of protection provided by the free speech clause, then, is not uniform throughout its coverage. It will rather vary depending on the type of speech or feature of the speech that the government is regulating, or perhaps some other characteristic of the way that government is regulating it. Speech content is shielded against government restriction by the nearly impermeable force field of strict scrutiny. The time, place, or manner of speech is shielded by the weaker force field of intermediate scrutiny. A few categories of speech content are also shielded more weakly than is most speech content: When government regulates advertising or other commercial speech, for example, it faces only intermediate scrutiny—not the strict scrutiny it normally faces when it restricts speech content. And, as noted earlier, the First Amendment's protection also abates when government takes on the role of a school administrator or employer—and regulates the speech of students or workers to assure the institution it runs can operate (Connick v. Myers, 1983, 147, 150–151; Garcetti v. Ceballos, 2006, 417–418; Hazelwood v. Kuhlmeier, 1969, 366; Tinker v. Des Moines School District, 1969, 737).

Core and Periphery of a Right's Coverage

The protection offered by a modern constitutional right is often at its strongest in circumstances that courts often describe as being at that right's "core." There are different ways that courts define a right's core—but often, it is a sphere where it is clearest that government has little justification to be in, or where the individual interests it protects are at their strongest. In First Amendment cases, the Court has often said that the kind of discussion most clearly (and strongly) insulated against government restriction is "core political speech" (See, e.g., Buckley v. American Constitutional Law Foundation, 1999, 639; McIntyre v. Ohio Elections Comm'n, 1995, 334). If there is any speech that government officials should be barred from restricting, it is criticism of government itself—and other speech integral to the deliberation necessary for democracy to function. The Court has also made clear that, even outside of political communications, the ideas we express cannot be limited simply because government finds them offensive or—motivated by paternalism—believes it should substitute other beliefs from the ones we have formed. "Content-based laws—those that target speech based on its communicative content" the Court has said, "are presumptively unconstitutional" (Reed v. Town of Gilbert, Arizona, 2015).

Of course, when courts define a certain subset of speech regulations as striking at a core, they implicitly classify others as being outside of this core—or in what we might call a "periphery." Here, government receives far more constitutional leeway to regulate speech: It cannot do so to impose ideological orthodoxy or shut off audiences from views or ideas the government believes they shouldn't read or hear. But it *can* do so when it puts aside any pretension to act (in Justice Jackson's words) exercising "guardianship of the public mind" (Thomas v. Collins, 1945, 545) and instead protects individuals from the physical harms or concrete disruptions that might accompany expression (when it occurs through burning objects or blocking traffic), or stem directly from it (in incitement or threats). It can do so when, instead of trying to control what people choose to say in public discourse or private conversation, it acts to manage, maintain order, and fulfill the institutional purposes of a public school, government workplace, or other organization defined by a particular mission. In all of those circumstances, the government must still act under constitutional rules that bar it from restricting speech without sufficiently strong reasons—and reasons of the right kind (Blitz, 2016a, 703–705). But the level of First Amendment protection is reduced.

One finds a similar division between core and peripheral realms of protection in the Fourth Amendment law. The Fourth Amendment of the United States Constitution protects individuals against "unreasonable searches and seizures." As the Supreme Court has said, its bar on "unreasonable searches" is designed to protect individuals against "a too-permeating police surveillance" (United States v. Di Re, 1948, 581, 595). But this protection against police surveillance is not equally strong everywhere. It is at its height in the home. As the Supreme Court said in Silverman v. United States, at the Fourth Amendment's "very core stands the right of a man to retreat into his own home and there be free from unreasonable governmental intrusion" (Silverman v. United States, 1961, 505, 511). Government officials can only search a home when they obtain a warrant from a neutral magistrate based on a showing of "probable cause" that there is evidence of criminal activity there. The Fourth Amendment extends the same staunch protection to other private spaces: Police must have a warrant to search through a person's purse or briefcase, or peruse the contents of her computer or cell phone (See Riley v. California, 2014, 386). But like the First Amendment, the Fourth Amendment's constitutional force field weakens as one moves away from this core: Law enforcement officers still need good reasons to search cars on public roadways or to pat down individuals on public streets under circumstances where an officer has reason to believe that "criminal activity may be a foot" (Terry v. Ohio, 1968, 30). But they don't need to obtain a warrant. The same schools, workplaces, and government organizations that are given greater leeway to regulate speech than lawmakers have to censor it also have greater leeway than police to surveil students, workers, or others who play certain roles in their institution in order to protect the community's safety and functioning: Schools and workplaces, for example, have been permitted by courts to randomly test certain students and workers for drugs so long as the procedures they use include adequate protections for Fourth Amendment privacy interests (See Board of Education of Independent School District, Pottawatomie Cty. v. Earls, 2002, 837–838; National Treasury Employees Union v. Von Raab, 1989, 679; Skinner v. Ry. Labor Executives Ass'n, 1989, 625; Vernonia School District v. Acton, 1995, 664–666).

One might argue that this spatial metaphor for how to understand a right adds little to what I previously said in the more general discussion of coverage and protection—namely that certain types of conduct within the coverage of a right receive greater protection from government interference than do others (or that certain types of government interference meet greater skepticism and resistance from courts). We can conceive of the subset with greater protection as a "core" surrounded by a "periphery" with less protection. But we need not conceive of this variable protection in spatial terms (especially as there may be varied levels of protection within the "core" or "periphery" of a right).

Still, the spatial metaphor is helpful in framing our understanding of these rights—largely because, the explanation for *why* a certain realm lies at the core of a right frequently depicts that core as being on the "inward" side of a constitutionally significant boundary line that, as Heyman describes it, divides the "outward realm of the state" from "the inward life of the individual" (Heyman, 2002, 657). As such, it is less the state's business than what is on the "outward" side of the line. Certain First Amendment scholars have drawn upon this kind of imagery to make better sense of how First Amendment law works. Burt Neuborne, for example, sees the First Amendment as beginning "in the interior precincts of the human spirit"—in its protection for religious liberty and conscience—and then extends its protection "outward, preserving the freedom to convey information and ideas to others," in protection for communication, and for freedom of the press (Neuborne,

2011, 18–19). Neil Richards makes use of a similar spatial metaphor for First Amendment protection. He conceives it as "a series of nested protections, with the most private area of our thoughts at the center, and gradually expanding outward to encompass our reading, our communications, and our expressive dealings with others" (Richards, 2008, 408). Interestingly, both of these visions of the First Amendment as a series of concentric circles place individuals' thoughts and beliefs at its core.

Envisioning free speech law as having a core and periphery also helps us make sense of scholarly arguments that aim to protect the strength of First Amendment protection by assuring that its core protections aren't confused with—or weakened to resemble—its periphery. Some writers do so by warning against defining the First Amendment's core too broadly—so that it includes even speech that government might intuitively have good grounds to restrict.

James Weinstein does this, for example, when he warns against stretching the First Amendment's core to cover speech beyond that which is necessary to sustain participatory democracy. In accordance with the courts' emphasis on political speech, he places democratic deliberation, not the exercise of intellectual autonomy, at this core. He asks readers to imagine a scientist invoking the First Amendment to protect dissemination of instructions or diagrams for producing a biological weapon—and notes that many will feel that government should have greater leeway to restrict such speech in order to protect public safety than strict scrutiny generally allows (Weinstein, 2011, 391). The problem, he points out, is that the same leeway for government will be out of place, and dangerous, if it is extended to allow government greater power to restrict core political speech. Consequently, he says, a sound theory of free speech law should "reserv[e] the most rigorous protection for the speech by which individuals participate in the democratic process, while at the same time providing meaningful but more flexible protection for other important free speech values, including important autonomy interests." (Id.)

Freedom of Thought as a Component of Other Constitutional Rights

This chapter began by describing freedom of thought as a distinct constitutional liberty—one that can stand on its own. Judges have sometimes seemed to do so as well. Stanley v. Georgia, for example, spoke of a First

Amendment right against mental manipulation. Drawing on this case, the Seventh Circuit Court of Appeals, in Doe v. Lafayette, emphasized that there is "no doubt" that government "runs afoul of the First Amendment when it punishes an individual for pure thought" (Doe v. City of Lafayette, 2004, 765).

But there is another way this freedom might be a part of American constitutional law: It might exist not as a free-standing right, but only as a component of other, more familiar constitutional rights. It may, for example, be a component of free speech protection. Or a component of the Fourth Amendment's protection of privacy. Or the privilege that criminal defendants have under the Fifth Amendment to remain silent—when the alternative would be compelled self-incrimination. Or our Fifth and Fourteenth Amendment "due process" rights to be free from state interference with our bodily autonomy or other types of personal freedom deeply rooted in American society.

In fact, one might argue, a principle of freedom of thought may help explain why some state measures that potentially run afoul of these constitutional rights are likely to be seen by courts as striking at the core of these rights' coverage. In the Fourth Amendment context, for example, when a government measure doesn't only intrude into our privacy, but also gives officials information about our unexpressed thoughts—this may be a reason for courts to treat this as striking at the core of our Fourth Amendment interests even if the government surveillance is occurring outside the home—and in a setting where we normally have lower privacy interests. Imagine that government officials develop ways to conduct brain scans that can provide detailed inferences about the thought content of students in a public school setting, drivers at a road checkpoint, or travelers in an airport.

These are areas where courts have held that we have Fourth Amendment privacy interests—and receive Fourth Amendment protection against unreasonable searches. But they have held that suspicionless searches of a kind that are unreasonable elsewhere are reasonable there. Public school students participating extracurricular activities may be subjected to random drug testing (Board of Education of Independent School District, Pottawatomie Cty. v. Earls, 2002, 837–838; Vernonia School District v. Acton, 1995, 668). Drivers can be subjected to warrantless breathalyzer tests to determine if they are driving under the influence of alcohol or drugs (Birchfield v. North Dakota, 2016, 2177–2178). They can also have their car searched for drugs by a dog trained to alert when it smells such drugs (Illinois v. Caballes, 2005, 408–409). And they can be stopped at a certain location and asked questions by law enforcement about whether they witnessed a recent accident at that location (Illinois v. Lidster, 2004, 427–428).

Airport security can and does conduct weapons searches on all individuals who enter an airport, not only those who they have reason to think might have weapons. But as I have written in earlier scholarship, that government can use suspicionless drug tests in schools and roadways, or use millimeter scanning devices to scan all travelers in airports, does *not* necessarily mean it could likewise use brain scanners to draw inferences about thoughts in the same situation (Blitz, 2017). We are normally *outside* the core of Fourth Amendment rights in these situations—even when government intrudes into our bodily privacy, as it does when it conducts random drug tests. But when government measures in these settings intrude upon our *mental* privacy, this arguably moves the government intrusion back into the core of Fourth Amendment protection—because it arguably gives government access to information that is more deeply private (and less the government's business) than information about whether we have a certain type of alcohol or another type of psychoactive drug in our blood (See Blitz, 2017; Farahany, 2012b, 1288–1289; Pustilnik, 2013, 12–15). A brain scan may require a warrant even in settings where government has generally been free to conduct warrantless (and suspicionless) searches of our clothing, a bag or package we are carrying, or our bodies (Blitz, 2017; See also Pardo and Patterson, 2013).

To be sure, this does not mean that *any* government collection about our mental state will necessarily trigger heightened Fourth Amendment scrutiny: A law enforcement officer who stops a driver on a roadway and asks her to touch her nose or who observes whether the driver is slurring her speech is, in doing so, gathering information which is intended to allow the officer to draw an inference about the driver's mental state—more specifically, whether the driver's concentration, decision-making ability, and awareness of her surroundings have been impaired by alcohol or some other drug. But where a law enforcement officer's investigatory methods entail a deeper intrusion into mental privacy—where they give the officer a window of sorts into thoughts that are normally not visible at all and perhaps are likely to be irrelevant for assuring road safety—then a court may well ratchet the degree of Fourth Amendment protection back up, and demand a warrant, or possibly even greater justification from the government, before allowing the search.

Some Fourth Amendment scholars have correctly pointed out that current Fourth Amendment case law on suspicionless searches does not give this kind of weight to mental privacy (See Farahany, 2012b, 1288–1289; Pustilnik, 2013, 12–15). But that courts have not yet emphasized mental privacy as a determinant of Fourth Amendment protection doesn't foreclose the possibility that they will do so. After all, Fourth Amendment protections in this area depend—according to the Court—on balancing of privacy interests and security interests—and where government is gathering information about our thoughts, the privacy interest may be much more significant (Blitz, 2017).

A principle of freedom of thought might have similar significance not only when government intrudes into our bodily privacy (in ways covered by the Fourth Amendment), but also when it constrains our bodily liberty (in ways also covered by the Fifth and Fourteenth Amendment). Just as government might face greater Fourth Amendment scrutiny from courts when it collects information from not just from our bloodstream but also from our brain, so it might face greater judicial scrutiny under the Fifth or Fourteenth Amendment when it exercises control over our body in a way that reshapes our mental operations. This is one lesson one might draw from a trio of Supreme Court cases addressing the question of when it is constitutional for government to compel prisoners or psychiatric patients to take anti-psychotic drugs against their will. In Washington v. Harper, in 1990, the Court asked whether a prison system could forcibly medicate a prisoner it deemed dangerous after a psychiatrist at the prison had authorized such treatment. The Court said in that case, that such compelled medication was permissible—but only where such a course of action served an important safety need and was found to be medically appropriate "by medical professionals rather than a judge" (Washington v. Harper, 1990, 231). In Riggins v. Nevada, in 1992, the Court found Nevada had acted unconstitutionally in compelling a prisoner to take anti-psychotic medications because it had failed to provide that "overriding justification and a determination of medical appropriateness" (Riggins v. Nevada, 1992, 135). In Sell v. United States, in 2003, it likewise found government officials had violated Sell's right to remain free from unwanted psychiatric medication when it compelled him to take this medication in order to make him competent to stand trial (Sell v. United States, 2003, 171–172).

The Court did not expressly state in these cases that individuals' constitutional rights to be free of compelled medication are stronger when such medication shapes a patients' mental operations, and not only their bodily freedom. In fact, in each of these cases, the Court majority seemed to go out of its way to avoid any discussion of freedom of thought. In Washington v. Harper, for example, it was only the dissenting opinion by Justice Stevens that emphasized the constitutional significance of compelled use of anti-psychotic medications for freedom of thought (Washington v. Harper, 1990, 237–238). The majority opinion, by contrast, characterized the constitutional concerns differently. In discussing the Harper's interest in being free from such medication, for example, it stressed a general interest in being free from compelled medical treatment and the numerous physical side effects that can arise from the drug Harper was administered—including "tardive dyskinesia," "a neurological disorder, irreversible in some cases, that is characterized by involuntary, uncontrollable movements of various muscles, especially around the face" (Washington v. Harper, 1990, 229–230). When striking down the measures in Riggins and Sell, the Court added another worry—which was that compelled psychiatric medication would undermine individuals' right to a fair trial by changing how they (and their counsel) might defend themselves. However, as reluctant as the Court was to frame these decisions in terms of freedom of thought, one might still argue that they are partly explained by a concern for such freedom: The Court is, in all these cases (even Harper, where it ultimately allowed forcible administration of psychoactive drugs), raising a constitutional bar against compelled use of psychotropic medications.

In both Fourth Amendment search and seizure law, and Fifth and Fourteenth Amendment due process law, freedom of thought concerns then would strengthen the constitutional protection that normally exists against intrusions into bodily privacy, under the Fourth Amendment, and bodily autonomy, under the Fifth and Fourteenth. If government monitoring of a person's brain chemistry or brain function compromises mental privacy, government may need a warrant—even if it normally wouldn't for similar intrusions into bodily privacy (such as random drug testing in schools or workplace). If government-compelled medical treatment not only causes unwanted effects to a prisoner's body, but also to her thinking processes, it may similarly face a higher level of scrutiny.

Other provisions of the Constitution may also protect freedom of thought—because they protect the thought we express in words or images, and may also extend to similar protection to the words and images we silently contemplate (but do not express). The Fifth Amendment, for example, gives to defendants in a criminal trial a privilege against having to testify against themselves—a privilege against self-incrimination. Even before a defendant has been charged with a crime, a defendant can exercise their "right to remain silent" in the face of police interrogation—and refuse to share information that might be used against them in a trial. Many scholars have argued that, if the government is constitutionally barred from forcing an individual to state incriminating words, it should likewise be barred from using brain scans or other technology to extract incriminating thoughts. Different scholars have given different versions of this argument. Some, like Dov Fox, Paul Root Wolpe, and Sarah Stoller have argued that forcing a criminal defendant to undergo *any kind* of neuroscience-enabled mind-reading would violate the privilege (Fox, 2009, 796; Stoller & Wolpe, 2007, 371). Others have understood the privilege to provide more limited protection against brain scans. Michael Pardo argues that the privilege bars compelled brain scanning that reveals "propositional content"—but not that which reveals psychological tendencies or characteristics (Pardo, 2006, 330). Nita Farahany argues that it bars government from obtaining evidence of unexpressed "utterances," but not most other kinds of unshared mental content (Farahany, 2012a, 366).

The First Amendment's protection for freedom of speech has likewise been understood by some scholars to protect freedom of thought. For some scholars, this is in part because individuals cannot be free to express themselves unless they are also free to engage in the thought that necessarily precedes such expression. As Neil Richards writes, one cannot protect "the marketplace of ideas" free speech is supposed to guarantee unless the law protects "the workshops" where "the ideas" in that marketplace "are crafted" (Richards, 2008, 396). For other scholars, protecting freedom of thought is not simply a necessary condition for freedom of speech—it is its central purpose. Rodney Smolla suggests that "the preferred position of freedom of speech" over other liberties can be traced to the fact that "speech is connected to thought in a manner that other forms of gratification are not" (Smolla, 1992, 11). Timothy Macklem writes that speech is integrally connected to thought and protected in part because of its role in shaping thought (Macklem, 2006, 11). (See also Blitz, 2010b, 1090–1094).

The most developed version of this "thinker-based" approach to free speech law comes from Seana Valentine Shiffrin. She argues that the central purpose of free speech protection is to secure "the individual agent's interest in the protection of the free development and operation of her mind" (Shiffrin, 2011, 287, 2014, 80–83). Freedom of thought, she writes, is—along with "freedom of communication"—one of "two related and mutually dependent freedoms" that it makes sense to place under the "the label, 'freedom of speech'" (Shiffrin, 2014, 79). The protection of freedom of thought, on this account, covers more than just protection for communication. It covers "mental contents" such as "non-discursive thoughts, images, sounds, and other perceptions and sensations as well as the workings of the imagination" (Id., 81, 113–114). Moreover, the interests supported by this account extend not only to generating particular thoughts or having certain mental experiences, but to developing an individual personality, and developing certain mental capacities (Id., at 87–88). Thus, "at the foundation of free speech protection" is not only a principle that forbids constraints on "interpersonal communication," but also "other measures that disrupt the free operation of the mind" (Id., at 94).

The Supreme Court has also sometimes treated freedom of thought as the underlying purpose of freedom of speech. It has said that freedom to speak is one component of a larger "freedom of mind" (Wooley v. Maynard, 1977, 714). "The right to think," it later said, "is the beginning of freedom," and we protect speech because "speech is the beginning of thought" (Ashcroft v. Free Speech Coalition, 2002, 253). It is perhaps not surprising then that Stanley v. Georgia's warning against allowing government to "contro[l] men's minds" came in a case that was, at least on the surface, about protecting the defendant's freedom of speech (Stanley v. Georgia, 1969, 366). If this is right, then a restriction of speech—or access to information—that intrudes more deeply into freedom of thought might, for that reason, be subject to more exacting judicial scrutiny.

Even for judges who do not believe that freedom of thought provides the underlying rationale for freedom of speech protection, protecting thought privacy may sometimes be necessary to protect speakers' expressive rights (or the derivative right of audiences to receive information). Imagine, for example, that government uses certain forms of brain-based "mind-reading" to identify individuals with dissenting views—and then exclude them from participating in certain public forums so that their views will not reach wide audiences. Or imagine that government shares certain types of historical information or scientific data *only* with individuals who can prove they have views of which the current administration approves. In these cases, government would be monitoring thoughts in order to identify and silence certain speakers, or deny information to certain readers. Its intrusion into mental privacy would likely be unconstitutional even if there were no constitutional right to mental privacy per se—because government would be using such mental surveillance to restrict speakers, or audiences seeking speech.

Treating a right to freedom of thought as a component—or variation—of another more familiar constitutional right would make it simpler for courts to spell out the constitutional implications of brain scans, cognitive enhancement technologies, or brain-computer interfaces. Rather than build a freedom of thought jurisprudence that doesn't yet exist, they could instead *refine* bodies of jurisprudence that *do*. And in making adjustments to search and seizure law or free speech law, for example, they might be guided by reasoning they have already used to adapt these areas of law to the destabilizing effects of other emerging technologies—such as cell phones, social media and other Internet communications, or GPS location tracking.

It might also, at least in the context of American constitutional law, provide a method of addressing the challenges with which the chapter opened—namely how might a jurisprudence of freedom of thought provide distinct protection against mental manipulation or mental privacy, or respond in different ways to government interference in our minds, or restraints on how we shape our own minds? Rather than answer such questions on a blank slate, one might argue, courts might instead ask whether and how each of the different protections for our mental freedom fit into an already-recognized constitutional right. Mental privacy, for example, might receive protection only to the extent it is protected by the Fourth or Fifth Amendment against law enforcement monitoring or extraction of our mental content, or protected by the First Amendment to protect the belief underlying speech. Our right to shape our own thought might sometimes fall within the coverage of the First Amendment's shield for formation and expression of beliefs and sometimes come within the different protection the Fifth and Fourteenth Amendment arguably offers to means by which we educate ourselves and shape our personal capacities.

But constitutional scholars should consider the possibility that a right to freedom of thought might exist as an *independent* right. The rise of new neuroscience-based technologies, or digital means of mental manipulation, might give rise to new threats to our mental autonomy in the twenty-first century for which the jurisprudence of the twentieth provides no answers. It is largely for this reason that Richard Glen Boire and Wrye Sententia proposed that courts protect not merely "freedom of thought" as traditionally conceived, but rather a broader right to "cognitive liberty." As Sententia defines it, this is a liberty that "updates notions of 'freedom of thought' for the twenty-first century by taking into account the power we now have, and increasingly will have, to monitor and manipulate cognitive function" (Sententia, 2004). Boire describes it as entailing a "right to control one's own consciousness" (Boire, 2000a)—and to do so using means that go beyond speech (including through pharmacological alteration of thinking patterns) and with protection from interference with thought that takes forms other than traditional censorship. Nita Farahany also argues for cognitive liberty protections—perhaps in the form of legislation rather than constitutional rights—to fill gaps that she identifies in the protection that current Fourth and Fifth Amendment law offer for mental privacy. (Farahany, 2012a, b).

As I have argued in earlier scholarship, First Amendment law might provide a "backstop" of sorts for some of these constitutional gaps (Blitz, 2017; See also Solove, 2007, 116–117)—especially when it is conceived broadly, as Shiffrin understands it, as entailing constitutional protection not only for communication, but for our "capacities for thought" and for "liv[ing] an autonomous life" (Shiffrin, 2014, 80). But even if such a right to freedom of thought finds a home in the First Amendment, it might be a right that is in many respects distinct from that of a right to free speech—and that requires a First Amendment jurisprudence extending substantially beyond that which courts have developed. This is especially true for what the chapter earlier called a "right to voluntary mind modification" with emerging technologies. Individuals, of course, have long-established First Amendment right to modify their thoughts by engaging in conversations, reading books, or watching movies. The Court has extended this right to video game play—and this extension arguably covers video games that reshape one's mental functioning with virtual reality interactions, neurofeedback, or other brain-computer interface technology (See Blitz, 2008, 2010a, 2018, 2021). But the right to modify one's thought with machines or other technologies (or consent to letting others engage in such modification, in psychotherapy, for example)

(See Blitz, 2016a; Haupt, 2016; Smolla, 2016) has received little analysis in existing constitutional jurisprudence.

Freedom of Thought as an Independent Right

A. **Freedom of Thought as an Absolute**

How then might we understand a right to freedom of thought as an independent constitutional right? How might courts define its coverage? Some cases appear clear-cut. Compelled neurosurgery aimed at reshaping our mental processes to government's liking would almost certainly implicate such a right[4] (See Kaimowitz v. Michigan Dep't of Mental Health, 1974; Winick, 1989, 19, 26). So too, if it were possible, would use of subliminal messaging to surreptitiously cause us to think or feel what the government wants us to think or feel (Bublitz & Merkel, 2014, 69–70; Scanlon, 1979). But other types of thought manipulation might raise more difficult questions: Would a person's right to freedom of thought be implicated by required education or training programs? Would a person be able to invoke such a right not only to insist that her mind remain free of manipulation by government, but also to modify her own mental functioning without government restriction (with drugs, BCI devices, or other technology)?[5] Moreover, even if a right to freedom of thought covered all of these circumstances, what level of protection would it offer?

There is perhaps more consensus on the level of protection that should come with a right to freedom of thought than there is about its scope or coverage. In short, freedom of thought is often described as an "absolute right" (See, e.g., Alegre, 2017; Richards, 2008, 2015). Frederick Schauer describes how we should understand such an "absolute" right in terms

[4] Any compelled surgery—of the brain or any part of the body—might violate American constitutional protections of physical liberty.

[5] In previous scholarship, I have explored different possible ways of understanding the "coverage" of a right to freedom of thought in American law. See Blitz (2008, 2010b, 2016a). Other legal scholarship has also explored similar questions. See also Shiffrin (2011, 2014), Bambauer (2014, 2018), and Kolber (2006).

of protection and coverage. "We may wish," he says, "to structure our rights such that protection is always absolute. The decision on coverage would be dispositive of protection, because no reason for restriction could outweigh the protection of the right" (Schauer, 1982, 90). If this were true of freedom of speech doctrine, for example, then any human conduct that counts as "speech" under the First Amendment would be *completely off-limits* to government regulation (or, alternatively, any government restriction that counts as the kind of restriction of speech to which the First Amendment presents a barrier would *always* face an *insuperable* barrier). Similarly, where an absolute right to freedom of thought applied (that is, where a certain activity or government restriction of it was within its coverage), government regulation would invariably be impermissible. To use other terminology above, the right would be all "core" and no "periphery."

As noted earlier, this is *not* true of modern First Amendment free speech doctrine. Certain writers have argued it should be—that free speech protection should be limited only to its core. In his 1961 article, *The First Amendment is an Absolute*, Alexander Meiklejohn made essentially this type of argument. Even though freedom of speech is not absolute in the sense of allowing anyone to say anything in any setting (or in any way they like)—it *is* absolute if one defines this freedom more narrowly. Freedom of speech, said Meiklejohn, is not the freedom to say anything anywhere (Meiklejohn, 1961). It is rather the freedom to engage in the discourse necessary to and in many ways, constitutive of—democratic self-government. It is in that latter sphere—in the content of that democracy-enabling discourse—which must remain *entirely* free of government officials' manipulation (Id.). But as is clear from earlier parts of the chapter, Meiklejohn's view of free speech coverage is not that of the modern court: Free speech law covers more than speech central to democratic deliberation. It covers commercial speech, for example, as well as the sometimes frivolous speech students engage in during or about school, and protects such speech, albeit more weakly than speech in public discourse (See, e.g., Mahanoy Area School District v. B.L. by and through Levy, 2021, 2048).

Why then might it make sense to define a right to freedom of thought as absolute when a right to freedom of speech is not? I've already noted one reason given by Justice Murphy: As a *practical matter*, he wrote, government simply *cannot* monitor and place limits on our unexpressed—and therefore unobservable—thoughts even it could find justification to do so. The internal nature of thought prevents it from doing so. The "[f]reedom to think," said Justice Murphy, "is *absolute of its own nature*" since even "the most tyrannical government is powerless to control the inward workings of the mind" (Jones v. Opelika, 1942, 618). Of course, if this remained true, courts would not need to build absolute protection for free thought into law because it would already be built into (and well-secured by) nature. The reason we may well need a constitutional right to freedom of thought is because the rise of brain imaging technology, brain-computer interfaces, and other development is eroding the natural protection that Murphy, in 1942, assumed would remain secure.

Why, then, if judges have to rebuild, in law, the freedom of thought protection we once found in nature should they make this protection absolute? The answer is that even if government officials *could* acquire the capacity to monitor or restrict our unexpressed thoughts, they might *always lack justification* to do so.

There are at least three possible reasons why a right to freedom of thought might be absolute in this normative sense. One was emphasized by the Justices in *Opelika*- both Murphy in dissent and the Justices in the majority—as they implicitly contrasted the absolute nature of freedom of thought with the more limited nature of freedom of speech. Our speech, the Justices in the majority observed, cannot be as completely shielded from government restriction as the "illimitable privileges of thought," because—unlike thought—speech affects others' rights and "ordinary requirements of civilized life compel [an] adjustment of interests" to balance expressive liberty with these rights. Speech, Justice Murphy likewise observed in his dissent, can produce "collisions with the rights of others" (Jones v. Opelika, 1942, 595, 618). Scholars have explained that it is in large part this possibility of collision with other rights that makes freedom of speech doctrine nuanced and complex—rather than a uniform shield of absolute strength. W. Bradley Wendel, for example, writes of free speech jurisprudence, "[t]he byzantine complexity of contemporary First Amendment law is [] the natural by-product of a recurring need to reconcile the basic political values of freedom and order" (Wendel, 2001, 359).

But if hidden and unexpressed thoughts do *not* threaten "order" in the same way, then freedom *of thought* jurisprudence will be free from the need for such repeated reconciliations, and the doctrinal complexity it generates. Free speech rights may have to leave room for government to protect individuals against defamatory or threatening speech, for example, or regulate speech that would otherwise disrupt the functioning of schools and workplaces. If unexpressed thoughts don't risk or cause similar disruption, then freedom of thought will not have to leave any such room for government intrusion. The shield it raises against government use of fMRI scanners or compelled memory dampening, for example, might remain impermeable.

There is a second reason that freedom of thought might be absolute in a way that freedom of speech is not: Even if unexpressed thoughts *do* create risks for public safety or otherwise threaten social order (see Schauer, 2015; Mendlow, 2018, 2021), it may be that—even so—constitutional law still cannot afford to let government officials regulate such risk-generating thought in the way it lets them address certain kinds of risk-generating speech (such as incitement). Our freedom to generate and hold our own beliefs, one might argue, is more central to our autonomy than our freedom to make certain statements. There is therefore less compromise that can occur in our right to freedom of thought and still leave us with minimal conditions necessary to live as free and autonomous individuals. A person who is required by government, or perhaps by other actors in society, to refrain from expressing her ideas in a certain circumstance can still silently hold those ideas in her mind—and perhaps give expression to them on another occasion. She can still silently explore and build upon those ideas. (Blitz, 2006). The realm of unexpressed thought, in other words, serves as a refuge of sorts where she can continue to exercise sovereignty over her life even when government officials (or other external actors) thoroughly control her external environment. If government actors extend their control into this refuge, by contrast, she will have nowhere to retreat (Christman, 1989).

Third, if the shield that nature once provided to our thought was absolute—because it completely blocked government from being able to monitor or control the invisible "inward workings of the mind" - then perhaps legal replacement for this eroded natural protection can only be considered effective if it comes with the *same degree* of protection that nature once provided. If we have come over decades to define our minimal mental freedom as entailing absolute protection against government observation of our unexpressed thoughts, then perhaps this gives us a claim to continue insisting on such absolute protection even when, thanks to the rise of fRMI and other technology nature can no longer provide it.

These arguments all have power with respect to at least certain aspects of freedom of thought. As Bublitz observes, our autonomy is more deeply threatened when government extends its power into a "forum internum"—our internal sphere—than it is when it exerts control over our external actions (including communication). As Alegre notes, the absolute nature we attribute to the right reflects its "profound importance for who we are as individuals and as societies" (Alegre, 2017). As McCarthy-Jones emphasizes in his contribution to this book, moreover, letting government exert control over some aspects of our thought (like speech we may consider less private than unexpressed thought) very likely gives it greater control over our thinking more generally (including our private contemplation). (McCarthy-Jones, 2021).

But it is hard to maintain that all of facets of our thought should receive such absolute protection. This is most clearly true of a right to voluntary mental modification. Thus, Bublitz, although arguing that the right to freedom of thought is absolute in certain respects, argues that a right to "re-configuration of one's own self"—although deserving of respect—"cannot be encompassed by the absolute protection of freedom of thought." Government, he says, cannot be required to ignore the "imminent dangers of psychoactive substances and social interests in the mental fabric of society" (Bublitz, 2014). This aspect of freedom of thought then may need to be reconciled with the public's need for order and safety. Some of those who write about the law and ethics of cognitive enhancement technologies have similarly stressed that the state may have *some* legitimate role in regulating their use—to protect the safety of those using the technologies. Henry Greely and his co-authors wrote in 2006, for example, that since "the risk of unintended side effects" from cognitive enhancement drugs is "both high and consequential," they should be available only under supervision from psychiatrists or other doctors (Greely et al., 2008, 704). Veljko Dubljevic has similarly discussed the possibility that cognition enhancement drugs might be available only to individuals who have been informed of their risks by a doctor or a mandatory course (Dubljevic, 2013, 179–187).

Moreover, if a right to freedom of thought is meant to recreate the mental privacy and integrity that nature once provided, then it may not automatically and invariably provide capacities for shaping our minds of a kind we lacked before the digital age and the rise of neurotechnology.

Bringing this kind of mind modification under the rubric of an absolute constitutional right to freedom of thought would present courts with a binary option: Either cognitive enhancement would be left entirely unprotected by the Constitution (leaving government free to regulate it any circumstances), or such cognition enhancement would be entirely shielded by it (and immune to state regulation).

The problem with such an approach, however, is that it is likely to limit the right's coverage so drastically that it will fail to protect an interest—in mental autonomy, for example—the moment that a plausible cause for government regulation exists on the other side of the balance. Anthony Amsterdam has warned about this problem in discussing Fourth Amendment search and seizure law. If we try to make the Fourth Amendment's protections against police surveillance strong—and keep them strong everywhere—such an "all-or-nothing" approach will probably cause courts to sharply limit the scope of the right so as not to constrain law enforcement too much (Amsterdam, 1974, 388). If Fourth Amendment protections come only in one super-strong form, courts will likely apply them only rarely—so they leave government with enough room to vigorously investigate and counter crime. The better solution perhaps is to abandon such an all-or-nothing approach, so that the right can still provide some protection for privacy even outside of the home (and other parts of the Fourth Amendment's core), but do so in a way that leaves government more room to function.

In fact, American courts have already limited the scope of the inchoate freedom of thought recognized in Stanley v. Georgia—and have arguably done so in part for this reason. In Stanley, the Supreme Court barred government from punishing someone (Robert Eli Stanley) solely for possessing an obscene film in his own home. But the Court subsequently made clear that this freedom of thought protection did *not* give adults a right to willingly view pornography in public places outside of the home (Paris Adult Theater v. Slaton I, 1973, 53–54), nor to transport obscene materials into their home for later viewing (United States v. Thirty-Seven (37) Photographs, 1971, 376; United States v. Orito, 1973, 140). Nor did it give them did not give them a right to possess or view child pornography, in their homes or anywhere else (Ohio v. Osborne, 1990, 108). In Doe v. City of Lafayette, the Seventh Circuit Court of Appeals—ruling en banc, that is with all judges of the court participating—said that Stanley's freedom of thought protection covers "*mere* thought, and not thought plus conduct." It thus rejected the argument of a convicted sex offender that he had been banned from public parks solely on the basis of the inappropriate sexual thoughts he had about children he observed in the park. Government, said the Court, wasn't regulating his thoughts, it was regulating his action (going to parks to observe children playing there). Giving constitutional protection to more than "pure thought" would leave government powerless to regulate all manner of conduct because "all regulation of conduct has some impact, albeit indirect, on thought" (Doe v. City of Lafayette, 2004, 765).

One might suggest, however, that even assuming the Seventh Circuit was correct in reaching that result, the jurisprudence of a right to freedom of thought need not be so narrow. Just as Fourth Amendment law provides some protection for privacy outside of the home, a right to freedom of thought might likewise continue to provide *some* protection for mental autonomy even outside of a core realm of the coverage of a right to freedom of thought—for example, when individuals are asserting a right to mind modification rather than mental integrity or privacy. It could continue to require that government officials provide *some* justification when they wish to intrude upon mental autonomy, even when their intrusion does not rise to the level that occurs in compelled psychosurgery or brainwashing, and even when government interests in regulation may have some force.

B. **The Right to Freedom of Thought as a Multifaceted Right**

As it does in the realm of freedom of speech, the level of protection that accompanies our right to freedom of thought may vary depending on exactly what it is being invoked to protect—or the type of threat that government action is presenting to our thought. Consider, for example, the possibility that a right to freedom of thought includes what Adam Kolber has described as "freedom of memory:" When the state tries to prevent us from using technology (like the drug, propranolol) to dampen or erase memories we no longer wish to retain, it may in doing so run afoul of a constitutional right we have to control the contents of our own minds (Kolber, 2006, 1622). But that doesn't mean that the state should be as powerless to preserve our memories as it to compel psychosurgery or engage in surreptitious belief manipulation. Even if we have a presumptive freedom to erase our own memories, one might argue it should remain illegal for us to do so where the justice system needs us to testify about those memories. As Kolber writes in analyzing the law and ethics of memory dampening in another chapter in this volume (and in previous scholarship), erasing one's memory might, at least in some circumstances, count as obstruction of justice (Kolber, 2006, 1589–1592, 2008, 145). When freedom of thought protects erasing and dampening memories, then, it may not protect it absolutely. Its level of protection might sometimes be weaker. It could conceivably be defined as a kind of strict scrutiny that gives way to the government's compelling interest in assuring that courts can have access to the evidence necessary to assure that justice is done. Or as a kind of intermediate scrutiny that would give government still more leeway to prevent us from reshaping our own memories in a way that undermines our ability to serve as witnesses. Or the freedom of memory might provide stronger protection to certain individuals (such as victims of a crime) than to others (like bystanders) (Kolber, 2021).

In this circumstance, our freedom of thought—like our freedom of speech—could produce a "collision with other rights" and require a jurisprudence that allows mental freedom to be reconciled with other interests (in this case, justice and safety) (Jones v. Opelika, 1942, 618). The right to shape our minds, then, might be subject to some of the same kind of analysis courts have done in free speech jurisprudence in reconciling speaker's autonomy interests with other important social interests (in this case, in the fair administration of justice).

In some cases, even rights to mental integrity or privacy may need to have certain exceptions that allow for government compulsion or interference in extraordinary circumstances. Consider, for example, circumstances where a particular person's unstated thoughts or intentions can have significant implications for others' safety: For example, when they are entrusted with piloting planes or with protecting key elements of national security. Thought surveillance may still be deeply concerning here. But some scholars have explored the question of whether government might not be absolutely prohibited from engaging in it, or from obligating certain officials (such as judges) to enhance certain mental capacities or to technologically-induce certain mental states (See, e.g., Chandler & Dodek, 2016; Lavazza, 2021). My general point here is that some activities may be covered by freedom of thought, but not protected by as strong a constitutional force field as we receive against paradigmatic examples of mental manipulation often viewed as entirely at odds with a free society (such as government-compelled psychosurgery).

A caveat is in order. It is sometimes difficult in both free speech law and search and seizure law—and might be also in a jurisprudence of freedom of thought—to distinguish types of regulations that are genuinely in what

I am calling a *periphery*, where the protection of a right is weaker, from regulations that are in a kind of *gray area*, where protection is uncertain—because the regulation may or may not be within the core of a right, or perhaps even within the coverage of the right, depending on how a court answers certain factual questions about the government's motives or other aspects of its regulation. Consider examples from free speech and search and seizure law. As I have already noted, government laws that restrict threats of violence may or may not face First Amendment free speech barriers—depending on whether government's targeting of the threat is selectively focused only on threats with certain ideologies (See Virginia v. Black, 1993, 359–360). Where government's selective restriction *is* ideological in nature, it will face strict scrutiny and thus be viewed as threatening a core area of First Amendment free speech protection (See R.A.V. v. St. Paul, 1992; Virginia v. Black, 1993, 359–360).

In search and seizure law, government typically faces no Fourth Amendment barriers when it obtains data collected by third parties, such as private communications companies (See Smith v. Maryland, 1979, 744–745; United States v. Miller, 1976, 443).[6] But matters are different if third parties have collected such data at the government's direction (United States v. Walther, 1981, 791–793; cf. Burdeau v. McDowell, 1921, 474–475). In that case, government will need to obtain a warrant based upon probable cause to see and use that data (since it is government, not voluntary private action, that produced the data). It may seem to some observers that free speech and Fourth Amendment protection are weaker in these circumstances (placing them at the periphery of each right), but it is more accurate to say that protection isn't weaker but rather indeterminate: It is not at some intermediate level of strength, but rather can be either of maximal strength (strict scrutiny in the First Amendment context, a probable cause-based warrant requirement in the Fourth) or non-existent, depending on what courts find when they analyze the government's actions. In this case, protection isn't simply weaker, coverage itself is in doubt.

[6] The Supreme Court has recently placed some limits on this "third party exception" to Fourth Amendment protections. Where a third party—like a cell phone company—is asked by the government to provide cell-site or other location data that would reveal an individual's whereabouts over an extended period of time, the government must obtain a warrant based on probable cause to obtain that information. See Carpenter v. United States (2018).

A right to freedom of thought might likewise offer protection that differs from absolute protection not in that it is weaker (as it would be if intermediate scrutiny applied) but more uncertain. This is what I have suggested earlier may be true when government bars individuals from using cognitive enhancement or memory-dampening technology to alter their own memories, emotional states, or thinking processes. Where government officials limit individuals' use of such technology for health and safety reasons, the right to freedom of thought may raise no barrier against their doing so. When they instead do so to prevent us from having certain mental experiences, they may face barriers as high as they do when they are within the core of the right. Of course, even where government officials *do* receive more leeway to restrict or otherwise regulate thought-shaping technology, the form such leeway takes may involve facing a lower barrier against regulation—rather than having a completely free hand. In the case of cognitive enhancement technology, for example, it may be the case that even where government has *valid health and safety interests* in limiting access to, or use of, such technology, it will not be entirely free from constitutional limits (See Blitz, 2010b, 2017, 2021). Even such health and safety interests will not give government an excuse to entirely ignore individuals' interests in exercising mental autonomy (or other freedom of thought interests). As between different means of achieving its health and safety interests, courts might find, government will be constitutionally required to choose the one that leaves individuals most free to exercise mental autonomy. The protection offered by a right to freedom of thought in this case would not be complete—government would still be able to restrict individuals' capacities to shape their own thought—but it would still exist in the form of a kind of "intermediate scrutiny" (See Blitz, 2016b, 301).

Whatever realm our right to freedom of thought may cover, how then would courts determine what level of protection exists for different areas within the coverage of such a multifaceted freedom of thought? How will they determine what belongs at the core and what belongs to the periphery of freedom of thought (and perhaps, draw more fine-grained distinctions between different levels of protection)? Answers to similar questions for free speech, Fourth Amendment privacy rights, or other rights are complex, nuanced, and marked by long-standing (and sometimes deep) disagreement between different judges, lawyers, and legal scholars. However, it is useful—in trying to imagine an emerging and future jurisprudence of freedom of thought—to take stock of two recurring themes in how courts have marked the boundaries of constitutional rights' coverage, and the (often blurry) lines that separate core from peripheral areas of that coverage.

Here is one high-level observation we might offer to explain what it is that makes protection so strong (sometimes, arguably approaching absolute protection) at the core of a right: This happens, perhaps, where the individual autonomy, privacy, or other individual interest protected by the right inevitably overwhelms any contrary interest government might invoke. And there are different variations on this kind of scenario.

A. **Balancing Individual Autonomy and the Public Interest**

First, a government interest might predictably be overridden because it is patently illegitimate—and thus cannot even get the government's case off the ground. Consider, for example, how the Court uses such an argument in *Lawrence v. Texas*, where the court found that laws criminalizing sodomy are unconstitutional. The government of Texas, the Court found, could have "no legitimate state interest" in controlling the private, consensual, and non-harmful sexual conduct of individual citizens. Our constitutional system doesn't accord any respect to government's desire to override individual autonomy in this context in order to bring our private lives into accord with the moral preferences of government (or the majority it represents) (Lawrence v. Texas, 2003, 577–578). The government has made a similar argument in rejecting paternalism as a valid government basis for restricting advertisements or other commercial speech. In Virginia Pharmacy Board v. Virginia Consumer Council, the Court emphasized that—while government has greater leeway to regulate commercial speech than other speech—it cannot do so on the ground that it needs to keep "the public in ignorance of the entirely lawful" drug price or other non-misleading commercial information (Virginia Pharmacy Board v. Virginia Consumer Council, 1976, 770). In this case, the type of speech in question—commercial speech—is normally *not* treated by courts by being at the core of the First Amendment. On the contrary, it is normally the kind of speech the government *does* have leeway to regulate. But not when the government's interest is patently at odds with constitutional principles about the proper role of government in regulating our affairs. Even though a commercial speaker's interest in expressing themselves free of government constraint is generally weaker than that of speakers engaging in public discourse, it will be strong enough to withstand a government justification that simply misconceives government's role in American constitutional democracy.

Arguably, the key statement that Stanley v. Georgia makes about freedom of thought fits this model. The Court there said that it is impermissible for government to "premise legislation on the desirability of controlling a person's private thoughts" (Stanley v. Georgia, 1969, 566). Like a government desire to impose its own morality on our private relationships, or to invoke paternalism in keeping us ignorant, a government desire to substitute its own preferred beliefs for those we hold, is simply an illegitimate goal. Even if some aspects of our thinking are arguably *not* crucial to our own autonomy, that doesn't mean government has any legitimate role in coercively controlling or shaping them.

Second, even where government's interest is legitimate—and perhaps quite powerful—it may still be one that courts cannot honor (or can, at best, rarely honor)—when doing so would require compromising a crucial autonomy, privacy, or other individual interest. For example, government might offer good reasons that it would be better equipped to protect public safety if it placed cameras inside of, and engaged in constant surveillance of, all individuals' in-home activities, or hacked into and surreptitiously surveilled every file every person stored on a home computer. The sacrifice of privacy this would entail, however, is not one the Fourth Amendment allows for—even when it would further powerful government interests. Similarly, courts have argued that even where government can plausibly argue that individuals' arguments might ultimately lead their audiences to adopt harmful views, the Court has held that this cannot generally justify government control of what individuals say. The First Amendment "core" or "bedrock" generally prevents government from controlling the content of individuals' ideas—even where it can explain how doing so might advance the interests of the public in some way (Texas v. Johnson, 1984, 414).

As I have noted earlier, the same argument can be offered about freedom of thought—and has provided one reason that freedom of thought has sometimes been regarded as absolute. If control over our own minds (or at least, freedom from others' control) is the last redoubt of freedom—an "inner citadel" where we can be guaranteed freedom even where we can find it nowhere else (Christman, 1989)—then courts could find that government must be excluded from this realm even where they can advance powerful interests by entering.

One clear basis to begin understanding what government regulations would cut at the core of the right to freedom of thought then would be to think more carefully about which such regulations threaten autonomy—or lack any legitimate government interest consistent with government's role in our constitutional system. Clear-cut violations of freedom of thought, such as compelled psychosurgery or hypothetical subliminal control, might count as such because they override our autonomy to a greater extent—or perhaps in a way that is harder for us to counteract—than measures that restrict only one tool we use for shaping our own cognition (such as a cognition enhancement drug or device) but not others (such as using speech or other cultural means to reshape our cognition). Bublitz and Merkel rely on an argument of this kind when they write that direct intervention into brain processes generally cause greater injury to autonomy than measures which influence our mind from the "outside" such as education (Bublitz & Merkel, 2014, 69–74). Even where our autonomy interests are low, however, government might still face a strong constitutional barrier against restricting or shaping thought when its motives are inconsistent with the role government may legitimately play. Restricting cognitive enhancement may be permissible but not where government engages in such restriction to prevent us from being better able to critically evaluate government policy. In short, like other rights, a right to freedom of thought is likely to be at its strongest when autonomy and privacy interests are high and public safety or other concerns are low.

As noted earlier, the courts might—in mapping this doctrine for a new right to freedom of unexpressed thought—find some guidance in the First Amendment jurisprudence that already protects expressed thought (that is, speech). It is worth considering, in particular, whether two aspects of free speech doctrine might provide a model for a future doctrine of freedom of thought. First, as noted earlier, government receives far more leeway to regulate speech when it steers clear of regulating the content of that speech and instead regulates it in a "content neutral" manner, for example, by regulating the use of a sound-truck to broadcast speech in a residential neighborhood or to reduce the threat certain means of expression can present to road traffic. Might government likewise have more leeway to regulate thought when it is doing so not to assure that a thinker has certain ideas rather than others but rather to help treat psychosis?

One might, of course, question whether a regulation aimed at banishing psychotic delusions is really "neutral" with respect to thought content—if the person wishes to retain those delusions. (See Stenlund, 2021).

A second-related question is whether—just as certain categories of speech content, like incitement or "true threats" of violence are unprotected by the First Amendment—certain types of thinking might likewise be unprotected by a right to freedom of thought. Even if the constitution protects rational thought, its protection for thinking patterns characteristic of insanity may be lower (or non-existent). As Gabriel Mendlow notes (citing Stephen Morse), treating psychosis "would appear to increase freedom of thought rather than to decrease it" (Mendlow, 2018, 2380 citing Morse, 2017, 15, 2021). On this account, an individual's freedom of thought may in some cases *require* government intervention in that person's thinking—in order to banish insanity and give them the minimal conditions for freedom of thought.[7]

Some First Amendment scholars have argued that, in some cases, the First Amendment gives government leeway to limit speakers' autonomy because doing so is necessary to protect the autonomy of listeners (or other speakers). Left unregulated, threats of violence might silence speakers at whom they are directed. One might likewise argue that even if government is barred from violating its citizens' mental privacy and integrity, it should be left with power—by a jurisprudence of freedom of thought—to protect its citizens' mental privacy and integrity from surveillance or manipulation by their employers, or other businesses, for example. One might, for example, argue that government should be able to limit when and how companies can ask individuals to consent to mental surveillance or manipulation in exchange for some commercial benefit: Individuals could plausibly argue that their freedom of thought gives them a right to consent to such shaping or sharing of their own mind—for example, in agreeing to do so while playing an online video game or surfing the Web. But others might plausibly argue that a more nuanced and complex freedom of thought regime would leave the government the power to protect them from being pressured or deceived into consenting to let a company (or another individual) control or observe their thinking—and to bar them, or at least nudge them away, from doing so in circumstances they might come to regret. (See Thaler &

[7] Mari Stenlunud explores this issue more fully in another chapter in this book (See Stenlund, 2021; See also Saks, 2002).

Sunstein, 2009; Niker, et al., 2021; Bublitz & Merkel, 2014; Bublitz, 2021).

B. **Social Convention**

When courts draw distinctions between the core and periphery of a right's coverage—when they distinguish between areas of higher and lower protection—there is also another important factor that involves reliance not solely on adherence to certain principles (such as protection for autonomy), but also on social convention. The distinctions that rights create might not always, and in all respects, track underlying inherent differences between the sides of the distinction. As David Strauss writes of constitutional doctrine, the solution it provides to a given legal problem sometimes merits courts' adherence (and that of citizens) not because it is the sole "right" solution—but because, there is a need for society to choose among multiple, equally valid options—and a tradition or social convention helps generate an agreed-upon solution. Strauss explains this in part by analogizing courts' adherence to the U.S. Constitution's text to the "focal points" that exist in what game theorists call "cooperation games":

> In a cooperative game with multiple equilibria, the solution will often depend on social conventions or other psychological facts. A simple example would be deciding whether traffic should keep to the left or the right, or who should call back if a telephone call is disconnected. These are games of pure cooperation, but even when there is some conflict of interest a "focal point"—a solution that, for cultural or psychological reasons, is more "salient" and therefore seems more natural—might be decisive. (Strauss, 1996, 910)

Strauss makes this point in arguing that allegiance to constitutional text as a whole operates as a kind of focal point. But this may also be true of the way that courts draw the boundary lines that mark the coverage of a right—or separate a right's core from its periphery. In short, modern societies need both spheres where the state can act vigorously serve the public's interest (investigating crimes, guaranteeing the free flow of traffic in streets, protecting individuals' health and safety) as well as some spheres where individuals can exercise autonomy free of state control, and find privacy free of state observation. The lines that demarcate such a sphere of autonomy or privacy may not necessarily correspond to spheres that are naturally more suited for autonomy or privacy than other spheres of our lives. In some circumstances, there may be no basis to say that one zone is more appropriate for privacy or autonomy than another. The territory of such a sphere may instead be a product of social convention, of "cultural or psychological reasons" that lead courts (or others) to regard it that way.

As I have noted in past scholarship, the primacy of the home in Fourth Amendment law is best understood as partly a product of such social conventions. "[E]ven if we can find places outside the home in which we feel more comfortable and safe from others' observation, these other places–for better or worse–do not have the same historically and legally significant pedigree that the private residence has acquired over the centuries" (Blitz, 2010a, 395). The point is that we need *some* space where we can have greater privacy from state observation than we do in most cases, and social convention has provided us with such a space in the home.

Even where the social conventions that partially define a right, or mark a core area within it, are to some extent the product of historical accident, this doesn't make them any less worthy of adherence. As I've written elsewhere in discussing freedom of thought and mental privacy, "a crucial roadway could just as easily have been built along a different path does not mean that we are not justified in preserving, maintaining, and using such a road where it has already been built. Similarly, the fact that we can imagine a counterfactual world that would justifiably protect intellectual privacy with different institutions, or by setting aside different spaces, does not mean we should abandon and cease to build upon, the intellectual-privacy traditions that we have" (Blitz, 2009, 20–21). To be sure, the status of the home in Fourth Amendment law is not wholly the product of social convention or accident: Its walls and other aspects of its architecture, as well our legal right to exclude others from it, help make it a place where it is easier to find privacy than it is in public space we share with others. But this underscores the extent to which boundary lines created by rights—and the "territory" that they most strongly insulate against state control—are the complex product of history, social conventions, and physical architecture, as well as of a commitment to protecting whatever interest underlies the right.

The same may be true in the development of a future jurisprudence of freedom of thought. The protection offered by a right to freedom of thought may come closest to being absolute when it protects realms that history, social convention, or the nature of the physical world has—in the past—placed beyond the reach of government. This may well be one reason why one of the threats to freedom of thought that seems most concerning to us are those in which government forcibly changes our brain—with surgery or with compelled medication. They are more "direct" attacks on freedom of thought than are attempts to influence us with words or perceptions. One might ask what makes this true of compelled use of psychoactive drugs, for example, since the state cannot currently, in carrying out such compulsion, simply implant or command a particular thought (in the manner in which mind control is sometimes carried out in science fiction). The government rather imposes certain physical or chemical changes on the brain with the hope it will cause an individual to act less violently, or to think more clearly. As Rodney J.S. Deaton says, talk therapy or intense propaganda by government may be more effective at instilling a particular idea than compelled drug use (Deaton, 2006, 214–221). Why then does the latter manner of attempting to shape thought feel any more "direct" than what occurs when the state confronts a person with words (e.g., in mandated talk therapy) or with images?

One possibility, perhaps, is that (as noted earlier) brain interventions impose changes on mental processing that are harder for the subject to resist than are similar changes produced by words or images, and other individuals have a well-established and constitutionally-protected right - a free speech right - to influence us with words that they do not have to influence us with "direct" manipulations (Bublitz & Merkel, 2014, 69–74). However, there is another reason that brain-based intervention may feel like a graver freedom of thought violation—one which feels like a violation of a *core* freedom of thought right. That is that our brain processes have ***historically*** been free from this kind of state manipulation. We have had a level of freedom from biological thought interventions that we have *not* had from the impact of others' words, or from the images they show us.

Courts therefore might apply freedom of thought in a way that involves what Orin Kerr, in the Fourth Amendment context, has called "equilibrium adjustment" (Kerr, 2011, 480). In 2001, for example, the Supreme Court ruled that police engage in a Fourth Amendment "search" requiring a warrant when they stand outside a home and use an infrared imager to view its interior. Even though police don't need a warrant to stand on a street outside a home and look at it with their natural vision, it is a search when they use technology (like an infrared imager) that lets them see parts of the home they could—in past times—have seen only by physically entering it. The Supreme Court, in other words, responded to technology which unsettled an equilibrium that existed between the state's surveillance power and individual's privacy: This technology contracted the sphere of privacy and expanded the sphere of surveillance by allowing police to gather (from the street) information from within a home that was previously inaccessible to them. The Court's response was to restore the equilibrium by replacing the natural barrier once created by the home's walls with a legal barrier that constitutionally restricted use of technology that can see through those walls.

The same approach might sometimes guide courts as they analyze government use of new technologies for monitoring or manipulating the brain. Even where a government-compelled brain scan recovers information that isn't all that private, courts might still find that it is violating a core mental privacy right because such a measure is giving government access to a realm we have long been able to regard as secured against government surveillance. The Court might understand a right to freedom of thought—or Fourth Amendment rights interpreted in light of freedom of thought interests—as assuring that we *retain* the mental of privacy we have long had. As I noted earlier, one might conceive of a right to freedom of thought as rebuilding—in law—the protection for the mental autonomy and privacy that fMRI scanners and other technologies have eroded in nature.

By contrast, other government measures that place limits on our thoughts might *not* extend government control in this way. Consider, for example, government measures that place limits on our ability to use cognitive enhancement technology—such as nootropic drugs, tDCS, deep brain stimulation, or brain implants. These measures place limits on our mental autonomy: They prevent us from shaping the content of our own minds. There is therefore a case, as Boire and others write, that they should fall within the coverage of freedom of thought or "cognitive liberty" (Boire, 2001a). At times, such a restriction might even have the same result as one that interferes with our thinking by direct intervention: A government measure that prevents us from endowing ourselves with a particular mental capacity (by forbidding us from using cognitive enhancement) might leave us in the same state we would be in if government used forcible medication or surgery to deprive us of such a mental capacity. (See Kolber, 2006; Blitz, 2010b).

However, in barring use of cognitive enhancement, government would not be depriving us of a kind of freedom or privacy we have long had. It would rather be regulating in areas where it has long regulated: The creation, sale, transfer, or use of drugs and medical devices, or delivery of professional psychological or psychiatric treatment (or medical treatment more generally). That doesn't mean that government should face no judicial scrutiny when it limits our use of such devices. But courts might leave government officials with more leeway in a realm where they have long had such leeway. As noted earlier, this is one reason why an absolute or near-absolute right against thought interference might not cover a right to voluntary mind modification.

There is, to be sure, a possible danger in relying on such an equilibrium adjustment model of rights in understanding freedom of thought: It risks freezing into the law an understanding of mental autonomy or privacy which, although appropriate for the twentieth century may not be a perfect fit for our lives in the twenty-first. The minimal conditions for autonomy in the pre-digital age may not be the same as those in a world where individuals have come to use computers not just as replacements for activities once carried out in the physical world, but to engage in new kinds of self-definition or personal action. Or where individuals have grown used to being able to dampen very painful memories and states of mind, or modify mental habits that interfere with their lives. Any judicial use of equilibrium adjustment must therefore take into account the possibility that technological and social change might not only alter the threats to our mental autonomy and privacy, but how we define that autonomy and privacy and understand its minimal conditions.

Bibliography

Books and Articles

Alegre, S. (2017). Rethinking freedom of thought for the digital age. *European Human Rights Law Review, 3*, 222–233.

Amsterdam, A. G. (1974). Perspectives on the Fourth Amendment. *Minnesota Law Review, 58*, 349–478.

Bambauer, J. (2014). Is data speech? *Stanford Law Review, 66*, 57–120.

Bambauer, J. (2018). The age of sensorship. In R. L. K. Collins, & D. Skover, D. (Eds.), *Robotica: Speech rights and artificial intelligence*. Cambridge University Press.

Blitz, M. J. (2006). Constitutional safeguard for silent experiments in living: Libraries, the right to read, and a First Amendment theory for an unaccompanied right to receive information. *UMKC Law Review, 74*, 799–882.

Blitz, M. J. (2008). Freedom of 3D thought: The First Amendment in virtual reality. *Cardozo Law Review, 30*, 1141–1242.

Blitz, M. J. (2009). The where and why of intellectual privacy. *Texas Law Review See Also, 87*, 15–23.

Blitz, M. J. (2010a). Stanley in cyberspace: Why the privacy protection of the First Amendment should be more like that of the fourth. *Hastings Law Journal, 62*, 357–400.

Blitz, M. J. (2010b). Freedom of thought for the extended mind: Cognitive enhancement and the constitution. *Wisconsin Law Review*, 1049–1117.

Blitz, M. J. (2016a). Free speech, occupational speech, and psychotherapy. *Hofstra Law Review, 44*, 681–780.

Blitz, M. J. (2016b). A constitutional right to thought enhancing technology. In V. Dubljevi, & F. Jotterand (Eds.) *Cognitive enhancement: Ethical and policy perspectives in international perspective*. Oxford University Press.

Blitz, M. J. (2017). *Searching minds by scanning brains: Neuroscience, technology, and constitutional privacy protection*. Palgrave Macmillan.

Blitz, M. J. (2018). The First Amendment. Video games, and virtual reality training. In W. Barfield, & M. J. Blitz (Eds.), *The law of augmented and virtual reality*. Edward Elgar.

Blitz, M. J. (2021). Cognitive enhancement and American constitutional law. In P. Riederer et al. (Eds.), *NeuroPsychopharmacotherapy*.

Boire, R. A. (2004). Neurocops: The politics of prohibition and the future of enforcing social policy from inside the body. *Journal of Law and Health, 19*, 215–256.

Boire, R. G. (2001a). Cognitive liberty Part I. *Journal of Cognitive Liberties, 1*(1), 1–3.

Boire, R. G. (2001b). Cognitive liberty Part II. *Journal of Cognitive Liberty, 1*(2), 1–6.

Bublitz, J. C. (2014). Freedom of thought in the age of neuroscience. *Archiv Rechts-Und Sozialphilosphie, 100*, 1–25.

Bublitz, J. C., & Merkel, R. (2014). Crimes against minds: On mental manipulations, harms, and human right to mental self-determination. *Criminal Law & Philosophy, 8*, 51–77.

Carter, J. A. (2021). Varieties of (extended) thought manipulation. In M. J. Blitz, & J. C. Bublitz (Eds.), *The law and ethics of freedom of thought: Neuroscience, autonomy and individual rights*. Palgrave Macmillan.

Carter, J. A., & Palermos, S. O. (2016). Is having your computer compromised a personal assault? The ethics of extended cognition. *Journal of the American Philosophical Association, 2*(4), 542–560.

Chandler, J., & Dodek, A. (2016). Cognitive enhancement in the courtroom: Ethical and policy implications in international perspectives. In V. Dubljevic, & F. Jotterand (Eds.), *Cognitive enhancement: Ethical and policy perspectives in international perspective*. Oxford University Press.

Christman, J. (1989). Introduction. In J. Christman (Ed.), *The inner citadel: Essays on individual autonomy*. Oxford University Press.

Clark, A., & Chalmers, D. J. (2008 [1998]). The extended mind. In A. Clark (Ed.), *Supersizing the mind: Embodiment, action, and the cognitive experience*. Oxford University Press.

Deaton, R. J. S. (2006). Neuroscience and the in corpore-ted First Amendment. *First Amendment Law Review, 4*, 181–221.

Dubljevic, V. (2013). Cognitive enhancement, rational choice and justification. *Neuroethics, 6*, 179–187.

Farahany, N. A. (2012a). Incriminating thoughts. *Stanford Law Review, 64*, 351–408.

Farahany, N. A. (2012b). Searching secrets. *Pennsylvania Law Review, 160*, 1239–1308.

Fox, D. (2008). Will memory detection technologies transform criminal justice in the United States? Brain imaging and the bill of rights. *American Journal of Bioethics, 8*(1), 1–4.

Fox, D. (2009). The right to silence as protecting mental control. *Akron Law Review, 42*, 763–801.

Fox, D., & Stein, A. (2015). Dualism and doctrine. *Indiana Law Journal, 90*, 975–1010.

Greely, H. et al. (2008). Towards responsible use of cognitive-enhancing drugs by the healthy. *Nature, 456*, 702–705.

Haupt, C. E. (2016). Professional speech. *Yale Law Journal, 125*, 1238–1303.

Heyman, S. J. (2002). Spheres of autonomy: Reforming the content neutral doctrine in First Amendment jurisprudence. *William & Mary Bill Rights Journal, 10*, 647–717.

Ienca, M., & Andorno, R. (2017). *Towards new human rights in the age of neuroscience and neurotechnology*. Society and Policy.

Kerr, O. S. (2011). An equilibrium-adjustment theory of the Fourth Amendment. *Harvard Law Review, 125*, 476–543.

Kolber, A. J. (2006). Therapeutic forgetting: The legal and ethical implications of memory dampening. *Vanderbilt Law Review, 59*, 1561–1626.

Kolber, A. J. (2008). Freedom of memory today. *Neuroethics, 1*, 145–148.

Kolber, A. J. (2021). *The ethics of memory dampening, in the law and ethics of freedom of thought: Neuroscience, autonomy and individual rights*. Palgrave Macmillan.

Lavazza, A. (2018). Freedom of thought and mental integrity: The moral requirements for any neural prosthesis. *Frontiers in Neuroscience, 12*, 82.

Lavazza, A. (2021). Technology against technology: A case for embedding mechanisms/restrictions/limits in new/neurodevices to protect our freedom of thought. In M. J. Blitz, & J. C. Bublitz (Eds.), *The law and ethics of freedom of thought: Neuroscience, autonomy and individual rights*. Palgrave Macmillan.

Levy, N. (2007). *Neuroethics: Challenges for the 21st century*. Cambridge University Press.

Macklem, T. (2006). Timothy. *Independence of Mind*, 1–13.

McCarthy-Jones, S. (2019). The autonomous mind: The right to freedom of thought in the twenty-first century. *Frontiers in Artificial Intelligence, 2*.

McCarthy-Jones, S. (2021). The who, what, and why of freedom of thought. In M. J. Blitz, & J. C. Bublitz (Eds.), *The law and ethics of freedom of thought: Neuroscience, autonomy and individual rights*. Palgrave Macmillan.

Meiklejohn, A. (1961). The First Amendment is an absolute. *Supreme Court Review, 245*.

Mendlow, G. S. (2018). Why is it wrong to punish thought? *Yale Law Journal, 127*, 2342–2386.

Mendlow, G. S. (2021). Why is it wrong to punish thought. In M. J. Blitz, & J. C. Bublitz (Eds.), *The law and ethics of freedom of thought: Neuroscience, autonomy and individual rights*. Palgrave Macmillan.

Morse, S. J. (2017). *Involuntary competence in United States criminal law* (University of Pennsylvania Law School, Public Law & Legal Theory Research Paper 975 No. 17–20), http://ssrn.com/abstract=2951966.

Neuborne, B. (2011). Madison's music: On reading the First Amendment.

Niker, F., Felsen, G., Nagel, S., & Reiner, P. (2021). Autonomy, evidence responsiveness, and the ethics of influence. In M. J. Blitz, & J. C. Bublitz (Eds.), *The law and ethics of freedom of thought: Neuroscience, autonomy and individual ights*. Palgrave Macmillan.

Pardo, M. (2006). Neuroscience evidence, legal culture, and criminal procedure. *American Journal of Criminal Law, 33*, 301–337.

Pardo, M., & Patterson, D. (2013). *Minds, brains and law: The conceptual foundations of law and neuroscience*. Oxford University Press.

Pustilnik, A. C. (2013). Neurotechnologies at the intersection of criminal procedure and constitution law. In S. Richardson, & J. Parry (Eds.), *The constitution and the future of criminal law*. Cambridge University Press.

Richards, N. (2008). Intellectual privacy. *Texas Law Review, 87*, 387–445.

Richards, N. (2015). *Intellectual privacy: Challenges for the 21st century*. Cambridge University Press.

Saks, E. R. (2002). *Refusing care: Forced treatment and the rights of mentally ill*. University of Chicago Press.

Scanlon, T.M. (1979). Freedom of expression and the categories of expression. *Pittsburgh Law Review, 40*, 519–550.

Schauer, F. (1982). *Free speech: A philosophical inquiry*. Cambridge University Press.

Schauer, F. (2015). On the distinction between speech and action. *Emory Law Journal, 65*, 427.

Schauer, F. (2020). Freedom of thought? *Social Philosophy and Policy, 37*(2), 72–89.

Sententia, W. (2004). Neuroethical considerations: Cognitive liberty and converging technologies for improving human cognition. *Annals of the New York Academy of Sciences, 1013* (1).

Shiffrin, S. V. (2011). A thinker-based approach to freedom of speech. *Constitutional Commentary, 27*, 283–307.

Shiffrin, S. V. (2014). *Speech matters: On lying, morality, and the law*. Princeton University Press.

Smolla, R. A. (1992). *Free speech in an open society*. Vintage.

Smolla, R. A. (2016). Professional speech and the First Amendment. *West Virginia Law Review, 119*, 67–112.

Solove, D. J. (2007). The First Amendment as criminal procedure. *New York University Law Review, 82*, 112–176.

Stenlund, M. (2021). Cognitive liberty of the person with a psychotic disorder. In M. J. Blitz, & J. C. Bublitz (Eds.), *The law and ethics of freedom of thought: Neuroscience, autonomy and individual rights*. Palgrave Macmillan.

Stoller, S. E., & Wolpe, P. R. (2007). Emerging technologies for lie detection and the Fifth Amendment. *American Journal of Law and Medicine, 33*(2/3), 359–374.

Strauss, D. A. (1996). Common Law Constitutional Interpretation. *University Chicago Law Review, 63*, 877–935.

Swain, L. (2021). Freedom of thought in political history. In M. J. Blitz, & J. C. Bublitz (Eds.), *The law and ethics of freedom of thought: Neuroscience, autonomy and individual rights*. Palgrave Macmillan.

Thaler, R., & Sunstein, C. (2009). *Nudge: improving decisions about health, wealth, and happiness* (Penguin Books).

Weinstein, J. (2011). Seana Shiffrin's thinker-based theory of free speech: Elegant and insightful, but will it work in practice? *Constitutional Commentary, 27*, 385–397.

Wendel, W. B. (2001). Free speech for lawyers. *Hastings Constitutional Law Quarterly, 28*, 305–444.

Winick, B. J. (1989). The right to refuse mental health treatment: A First Amendment perspective. *University Miami Law Review, 44*, 1–103.

Cases

American Civil Liberties Union v. Alvarez, 679 F.3d 583 (7th Cir. 2012).
Ashcroft v. Free Speech Coalition, Ashcroft v. Free Speech Coal., 535 U.S. 234 (2002).
Bethel Sch. Dist. No. 403 v. Fraser, 478 U.S. 675 (1986).
Birchfield v. North Dakota, 136 S.Ct. 2160 (2016).
Buckley v. American Constitutional Law Foundation, 525 U.S. 182 (1999).
Burdeau v. McDowell, 256 U.S. 465 (1921).
Carpenter v. United States, 138 S.Ct. 2206 (2018).
Clark v. Community for Creative Non-Violence, 468 U.S. 288 (1989).
Connick v. Myers, 461 U.S. 138 (1983).
Doe v. City of Lafayette, Indiana, 377 F.3d 757 (7th Cir. 2004).
Delaware v. Prouse, 440 U.S. 648 (1979).
Garcetti v. Ceballos, 547 U.S. 410 (2006).
Hazelwood v. Kuhlmeier, 484 U.S. 260 (1988).
Illinois v. Caballes, 543 U.S. 405 (2005).
Illinois v. Lidster, 540 U.S. 419 (2004).
Jones v. Opelika, 316 U.S. 584 (1942).
Kaimowitz v. Michigan Dep't of Mental Health, 42 U.S.L.W. 2063 (Cir. Ct. Wayne Cty., Mich., 1973).
King v. Governor of N.J., 767 F.3d 216 (3d Cir. 2014).
Kyllo v. United States, 533 U.S. 27 (2001).
Lawrence v. Texas, 539 U.S. 558 (2003).
Mahanoy Area Sch. Dist. v. B. L. by & through Levy, 141 S. Ct. 2038 (2021).
Masterpiece Cakeshop v. Colorado Civil Rights Commission, 138 S. Ct. 1719 (2018).
McIntyre v. Ohio Elections Comm'n, 514 U.S. 334 (1995).
Meyer v. Grant, 486 U.S. 414 (1988).
Miller v. United States, 425 U.S. 435 (1976).
Morse v. Frederick, 551 U.S. 393 (2007).
Nat'l Ass'n for Advancement of Psychoanalysis v. Cal. Bd. of Psychology, 228 F.3d 1043 (2000).
National Treasury Employees Union v. Von Raab, 489 U.S. 656 (1989).
Occupy Fresno v. County of Fresno, 835 F. Supp. 2d 849 (E.D. Cal. 2011).
Osborne v. Ohio, 495 U.S. 103 (1990).
Palko v. Connecticut, 302 U.S. 319 (1937).
Paris Adult Theatre I v. Slaton, 413 U.S. 49 (1973).
Pickup v. Brown, 740 F.3d 1208 (9th Cir. 2014).
Board of Education of Independent School District, Pottawatomie Cty. v. Earls, 536 U.S. 822 (2002).
R.A.V. v. St. Paul, 505 U.S. 377 (1992).
Reed v. Town of Gilbert, Arizona, 576 U.S. 155 (2015).

Riggins v. Nevada, 504 U.S. 127 (1992).
Riley v. California, 573 U.S. 373 (2014).
Sellv. United States, 539 U.S. 166 (2003).
Silverman v. United States, 365 U.S. 505 (1961).
Smith v. Maryland, 442 U.S. 735 (1979).
Skinner v. Ry. Labor Executives' Ass'n, 489 U.S. 602 (1989).
Stanley v. Georgia, 394 U.S. 557 (1969).
Terminiello v. Chicago, 337 U.S. 1 (1949).
Terry v. Ohio, 1968, 392 U.S. 1 (1968).
Texas v. Johnson, 491 U.S. 397 (1989).
Thomas v. Collins, 323 U.S. 516 (1945).
Tinker v. Des Moines Indep. Cmty. Sch. Dist., 393 U.S. 503 (1969).
Universal City Studios v. Corley, 273 F.3d 429 (2d Cir. 2001).
United States v. Orito, 413 U.S. 139 (1973).
United States v. Thirty-Seven (37) Photographs, 402 U.S. 363, (1971).
United States v. Di Re, 332 U.S. 581 (1948).
United States v. Playboy Entertainment Group, 529 U.S. 803 (2000).
Vernonia School District v. Acton, 515 U.S. 646 (1995).
Virginia v. Black, 538 U.S. 343 (1993).
Ward v. Rock Against Racism, 491 U.S. 781 (1989).
Washington v. Harper, 494 U.S. 210 (1990).
West Virginia State Bd. of Educ. v. Barnette, 319 U.S. 624 (1943).
Wooley v. Maynard, 430 U.S. 705 (1977).

CHAPTER 5

Why is It Wrong to Punish Thought?

Gabriel S. Mendlow

Introduction

It's a venerable maxim of criminal jurisprudence that the state must never punish people for their mere thoughts—for their beliefs, desires, fantasies, and unexecuted intentions. This maxim is all but unquestioned, yet its true justification is something of a mystery. Jurists often say that mere thoughts are unpunishable because they're harmless, innocent, and unprovable. But, as I'll argue, certain thoughts are every bit as dangerous, wrongful, and provable as actions we readily criminalize. If mere thoughts are unpunishable, it's instead because they're *immune* from punishment despite deserving it. Unlike various legal immunities,

---

This chapter is adapted from Gabriel S. Mendlow, Why Is It Wrong to Punish Thought, 127 *Yale Law Journal* 2342 (2018). The original essay is considerably longer, with more detailed citations and acknowledgments.

G. S. Mendlow (✉)
University of Michigan, Ann Arbor, MI, USA
e-mail: mendlow@umich.edu

M. J. Blitz et al. (eds.), *The Law and Ethics of Freedom of Thought, Volume 1*, Palgrave Studies in Law, Neuroscience, and Human Behavior,
https://doi.org/10.1007/978-3-030-84494-3_5

however, the immunity of thought can't rest on a pragmatic foundation. Although the specter of intrusively oppressive policing may give us reason to *treat* thoughts as immune from punishment, it doesn't establish that they actually are. It doesn't establish that every act of punishment for thought involves an intrinsic (i.e., consequence-independent) injustice to the person punished: that every such act necessarily *wrongs the thinker*.

In place of these flawed rationales, the essay proposes that punishment for thought is intrinsically unjust because it's a form of indirect mind control. The proposed rationale captures the widely shared intuition that punishment for thought isn't simply disfavored by the balance of reasons but is morally wrongful in itself, an intrinsic injustice to the person punished. The proposed rationale also shows how thought's immunity from punishment relates to a principle of freedom of mind, a linkage often assumed but never explained. In explaining it here, I argue that thought's penal immunity springs from the interaction of two principles of broad significance: one familiar but poorly understood, the other seemingly unnoticed. The familiar principle is that persons possess a *right of mental integrity*, a right to be free from the direct and forcible manipulation of their minds. We'll see that this right undergirds a set of important principles governing the relationship between the mind and the state (principles concerning such things as education, brainwashing, and forced medication), of which the ban on thought crime is merely one. The seemingly unnoticed principle is that the state's authority to punish transgressions of a given type extends no further than its authority to thwart or disrupt such transgressions using direct compulsive force. This principle, which I call the *Enforceability Constraint*, holds that the state may ensure compliance with a given norm through criminal punishment only when the state may, in principle, force compliance with that norm directly.

Heretofore unexamined, the Enforceability Constraint is in fact a signal feature of our system of criminal administration, governing the scope and limits of the criminal law. When conjoined with the principle that persons possess a right of mental integrity, the Enforceability Constraint entails that punishment for thought is intrinsically unjust: if using mind control to force compliance with a thought-proscribing norm would violate a potential norm-breaker's right to mental integrity, then so too would exposing the norm-breaker to punishment. That is why it's wrong to punish thought.

i. Inadequate rationales for the ban on thought crime

Theorists often claim that criminalizing mere thought would unleash the worst sort of tyranny and oppression. According to James Fitzjames Stephen, if we criminalized every improper thought, "all mankind would be criminals, and most of their lives would be passed in trying and punishing each other for offenses which could never be proved" (Stephen, 1883, 78). H.L.A. Hart adds: "Not only would it be a matter of extreme difficulty to ferret out those who were guilty of harboring, but not executing, mere intentions to commit crimes, but the effort to do so would involve vast incursions into individual privacy and liberty" (Hart, 2008, 127). Quoting Stephen, Hart concludes: "[T]o punish bare intention 'would be utterly intolerable'" (ibid., 78).

These assertions are facile. To be sure, life would be intolerable under a regime that punished every improper mental state—every sadistic fantasy, evil desire, and hateful belief. But life also would be intolerable under a regime that punished every improper act—every unkindness and petty betrayal, no matter how harmless, innocent, or difficult to prove. That's an excellent reason not to punish every improper act. It's a terrible reason not to punish *any* act. In punishing acts, legal systems can and do discriminate between the grave and the paltry. If a legal system elected to punish thoughts (the word I'll often use to denote the entire class of mental states), the state could exercise like discretion, punishing only the rare thought that's dangerous, depraved, and provable. The key question is whether any such thought exists, and it's a question that Stephen and Hart evade.

I'll argue that the answer is yes. Contrary to the received wisdom, certain thoughts are dangerous, depraved, and provable. Thus, the ban on punishing mere thoughts can't be justified by any of the leading rationales: the harm principle, the requirement that criminal transgressions be culpable wrongs, or the requirement that criminal transgressions be proved beyond a reasonable doubt.

I'll consider these rationales in turn.

A. The Harm Principle

Reporting a common view, P.J. Fitzgerald notes that "[t]he comparative harmlessness of mere thoughts and intentions by themselves is considered sufficient reason for not punishing them. The small degree of harm likely to result from such intentions is not thought to justify the interference with liberty which punishment would involve" (Fitzgerald, 1962, 97).

If thoughts aren't more than minimally harmful, then criminalizing them violates John Stuart Mill's harm principle. According to the harm principle, "the only purpose for which power can be rightfully exercised over any member of a civilized community, against his will, is to prevent harm to others" (Mill, 1859, 9). But is Fitzgerald right that thoughts never risk more than a "small degree of harm"?

Consider a person's intention to kill, particularly when formed after extensive reflection and deliberation. Is such an intention really less likely to cause harm than driving recklessly or possessing volatile explosives—activities that we don't hesitate to criminalize on account of their dangerousness? If no lethal intention were more than minimally dangerous, it would be irrational for me to fear you simply because you intended to kill me. But it's difficult to accept that such fear is irrational. There would be little point to forming intentions if intentions didn't generally increase the likelihood of actions. It's one thing for you to *want* to kill your enemy, or to *believe* that killing him has something to be said for it. Wanting and believing these things are common enough occurrences, which don't necessarily indicate a propensity to violence. It's another thing entirely for you to *intend* to kill your enemy, to make killing him your goal. To make killing someone your goal is to embrace a distinctive and unusual set of rational commitments. It's to commit to watching for an opportunity to kill him, to seizing such an opportunity when practicable, and to refraining from conduct that would make performance impossible. Rational commitments of this sort are what distinguish intending to kill, which is rare, from desiring to kill, which is sometimes said to be common. Rationality doesn't demand of one who desires to kill that she abandon all contrary intentions. Rationality doesn't even demand that she abandon all contrary desires. But rationality does demand that an intending killer *kill*, or else abandon her intention.

It's true that intentions can be rescinded, decisions rethought, and plans discarded, but it doesn't follow that your intending to do something never increases the likelihood that you'll do it. A characteristic effect of forming an intention is to place yourself under rational and psychological pressure to follow through, pressure compounded by a range of familiar cognitive biases that further reduce the likelihood you'll change your mind. The more invested we feel in a decision, the less likely we are to reconsider it (Arkes & Blumer, 1985, 124). We also tend to remember our past decisions as being more justified than they actually were (Brehm, 1956, 384, 386; Mather et al., 2000, 132). When confronted with new evidence, we tend to revise our opinions insufficiently (Edwards, 1968, 17, 18). And we generally tend to place more credence in evidence that confirms our beliefs than in evidence that contradicts them (Nickerson, 1998, 175). Evidence that contradicts our beliefs sometimes perversely strengthens them (Sanna, Schwarz, & Stocker, 2002, 497).

If the rational pressures intrinsic to intention and the cognitive biases that reinforce those pressures all increase the odds that you'll do what you intend to do, forming a lethal intention creates a risk of death. If you're a competent person with the means to kill, the danger posed by your lethal intention could be at least as great as that posed by many risky activities we seldom think twice about punishing, such as driving recklessly and possessing volatile explosives.

B. The Requirement That Criminal Transgressions Be Culpable Wrongs

Not only can lethal intentions be dangerous, but for that very reason they also can be culpably wrongful, at least potentially. If it's sometimes culpably wrongful to create a risk of nondeadly injury inadvertently, then presumably it's sometimes culpably wrongful to create a risk of deadly injury knowingly—which is what you do when you form the intention to kill, assuming you're a competent person with the necessary means. Knowingly creating a risk of death is a serious wrong, a wrong the public seemingly has standing to condemn. It's hard to accept that the public could lack standing to complain of some risk just because the risk originates inside a person's head rather than on the outside. The site of the risk seems to lack independent moral significance.

When a person forms the intention to kill, she culpably creates in herself a psychological condition the purpose and possible effect of which is to cause a death. Although she can eliminate the risk of death by abandoning the intention, we shouldn't pretend that abandoning an intention is as easy as flipping a mental switch. As I noted a moment ago, intentions carry substantial mental inertia. When a person forms the intention to kill, she sets herself on a path that makes someone's death at least a little bit more likely—just as a person may do when she acquires a safely stored but very deadly weapon or appropriates the nuclear launch codes.

Like forming a lethal intention, these activities may properly be subjected to public condemnation even though the risks they create remain exclusively within the actor's control. It's everyone's business when someone knowingly creates an impermissible risk, wherever and by whatever means.

But riskiness is only part of what makes lethal intentions wrongful, and probably not even the largest part. If, thanks to fortuity or incompetence, your intention to kill me creates no appreciable risk that I'll die, you wrong me nonetheless, just by aiming at my death. The wrongfulness of your intention derives not only from the risk it creates, but also—and perhaps more fundamentally—from the wrongfulness of the action toward which it aims. Ordinarily, you have a conclusive moral reason not to kill me, which is virtually always[1] also a conclusive moral reason not to *try* to kill me, *prepare* to kill me, *plot* to kill me, *plan* to kill me, or *intend* to kill me (Duff, 2012, 121, 135–36). When you form the intention to kill me, you therefore do something you have a conclusive moral reason not to do. And when you do something you have a conclusive moral reason not to do, you do something wrongful—even if all you do is form a mental state.

It's therefore unsurprising that the wrongfulness of malevolent intentions is presupposed by a range of moral judgments and emotional reactions both natural and inevitable. Consider the host of attitudes and demands we'd have to disclaim if your unexecuted intention to kill me weren't a culpable wrong. I couldn't resent you for your intention. I couldn't demand that you abandon it. I couldn't even demand that you apologize for it. I could think the worse of you on account of your intention, but I couldn't say, "How dare you intend to kill me?" If you've done me no wrong, I lack the standing to condemn you. Although I could view your intention as a moral failing—a character flaw—I couldn't view it as a moral transgression. I couldn't view it as a moral transgression even if you unquestionably formed it voluntarily. And it seems clear that at least

[1] In a bizarre scenario like Gregory Kavka's toxin puzzle (Kavka, 1983, 33–35), your conclusive moral reason not to kill me might be no more than a nonconclusive (i.e., defeated or outweighed) moral reason not to intend to kill me. Suppose an eccentric billionaire offers to pay you a million dollars if, at midnight tonight, you intend to kill me tomorrow afternoon. He emphasizes that the money will be in your bank account by 10 a.m. tomorrow morning, so you don't actually have to go through with the killing. You just have to *intend* to. In this scenario, you've got a conclusive moral reason not to kill me, but a defeated moral reason not to intend to kill me. I assume that scenarios with these rational implications are exceedingly rare.

some intentions are subject to a person's voluntary control, particularly intentions that a person forms after reflection and deliberation.

C. The Requirement That Criminal Transgressions Be Proved Beyond a Reasonable Doubt

Even if dangerous and wrongful, lethal intentions would be inapt for punishment if, as Stephen asserts, they "could never be proved." (Stephen, 1883, 78). It is sometimes said that, in the absence of evidence that a person has taken steps to fulfill his purported intention, we can never know whether the purported intention is anything more than a mere desire or fantasy. Our everyday experience says otherwise, however. We routinely rely on what a person says she intends to do, well before she's begun to act on her stated intention. Examples range from the mundane to the vital: from the friend who says she'll meet you for lunch at noon to the gangster who, without budging, tells you to get lost or he'll kill you.

That unexecuted intentions are sometimes provable doesn't mean that proving them is always easy. Proving them is often difficult and that difficulty alone is reason not to criminalize them, especially if the difficult can be met only by intrusive investigations. The dangers of such investigative methods surely give us some reason not to punish thought. Indeed, they give us some reason to treat punishment for thought as though it were morally forbidden. But they don't establish that punishment for thought is morally forbidden in fact. The risk of intrusively oppressive policing doesn't establish that there's an intrinsic (i.e., consequence-independent) injustice in every act of punishment for thought, any more than the unreliability of coerced confessions establishes that there's an intrinsic injustice in every act of interrogational torture. Although it might be politically expedient to oppose torture on instrumental grounds, the basic moral reason to refrain from torture isn't that torture produces unreliable information, or that torturing our adversaries encourages them to torture us when we fall into their hands, or that engaging in torture tends to undermine other legal norms against state brutality. All of these things are probably true, and all of them give us good reason to conduct ourselves as though torture were morally forbidden. But none of them shows that torture actually is forbidden in itself—that each act of torture, irrespective of its consequences, is unjust.

ii. The ban on thought crime as a categorical moral immunity

Even if malevolent criminal intentions can be dangerous, culpable, provable wrongs, we should hesitate to let go of the idea that it's always unjust to punish thought. The revulsion many commentators—indeed, most people—express at the prospect of punishment for mere mental states seems to emanate from a source firmer than the dubious assumption that no single mental state is culpably wrongful. Commentators vehemently assert that punishing mere mental states transgresses a principle of "natural justice" (Endlich, 1891, 831–832) founded in "the inviolability of thoughts" (Dan-Cohen, 1999, 379), a principle whose disregard constitutes a "monstrous" (Yaffe, 2014, 101) intrusion into a person's "private world" (Ashworth, 2011, 126, 134), and an invasion of her "essential . . . human right to freedom of thought" (Calvert, 2005, 125). These remarks describe a supposed injustice both narrower and deeper than that of punishing someone for a transgression undeserving of punishment. The supposed injustice is narrower in that it's peculiar to the mind; it is deeper in that it transcends the injustice of punishing someone for a transgression that isn't culpably wrongful. If punishing someone for a mental state is a "monstrous" intrusion into her "private world," it presumably remains so even when the mental state in question is a dangerous, culpable wrong.

Now, even if none of the conventional rationales suffices on its own to ground a categorical ban on thought crime, the collective weight of these considerations might well support fidelity to a categorical ban. If it's simply too costly, too risky, and too oppressive to try to distinguish the few mental states that merit punishment from the many that don't, then, on balance, we shouldn't criminalize any. But to adopt a categorical ban on these grounds alone is to give up on the idea that there's an intrinsic (consequence-independent) injustice in each act of punishment for thought. It's to dismiss as hyperbole commentators' assertions about "the inviolability of thoughts" (Dan-Cohen, 1999, 379) and the "monstrous" intrusion (Yaffe, 2014, 101) into a person's "private world" (Ashworth, 2011, 134) that occurs when her thoughts are made the object of punishment. To give up on these ideas and to dismiss the associated rhetoric as hyperbole is akin to giving up on the idea that there's an intrinsic injustice in torture, the idea that torture's injustice isn't solely a function of its downstream consequences.

To view torture's injustice as intrinsic isn't necessarily to see the moral ban on torture as *absolute*. It's instead to see every act of torture as involving a grievous moral sacrifice, even in the hypothetical circumstance in which the state's vital ends supposedly justify its torturous means. I submit that any purported justification of the ban on torture is morally deformed if it gives no account of this moral sacrifice, if it makes no effort to elucidate torture's intrinsic injustice and speaks instead only of torture's instrumental shortcomings. The basic moral reason not to torture is that torturing a person does an injustice *to that person*. The torture victim's signal complaint is that she herself has been wronged, not that the practice to which she's been subjected engenders various other abuses. A person punished for her thoughts is prone to lodge a similar complaint, to complain that she herself has been wronged. This complaint is sound if, but only if, there's an intrinsic injustice in every act of punishment for thought. What's needed is an explanation of why it's intrinsically unjust to punish mental states that are provable, dangerous, and culpably wrongful: mental states that bear the chief hallmarks of paradigmatic punishable actions.

In itself, there's nothing especially puzzling about the idea that a class of dangerous and culpably wrongful transgressions is immune from punishment. Criminal law contains a miscellaneous assortment of what Paul Robinson calls "nonexculpatory defenses," defenses like diplomatic, judicial, legislative, and executive immunity, all of which preclude liability "where the actor by all measures deserves condemnation and punishment" (Robinson, 1984, §201). These defenses provide a poor analogy to the prohibition on punishing thought, however, because none of them takes its primary justification from the notion that withholding the defense would perpetrate an intrinsic injustice *on defendants*. Rather, as Robinson explains, "[n]onexculpatory defenses arise where an important public policy other than that of convicting culpable offenders, is protected or furthered by foregoing trial or conviction and punishment" (ibid.).

Certainly, the ban on thought crime furthers important public policies—as does the ban's closest counterpart, the ban on punishing speech and other forms of expression. In fact, the most famous of all arguments for freedom of expression, Mill's marketplace-of-ideas argument in Chapter 2 of *On Liberty*, is a classic example of what lawyers call a "policy argument." Mill writes,

> the peculiar evil of silencing the expression of an opinion is, that it is robbing the human race; posterity as well as the existing generation; those who dissent from the opinion, still more than those who hold it. If the opinion is right, they are deprived of the opportunity of exchanging error for truth; if wrong, they lose, what is almost as great a benefit, the clearer perception and livelier impression of truth, produced by its collision with error (Mill, 1859, 16).

No part of Mill's argument credits the idea that suppressing speech is wrong *because it wrongs the speaker*. If we're to vindicate the notion that punishing pure thought is wrong *because it wrongs the thinker*, we can't rely on any sort of policy argument. We need an argument that depicts thought's immunity from punishment not as an immunity based in good public policy but as an immunity based in the thinker's status as a moral being.

iii. Mental immunity and freedom of mind

We've yet to uncover a principled basis for the idea that punishing thought is categorically impermissible. So it remains a mystery what commentators are actually describing when they speak of "the inviolability of thoughts" (Dan-Cohen, 1999, 379), or when they call punishment for mere mental states a "monstrous" (Yaffe, 2014, 101), intrusion into a person's "private world" (Ashworth, 2011, 134), and an invasion of her "essential ... human right to freedom of thought" (Calvert, 2005, 125).

I aim in what follows to mine the foundations of this rhetoric and lay bare the premises of an argument of my own. The argument gives analytical clarity to the attractive but heretofore unexplained idea that thought's immunity from punishment relates to a principle of freedom of mind. Although I hope to render the argument's premises plausible, my primary objective is to show that our legal order presupposes these premises, and thus to explain why the conclusion they entail seems so intuitive.

A. The Basic Idea

Given how often and how fervently theorists associate the ban on thought crime with a principle of freedom of mind, it's somewhat surprising that no one has bothered to show how the second principle might undergird the first. Theorists may think the linkage is just obvious. Or they may assume there is so little conceptual space between the two principles that any demonstration of the linkage would be uninteresting. As we'll see, the linkage is both interesting and unobvious.

In brief, I propose that the injustice of punishment for mere mental states takes its character from the injustice of a more literal breach of the "inviolability of thoughts": namely, a direct and forcible intrusion into the mind.

This more literal breach of the "inviolability of thoughts" is the sort of intrusion that the state would perpetrate if it exposed you to a mind-altering drug in order to disrupt your criminal intentions. It's natural to suppose that this sort of direct and forcible mind control is unjust insofar as it violates your *right of mental integrity*, your right to be free from unwanted mental interference or manipulation. I'll say more about the contours and limits of this right in a later section. For now, an example will convey the basic idea. Suppose you're an intending criminal. Without invading your right to mental integrity, the government may question you about your criminal intention, try to persuade you to abandon it, surveil you, tail you, and stand ready to thwart you if you attempt to carry your intention out. But the government will invade your right to mental integrity if it causes you to abandon your intention by forcing you to ingest mind-altering drugs, by exposing you to psychotropic gas, or by employing some other form of forcible mind control.[2]

To be sure, many of these intrusions also may invade your right to *bodily* integrity. Forcing you to ingest or inhale an unwanted substance is a classic battery. But if you possess a right to mental integrity, none of these actions is just a battery. Each is also an attempt at forcible mind control, which is a distinctive rights invasion. It's this rights invasion that forms the gravamen of the wrong that the state perpetrates when it forces you to ingest or inhale something mind-altering—the physical battery being slight and potentially harmless. If the government could control

[2] I must emphasize that I am using the label "right of mental integrity" to designate a relatively narrow right against the direct and forcible manipulation of a person's mind (e.g., through the forced administration of intoxicants or psychotropic medications). If I used the label to designate a broader right against all forms of impermissible mental manipulation, including the form of indirect mental coercion that the state perpetrates when it criminalizes mere thought, then my explanation of why it's wrong to punish thought would be all but circular, and it would obscure rather than illuminate the connection between direct and indirect mind control.

your mind without battering you at all (say, by using light and sound to hypnotize you involuntarily), the intrusion still would wrong you, and it would wrong you because it would violate your right to mental integrity.

The claim I'll defend over the next two sections is that punishment for mere mental states is intrinsically unjust because it's a form of *indirect* mind control.

Not only does this claim promise to give content to the picturesque but imprecise assertion that punishment for mere mental states transgresses the "inviolability of thoughts," but it also captures the essence of relevant American legal doctrine. Consider *Stanley v. Georgia* and *Ashcroft v. Free Speech Coalition*, two well-known cases in which the Supreme Court cited a constitutional prohibition on mind control to justify striking down statutes the enforcement of which had no direct effect on a person's mind. In *Stanley*, the Supreme Court struck down a state statute "forbidding mere private possession of [obscene] material" (Stanley v. Georgia, 1969, 564). The Court rejected the government's claim to a "right to control the moral content of a person's thoughts" (ibid., 565), noting that "[o]ur whole constitutional heritage rebels at the thought of giving government the power to control men's minds" (ibid.). Decades later, in *Free Speech Coalition*, the Court gave the same justification for striking down a federal statute prohibiting visual depictions of "an actor [who] 'appears to be' a minor engaging in 'actual or simulated . . . sexual intercourse'" (Ashcroft v. Free Speech Coalition, 2002, 241). The Court in *Free Speech Coalition* had to distinguish an earlier decision in which it had permitted the government to ban pornography involving real children on account of the harm done to the children depicted (ibid., 240; New York v. Ferber, 1982). Unlike real child pornography, explained the Court in *Free Speech Coalition*, *simulated* child pornography is anathema for one reason alone: its effect on a viewer's mind. The Court deemed this reason an impermissible basis for criminal legislation. "The [g]overnment submits . . . that virtual child pornography whets the appetites of pedophiles and encourages them to engage in illegal conduct. This rationale cannot sustain the provision in question. The mere tendency of speech to encourage unlawful acts is not a sufficient reason for banning it" (Ashcroft v. Free Speech Coalition, 2002, 240, 241). Quoting *Stanley*, the Court concluded: "The government 'cannot constitutionally premise legislation on the desirability of controlling a person's private thoughts'" (ibid., 253; Stanley v. Georgia, 1969, 564). In *Free Speech Coalition*, as in *Stanley*, the Court based its analysis on a constitutional prohibition on mind control even though the statute it found unconstitutional did not affect the mind directly: enforcing statutory bans on obscenity and simulated child pornography is a far cry from administering unwanted mind-altering drugs. The Court's position seems to have been that, because forcible mind control is impermissible, so too are certain governmental efforts designed to achieve the same end by indirect means.

The indirect method of mind control that the Court deemed impermissible in *Stanley* and *Free Speech Coalition* was the state's practice of punishing people for conduct believed likely to produce undesirable thoughts. A more blatant method of indirect mind control, which I presume the Court would disapprove of for the same reason, is the practice of punishing people for their undesirable thoughts themselves. The basic idea is easy to state: it's *because* the state mustn't control thoughts that the state mustn't punish them.

In what follows, I'll show how this idea follows from two interlocking propositions presupposed by our legal order—propositions that I won't be able to defend fully, but that I'll do my best to render plausible. The first proposition—the *Enforceability Constraint*—is that it's wrong for the state to punish offenses of a given type if it's always wrong in principle for the state to forcibly disrupt such offenses merely on the ground that they're censurable transgressions (transgressions that are dangerous or wrongful and for this reason worthy of condemnation). The second proposition—grounded in the right of mental integrity—is that it's always wrong in principle for the state to forcibly disrupt a given mental state merely on the ground that it's a censurable transgression (although the state sometimes may disrupt a mental state on more exigent grounds). I'll defend these propositions in turn.

B. The Enforceability Constraint

In our system of criminal administration, the state may ensure compliance with penal norms not only indirectly through punishment, but also through direct compulsive force. When you're selling loose cigarettes, the police may take them from your hand. When you're making a bomb, the police may escort you from your laboratory. When you're absconding with stolen goods, the police may stop you and seize them.

An unexamined but signal feature of our system is that the direct and indirect enforcement authorities are linked in a particular way: in practice, and seemingly not by accident, the state may enforce a given penal norm indirectly only when it also may enforce that norm directly. In other words, the state may punish someone for transgressions of a given type only when the state may in principle use reasonable force to thwart such transgressions merely on the ground that they're criminally wrongful, that is, without supplying any additional justification. If the state may not even in principle use force to thwart instances of a given transgression on the ground that they're criminally wrongful, then the state also may not make that type of transgression an object of punishment.

Why may the state ensure compliance with a given legal norm through punishment only when the state may ensure contemporaneous compliance with that norm through direct compulsive force? My answer, in brief, is this: if ensuring compliance with a given norm through direct compulsive force would violate your rights, so too would ensuring compliance with that norm through the threat and imposition of the severest form of sanction and censure. I'll establish this proposition more firmly by means of an informal conditional proof, starting with the supposition that some supposed transgression is off limits to forcible disruption, and reasoning from that supposition to the conclusion that the transgression is off limits to punishment.

Suppose, as our starting point, that the state would wrong you if it forcibly disrupted some supposed transgression of yours, T, merely on the ground that T is a censurable transgression. Suppose, further, that the wrong the state would perpetrate against you if it disrupted your T-ing is a wrong *intrinsic* to the disruption—a wrong that consists at least partly in the disruption of T itself, rather than consisting entirely in the fact (if it is one) that the method of disruption injures you in some other way.

Now, if it's the case that the state would wrong you intrinsically if it disrupted your T-ing merely on the ground that T is a censurable transgression, then there must be some reason *why* this is so. And the reason can't be that the method of disruption injures you in some other way, because we've supposed that the wrong is intrinsic—that it consists at least partly in the disruption of T itself. Why, then, does the state wrong you intrinsically when it disrupts your T-ing merely on the ground that T is a censurable transgression?

One possibility is that T is perfectly innocent and innocuous (like consensual sexual conduct between adults) or is at least less wrongful and less harmful than any censurable transgression that the state legitimately may criminalize. In either case, it follows straightforwardly that the state would wrong you if it punished you for T-ing.

But some transgressions may be immune from disruption on grounds of censurability even though they're wrongful and arguably dangerous. (Certain speech acts fall into this category, and so may certain thoughts, as I'll argue in the next section. When the state prevents you from performing these speech acts or from thinking these thoughts, the state wrongs you. And it wrongs you intrinsically—which is to say, it wrongs you even if it uses means of prevention so delicate and precise that they cause you no injury.)

Suppose, then, that T is as wrongful and harmful as other censurable transgressions that the state may criminalize, yet the state nevertheless would wrong you intrinsically if it disrupted your T-ing merely on the ground that T is a censurable transgression.

If the state would wrong you intrinsically if it disrupted your T-ing on this ground alone, yet your T-ing is dangerous and wrongful, then a likely explanation—perhaps the only possible explanation—is that you've got a *right* to perform T, a right that the state would violate if it forcibly disrupted your T-ing merely on the ground that T is a censurable transgression.

Now, if the state would violate your right if it forcibly disrupted your T-ing merely on the ground that T is a censurable transgression, then I suggest that the state also would violate your right if it disrupted your T-ing in a particular *indirect* fashion: by imposing terrible consequences on you for T-ing, merely on the ground that T is a censurable transgression.

But when the state *punishes* you for T-ing, it thereby imposes terrible consequences on you for T-ing, and it does so on no ground other than that T is a censurable transgression. (Ordinarily, to justify punishing someone, the state need only show that the person committed a criminal wrong.) So we may conclude that when the state punishes you for T-ing, it violates your rights. It wrongs you.

We've arrived at the following conditional claim: whether T is innocent and innocuous or wrongful and dangerous, if the state would wrong you if it forcibly disrupted your T-ing on the ground that T is a censurable transgression (our initial supposition), then so too would the state wrong you if it punished you for T-ing (our conclusion). This conditional claim is none other than the Enforceability Constraint.[3]

Justifying the Enforceability Constraint more fully is beyond the scope of this essay. My present goal is more modest. It's to show how abnormal it would be to treat any type of transgression as an exception to the Enforceability Constraint. Deeming mental transgressions an exception would yield an anomaly: a type of crime that the state may punish but never forcibly disrupt on grounds of criminality alone.

No such type of crime exists, nor does any recognized limit to the state's enforcement power belie the gist of the Enforceability Constraint. In fact, no recognized limit on the state's enforcement power does more than restrict when, how, or pursuant to what procedures given instances of an offense may be forcibly disrupted.

The most salient limit on the state's enforcement power is the principle of reasonable force (Graham v. Connor, 1989, 395). This principle governs *how much* force the state may deploy to make someone comply with a given penal norm on a given occasion, not whether such force may be deployed at all. In the typical case, the state may deploy an amount of force sufficient but not greater than necessary to stop the relevant norm violation. If you're selling loose cigarettes, the police may pull them from your hand, but they may not put you in a chokehold (Rahman & Barr, 2014).

Of course circumstances sometimes arise where the amount of force necessary and sufficient to stop a given transgression is unreasonably great. Suppose a narcochemist is manufacturing methamphetamine in a treehouse and the only way the police can stop him is by cutting the

[3] Although I've presented these considerations as an argument for the Enforceability Constraint, they may in fact justify both more and less than the Enforceability Constraint. Insofar as certain forms of what we regard as punishment might fall short of imposing *terrible* consequences on an offender, the argument in the text won't establish that the state is always forbidden to punish what it may not disrupt directly merely on grounds of wrongfulness. Certain "lighter" forms of punishment might still be permissible—just as *nonpenal* sanctions are often permissible even when direct enforcement of the relevant (nonpenal) norm is forbidden, the way it's often permissible to award damages as a sanction for conduct that a court couldn't enjoin and that a plaintiff couldn't lawfully disrupt through self-defensive force. Furthermore, insofar as punishing someone for T-ing is but one way of indirectly violating his right to T, the argument in the text may in fact justify principles *beyond* the Enforceability Constraint, including a principle forbidding the state from preventively but nonpunitively detaining people for T-ing. I return to this possibility in the next section.

tree down, paralyzing him in the process. May the police cut down the tree? Clearly not, and the Enforceability Constraint agrees. What the state may punish, the state in principle may impede—but only with reasonable force. Unreasonable force wrongs the narcochemist.

It wrongs him because he has a right not to be paralyzed absent truly exigent circumstances—not because he has a right to make methamphetamine. And that's important. The Enforceability Constraint permits the state to subject the narcochemist to punishment, even as the principle of reasonable force forbids the state to thwart his meth-making. In a world where no single instance of a given offense is disruptable through reasonable force—a world where every narcochemist operates from a fortified treehouse—the Enforceability Constraint still permits offenders to be punished. The Enforceability Constraint says that an offense is unpunishable if it's always wrong *in principle* to disrupt instances of that offense merely on grounds of wrongfulness. In a world of fortified treehouse meth labs, it's always wrong to disrupt meth-making in practice, but it isn't always (or perhaps ever) wrong to do so in principle.

Other limits to the state's enforcement power concern *when* and pursuant to *what procedures* the state may use force to stop a given transgression. Like the principle of reasonable force, these limits are fully consistent with the Enforceability Constraint. Consider the First Amendment doctrine of prior restraint, which holds that certain expressive acts that are punishable after the fact may not be blocked in advance by a judicial order or administrative ruling (Near v. Minnesota, 1931, 713–714; Neb. Press Ass'n v. Stuart, 1976, 559). The doctrine's primary rationales are evidentiary and institutional. "It is always difficult to know in advance what an individual will say," the Supreme Court notes, "and the line between legitimate and illegitimate speech is often so finely drawn that the risks of freewheeling censorship are formidable" (Southeast Promotions, Ltd. v. Conrad, 1975, 559). Moreover, as the Court observes elsewhere, "[a] criminal penalty . . . is subject to the whole panoply of protections afforded by deferring the impact of the judgment until all avenues of appellate review have been exhausted. . . . A prior restraint, by contrast and by definition, has an immediate and irreversible sanction" (Neb. Press Ass'n v. Stuart, 1976, 559). If the Court is correct, these evidentiary and institutional considerations support the view that norms prohibiting certain types of speech may not be enforced at particular times (e.g., prior to a jury trial) or in particular ways (e.g., by a bureaucrat's edict).

What these considerations don't support (and have never been interpreted as entailing) is the view that certain penal norms may not be enforced at all except by criminal punishment. It's widely accepted, for example, that an expressive act immune from pretrial injunction may be blocked by a judicial order once the act has been formally adjudicated as unlawful. As the California Supreme Court explains, "[p]rohibiting a person from making a statement or publishing a writing *before* that statement is spoken or the writing is published is far different from prohibiting a defendant from *repeating* a statement or *republishing* a writing that has been determined at trial to be defamatory and, thus, unlawful" (Balboa Island Vill. Inn, Inc. v. Lemen, 156 P.3d 339 [Cal. 2007]). The doctrine of prior restraint therefore isn't a counterexample to the Enforceability Constraint; to the contrary, it assumes the Constraint's soundness. The doctrine maintains only that criminal norms prohibiting speech acts are unenforceable at certain times and pursuant to certain procedures. The doctrine doesn't maintain that these norms are unenforceable in principle.

Now, what's enforceable in principle might not always be justifiably enforced in practice. It's conceivable that the above-mentioned limits on the state's enforcement power, if applied to penal norms prohibiting mere thought, would render such norms practically unenforceable except by retrospective criminal punishment. For one thing, it's possible that any direct effort by the state to disrupt the commission of a purely mental transgression would flout limits of timing and procedure. Given the relative inscrutability of the mind, in the absence of a judicial inquest the risks of erroneous intrusion might be too great to bear (Moore, 1993, 48). It's also possible that any amount of force would be excessive if deployed to disrupt a person's mere mental states. Given the crude technologies of mind control currently available, forcible intrusion into the mind might inevitably cause serious physical injuries or deleterious changes to a person's personality or mental well-being. Even if all these things are true, however, limits of timing, procedure, and proportionality still don't entail that mental intrusion is objectionable *in principle*. They don't entail that mental intrusion would be objectionable even if it could be carried out flawlessly: by a device that could detect malevolent intentions with high reliability and psycho-surgically remove them without doing other damage.

If such intrusion isn't objectionable in principle, then the Enforceability Constraint doesn't yield the conclusion that punishing thought is intrinsically unjust. So the question is whether psycho-surgical policing is actually objectionable in principle. May the state thwart your mental states merely on the ground that they're censurable transgressions?

C. The Right of Mental Integrity

My contention is that psycho-surgical policing is indeed objectionable in principle, and it's objectionable in principle because it violates the right to mental integrity, the right to be free from unwanted mental interference or manipulation of a direct and forcible sort.

A commitment to this right, like a commitment to the Enforceability Constraint, seems a basic feature of our system of criminal administration. The right to mental integrity figures not only in the reasoning of *Stanley* and *Free Speech Coalition* but also in the legal principles governing when the state may forcibly medicate a defendant to render him competent to stand trial (Sell v. United States, 2003, 179; Riggins v. Nevada, 1992) and when the state may forcibly medicate a mentally ill prisoner to ensure public safety (Washington v. Harper, 1990, 210).[4]The right to mental integrity also applies in a decidedly nonpenal context, undergirding a civilly committed person's right to refuse involuntary psychiatric treatment. As one court explained, "[t]he [constitutional] right of privacy is broad enough to include the right to protect one's mental processes from governmental interference" (Rennie v. Klein, 1978, 1144). Legal principles aside, we generally blanch at the idea of brainwashing—the idea of one person controlling the thoughts of another through forcible conditioning—whether the controller is a cult leader or a totalitarian government.

The main obstacle to appreciating that our legal and moral order presupposes a right to mental integrity is the mistaken view that, if such a right existed, it would be unqualified or absolute. If the right to mental integrity were absolute, forcible manipulation of a person's mind would be absolutely forbidden. But forcible manipulation of a person's mind doesn't seem absolutely forbidden. For example, it might be permissible for the state to force a mentally ill prisoner to ingest psychiatric medication, as the Supreme Court recognized in *Washington v. Harper*

[4] I acknowledge that the Supreme Court itself (as well as some interpreters of its jurisprudence) might deny that these decisions are grounded most fundamentally in a right of mental integrity, as opposed to a broader due process right against all coercive medical interventions, including but not limited to interventions that intrude on the mind.

(Washington v. Harper, 1990, 210). If this sort of mental intrusion is justifiable, that might be thought to entail that there's no right to mental integrity after all—no right to be free from forcible mind control. But the justifiability of mental intrusion entails merely that the right to mental integrity, if it exists, is qualified or non-absolute.

In fact, the Court's willingness to permit forced medication in *Harper* actually seems to rest on an acknowledgment that people possess a *qualified* right to mental integrity rather than on a denial that any such right exists. In *Harper*, a mentally ill prisoner claimed that the state should be barred from forcing him to ingest antipsychotic drugs unless it could prove that he would consent to such treatment if competent (ibid., 222). The Supreme Court denied the prisoner's claim, holding that the state may force a seriously mentally ill prisoner to ingest antipsychotic medication against his will as long as the state first establishes that he's "dangerous to himself or others" and that such treatment is in his "medical interest" (ibid., 227). If this holding is correct—as a matter of political philosophy, whether or not as a matter of constitutional law—then the government doesn't violate (i.e., unjustifiably invade) an inmate's right to mental integrity by interfering directly with his thoughts if doing so is practically necessary to ensure public safety and is in the person's "medical interest." It doesn't follow, however, that the proposed right of mental integrity is illusory. Nor does it follow that public necessity temporarily extinguishes the inmate's right to mental integrity, such that the right exerts no moral force in the covered circumstance. Rather, the best explanation of the Court's holding is that public necessity *overrides* the inmate's right without extinguishing it. If the right persists even when justifiably overridden, then the right continues to exert moral force. That explains why the unwanted psychiatric intervention must end as soon as possible, why the intervention must be no more intrusive than necessary to serve its purpose, and why the very question of the intervention's permissibility is so momentous in the first place (Sell v. United States, 2003, 179).

As my analysis of *Harper* shows, we can allow that the state may manipulate your mental states on grounds of public necessity without thereby denying the existence of a right to mental integrity. Just as important, we can allow that the state may manipulate your mental states on grounds of public necessity without thereby conceding that the state may infringe your right to mental integrity on grounds *other* than public necessity—such as the ground that the targeted mental state is a censurable transgression, a ground on which (per the Enforceability Constraint) the state would have to be allowed to invade the right if it were allowed to make mere thought an object of punishment.

Public necessity may justify many kinds of rights invasion that would be impermissible if undertaken on other grounds. For example, the state may subject you to excruciating pain as a way of preventing you from killing someone, but not as a way of punishing you for a criminal offense. Your right not to be subjected to excruciating pain prohibits the state from performing certain actions for certain reasons without forbidding the state from performing those actions altogether. Thus, your right not to be subjected to excruciating pain forbids the state from causing you excruciating pain on the ground that doing so will serve as an unpleasant sanction that expresses the state's disapproval of your past wrongdoing (punishment)—but the state violates no right of yours when it subjects you to the exact same measure of excruciating pain on the ground that doing so will make you drop the gun you're threatening to fire at an innocent child (contemporaneous disruption).

Similarly, your right to mental integrity forbids the state from forcibly disrupting your mental states on the ground that they're censurable transgressions—but, if the holding of *Harper* is sound, the state doesn't violate your right to mental integrity when it forcibly disrupts your mental states on the ground that doing so is necessary to protect the public and is in your "medical interest" anyway. Indeed, mental intrusion on grounds of public necessity seems permissible even when it's not in your "medical interest." Imagine that a terrorist intends to detonate a bomb and the police have only three ways of stopping him: they can incapacitate him (e.g., shoot him), restrain him physically (e.g., handcuff him), or restrain him psychically (e.g., deploy a stun grenade). If the police aren't close enough to the terrorist to restrain him physically, they're left with two options: incapacitation and psychical restraint. Because the threat to public safety is grave—and because temporary psychical restraint is a mild invasion of a person's mental integrity, whereas permanent physical incapacitation is a grievous invasion of his bodily integrity—I presume that the government may forcibly disrupt the terrorist's intention (e.g., with a stun grenade) on the ground that doing so is necessary to prevent the terrorist from detonating the bomb.

In fact, I don't see any barrier in principle to the state *preventively detaining* people on the basis of their thoughts alone. But consider how heavy a burden the state would have to bear in practice if it sought to justify such a measure by appeal to the considerations generally thought necessary to justify direct mental intrusion. To forcibly medicate a prisoner, for example, the state must show that the prisoner is dangerous and that less intrusive alternatives to forced medication are unavailable. If the state could make a similar showing in regard to detaining a person on the basis of a given thought—if it could show that doing so were necessary to protect the public, less intrusive alternatives being unavailable—then I'd be willing to concede that it isn't always wrong to preventively detain people on the basis of that particular thought. I simply doubt whether the state could ever make the requisite showing. It isn't enough for the state to show that certain thoughts present an exceptional danger. It's also necessary for the state to show that the danger can be allayed in one way only: by preventively detaining people on the basis of those thoughts alone. No actual jurisdiction takes the possibility seriously. Several American states have laws permitting the preventive detention of "sexually violent predators," but these laws require proof of previous violent conduct, rather than mere proclivity (People v. Field, 2016, 553).

Yet there's one strain of American law that might seem to lower the barrier to mental intrusion: the doctrine permitting the government to administer involuntary medication *without* a showing of public necessity when the purpose is to render a psychotic defendant fit for trial. Under current Supreme Court precedent, the government may administer involuntary medication for this purpose if "the treatment is medically appropriate, is substantially unlikely to have side effects that may undermine the fairness of the trial, and, taking account of less intrusive alternatives, is necessary significantly to further important governmental trial-related interests" (Sell v. United States, 2003, 179). Stephen Morse rationalizes this doctrine on the ground that the state's "interest in adjudicating guilt and innocence and achieving finality in the criminal process is . . . 'essential' or important" (Morse, 2017, 17), whereas the defendant's interest in freedom from unwanted mental intrusion is minimal under the circumstances. Forcibly medicating a psychotic defendant, Morse argues, "would appear to increase freedom of thought rather than to decrease it. . . . [T]he 'freedom' to be psychotic does not seem to be a freedom worth having or freedom at all" (ibid., 15).

If this reasoning and the doctrine it supports are sound, it's natural to ask whether the need to prevent people from having culpably wrongful thoughts couldn't sometimes be at least as pressing as the need to rid defendants of delusions pretrial. I'm not certain that the doctrine is sound, however, so I'm neutral between the following possibilities:

(1) The need to rid defendants of delusions pretrial is more pressing than the need to prevent people from having culpably wrongful thoughts. Accordingly, although the state may forcibly medicate defendants pretrial, it may not punish people for their thoughts (thanks to the Enforceability Constraint).
(2) The need to rid defendants of delusions pretrial *isn't* more pressing than the need to prevent people from having culpably wrongful thoughts, and each of these needs is insufficient to justify mental intrusion. Accordingly, the state may not forcibly medicate defendants pretrial, nor (thanks to the Enforceability Constraint) may the state punish people for their thoughts.

My claim is simply that (1) is coherent. It nevertheless might be false. The better view might be (2): it might be that mental states are unpunishable only if forcibly medicating defendants pretrial is unjustifiable. This possibility doesn't seem a *reductio ad absurdum* of the proposition that mental states are unpunishable. We shouldn't unquestioningly accept that the government's trial-related interests truly justify infringing the mental autonomy of psychotic defendants.

The one possibility I've rejected is this:

(3) The need to rid defendants of delusions pretrial isn't more pressing than the need to prevent people from having culpably wrongful thoughts, yet each of these needs is *sufficient* to justify mental intrusion. Accordingly, the state not only may forcibly medicate defendants pretrial but it also may punish people for their thoughts.

I've rejected this possibility out of hand—precipitously, some might say. Although our legal order presupposes a right to mental integrity that applies across a range of penal and nonpenal contexts, in many of these contexts the right gives way to competing values. As conceived in law, the right to mental integrity clearly isn't absolute. This raises a basic question. If the right to mental integrity can be overridden on grounds of public necessity, and maybe also on grounds of judicial finality, why can't the right to mental integrity ever be overridden on the ground that it's being exercised wrongfully? If mental intrusion can be justified by the imperatives of public safety and criminal adjudication, why can't it also be justified by the imperative of law enforcement? Why can't the state at least sometimes manipulate a person's mind on the ground that his mental states are censurable transgressions?

I think this line of rhetorical questions gets things backward. Part of what it means to have a right is that any proposed invasion of the sphere that the right protects requires affirmative justification. Absent such justification, we can repel a proposed invasion just by asserting the right. Thus, if there's a right to mental integrity—as our legal order presupposes, and as intuitively seems to be the case—then the question we must ask of any proposed invasion of the right isn't "why *shouldn't* it be permitted?" but "why *should* it?". The burden is on the intruder to justify the intrusion, not on the right-bearer to defeat it.

Now, I don't mean to imply that such justification is unimaginable. We simply know too little about the foundations of either the state's enforcement power or the right to mental integrity to assert confidently that mental intrusion can never be justified merely on the ground that a person's mental states are censurable transgressions. Thus, we can't yet say whether the imperative of law enforcement is more or less compelling than the imperative of criminal adjudication—although I do think we can assume that the countervailing individual interests in the adjudication context are probably somewhat weaker. As Morse suggests, "the 'freedom' to be psychotic [may not] be a freedom worth having or freedom at all" (ibid., 15).

I also think we can assume that the countervailing individual interests are weaker when the right in question is that of *bodily* integrity. I've assumed, as everyone does, that the right of bodily integrity routinely gives way to the imperative of law enforcement: that proposed invasions of the right to bodily integrity can be justified on the mere ground that the right-bearer is committing a censurable transgression. The police may take loose cigarettes from your hand, escort you from your bomb-making laboratory, and seize your stolen goods—all without violating your right to bodily integrity.

But why? If, as I've said, the burden is always on the potential right-intruder to justify an intrusion, not on the right-bearer to defeat it, then why does the imperative of law enforcement—the state's imperative to disrupt censurable transgressions merely on the ground that they're censurable transgressions—justify invading the body if it doesn't justify invading the mind?

To this important question, I can offer only the beginning of an answer. My suspicion is that the right to mental integrity may derive (in a way that the right to bodily integrity does not) from the nature and moral significance of personhood. At the root of the normative asymmetry between mind and body may be the fact that one's mental states, far more than one's actions, determine who one is as a person. As Seana Shiffrin writes, "what makes one a distinctive individual *qua person* is *largely* a matter of the contents of one's mind" (Shiffrin, 2011, 291). Thus, if one has an interest in controlling one's identity as a "distinctive individual"—an interest in controlling who one is as a person—then one has an interest in controlling the contents of one's mind. I assume that this fundamental interest grounds the right to mental integrity and that this right, unlike the right to bodily integrity, therefore serves as a decisive counterweight to the imperative of law enforcement.

In making these assumptions—in assuming that the state necessarily violates your right to mental integrity when it forcibly disrupts your thoughts on the ground that they're censurable transgressions—I've not simply assumed what I set out to prove: that thought is unpunishable. Grounding thought's immunity from punishment in its immunity from direct manipulation has required me to defend an unexamined but signal feature of our system of criminal administration: that the state's authority to punish transgressions of a given type extends no further than its authority to disrupt transgressions of that type using direct compulsive force. If sound, the Enforceability Constraint isn't a conceptual or semantic truth; it's a normative one. And it's a normative truth that doesn't hold for nonpenal law, where retrospective sanction is often permissible even when contemporaneous compulsion is not.

* * *

I've argued in this Part that the intrinsic injustice of punishment for thought has the following origins:

(1) It's wrong for the state to punish you for your thoughts if it's always wrong in principle for the state to use force to thwart or disrupt your thoughts merely on the ground that they're censurable transgressions.

(2) It's always wrong in principle for the state to use force to thwart or disrupt your thoughts merely on the ground that they're censurable transgressions.
(3) Therefore, it's wrong for the state to punish you for your thoughts.

The first of these propositions draws support from the Enforceability Constraint, and the second from the right to mental integrity—two ideas to which our legal order seems resolutely committed. In explaining these commitments, I did my best to make both seem reasonable. I didn't pretend to offer a full justification of either. It's unlikely that any such justification would be beyond controversy, anyway. It would be surprising indeed if a somewhat controversial proposition—that there's an *intrinsic* injustice in punishment for mere mental states—followed straightforwardly from propositions that were themselves uncontentious.

Concluding Remarks

The state's enforcement power and the mind's inviolability are rich topics worthy of further inquiry (for a general discussion of the right to mental integrity, see Bublitz & Merkel 2014). Especially ripe for study is their point of intersection. Positing a right to mental integrity raises difficult questions about the limits of the state's enforcement power, foremost among them the question of the right's precise scope vis-à-vis the state.

It can't be that the state violates your right to mental integrity every time it tries to influence your thoughts. The state violates no one's right to mental integrity when it pleads with a hostage taker, requires children to be educated, or simply attempts to communicate with its citizens. A police officer doesn't violate your right to mental integrity when she approaches you and begins talking, even though by doing so she causes you to experience certain perceptions and beliefs that you might not want to experience.

As these examples show, distinguishing between permissible and impermissible modes of interference with a person's mental life presents no small task. Why does the police officer's communicative act not violate your right to be free from unwanted mental intrusion? Is it because the means of interference (stimulating your perceptive faculties) isn't forcible? Is it because you implicitly consent to this type of mental intrusion just by going around in the world with open eyes and ears? Is it because your right to mental integrity simply doesn't cover perceptions and perceptual beliefs, the right being limited to other sorts of mental state? Or is it because of the purpose for which the intrusion is undertaken?

A complete theory of mental integrity would answer these questions by yielding an analytical framework for distinguishing in a principled way between modes of state interference that respect the right to mental integrity and modes that constitute impermissible mental intrusions. Like any moral or legal right, the right to mental integrity can be analyzed in terms of three aspects: (i) the domain over which the right ranges; (ii) the type of mental intrusions that qualify as invasions of the right; and (iii) the kind of circumstances (including state motivations) that make an invasion a *violation*, an invasion that's impermissible.

By distinguishing these three aspects of the right to mental integrity, we might begin to make progress on questions like those above. Why doesn't the state violate your right to mental integrity when a police officer accosts you and asks you questions? Plausibly, the perceptions and perceptual beliefs that the police officer causes you to experience don't fall within the domain over which the right ranges (see [i]). Why doesn't a liberal state violate a child's right to mental integrity when it compels her to receive an education of one sort or another? A possible answer is that, even though the beliefs and dispositions that a liberal education instills all fall within the domain that the right protects (see [i]), a liberal education engages directly with a child's rational faculties, instead of bypassing those faculties in the fashion of brainwashing or indoctrination. Thus, compulsory education may not qualify as a rights invasion (see [ii]). Why doesn't the state violate a mentally ill inmate's right to mental integrity when it forces her to ingest psychiatric medication as a means of ensuring community safety? Plausibly, the circumstances and intended effect of the intrusion render the rights invasion permissible (see [iii]).

Each of these tentative answers alludes to some general operating principle that differentiates impermissible mind control from softer modes of influence that leave people's mental integrity tolerably intact. Some such principles *must* exist, or else the state would be altogether forbidden from influencing people's beliefs and desires—an implausible position. The operating principle that this essay has aimed to vindicate is the age-old maxim of criminal jurisprudence *cogitationis poenam nemo patitur* ("no one may be punished merely for thinking"). But this operating principle is potentially just one among many.

References

Articles and Books

Arkes, H. R., & Blumer, C. (1985). The psychology of sunk cost, 35 *Organizational Behavior and Human Decision Processes* 124.

Ashworth, A. (2011). Attempts. In J. Deigh, & D. Dolinko (eds.), *The Oxford Handbook of Philosophy of Criminal Law* 126.

Brehm, J. W. (1956). Postdecision changes in the desirability of alternatives, 52 *Journal of Abnormal & Social Psychology* 384.

Bublitz, J., & Merkel, R. (2014). Crimes against minds: On mental manipulations, harms and a human right to mental selfdetermination, 8 *Criminal Law & Philosophy* 51.

Calvert, C. (2005). Freedom of thought, offensive fantasies and the fundamental human right to hold deviant ideas: Why the seventh circuit got it wrong in Doe v. City of Lafayette, Indiana, 3 *Pierce Law Review* 125.

Dan-Cohen, M. (1999). Harmful thoughts, 18 *Law & Philosophy* 379.

Duff, R. A. (2012). Risks, culpability and criminal liability. In G. R. Sullivan & Ian Dennis (eds.), *Seeking security: Pre-empting the commission of criminal harms.*

Edwards, W. (1968). Conservatism in human information processing. In Benjamin Kleinmuntz (ed.), *Formal representation of human judgment.*

Endlich, G. A. (1891). The doctrine of mens rea, 13 *Criminal Law Magazine & Reporter* 831.

Fitzgerald, P. J. (1962). Criminal law and punishment.

Hart, H. L. A. (2008). Punishment and responsibility: Essays in the philosophy of law (2nd edition).

Kavka, G. S. (1983). The toxin puzzle, 43 *Analysis* 33.

Mather, M., Shafir, E., & Johnson, M. K. (2000). Misremembrance of options past: Source monitoring and choice, 11 *Psychological Science* 132.

Morse, S. (2017). Involuntary competence in United States criminal law (University of Pennsylvania Law School, Public Law & Legal Theory Research Paper No. 17–20), http://ssrn.com/abstract=2951966 [http://perma.cc/BUX3-642U].

Mill, J. S. (1978 [1859]). On Liberty, Elizabeth Rapaport (ed.), Hackett Publishing Co.

Nickerson, R. S. (1998). Confirmation bias: A ubiquitous phenomenon in many guises, 2 *Review of General Psychology* 175.

Moore, M. (1993). Act and crime: The philosophy of action and its implications for criminal law.

Rahman, S., & Barr, S. (2014, December 5). Opinion, Eric Garner and the legal rules that enable police violence, NY Times, https://www.nytimes.com/2014/12/06/opinion/eric-garner-and-the-legal-rules-that-enable-police-violence.html [http://perma.cc/43AC-X75R].

Robinson, P. H. (1984). Criminal law defenses.

Sanna, L. J., Schwarz, N., & Stocker, S. L. (2002). When debiasing backfires: Accessible content and accessibility experiences in debiasing hindsight, 28 *Journal of Experimental Psychology: Learning, Memory & Cognition* 497.

Shiffrin, S. V. (2011). A thinker-based approach to freedom of speech, 27 *Constitutional Commentary* 283.

Stephen, J. F. (1883). A history of the criminal law of England: Vol. 2.

Yaffe, G. (2014). Criminal attempts, 124 *The Yale Law Journal* 92.

Cases and Statutes

Ashcroft v. Free Speech Coalition, 535 U.S. 234 (2002).

Balboa Island Vill. Inn, Inc. v. Lemen, 156 P.3d 339 (Cal. 2007).

Graham v. Connor, 490 U.S. 386 (1989).

Near v. Minnesota, 283 U.S. 697 (1931).

Nebraska Press Ass'n v. Stuart, 427 U.S. 539 (1976).

New York v. Ferber, 458 U.S. 747 (1982).

People v. Field, 204 Cal. Rptr. 3d 548 (Cal. Ct. App. 2016).

Rennie v. Klein, 462 F. Supp. 1131 (D.N.J. 1978).

Riggins v. Nevada, 504 U.S. 127 (1992).

Sell v. United States, 539 U.S. 166 (2003).

Southeast Promotions, Ltd. v. Conrad, 420 U.S. 546 (1975).

Stanley v. Georgia, 394 U.S. 557 (1969).

Washington v. Harper, 494 U.S. 210 (1990).

CHAPTER 6

Autonomy, Evidence-Responsiveness, and the Ethics of Influence

Fay Niker, Gidon Felsen, Saskia K. Nagel, and Peter B. Reiner

Introduction

It is uncontroversial that the rise of the cognitive sciences, broadly construed, has had a significant impact on how we understand how humans think and behave. Robust sets of neurobiological and psychological findings concerning human cognitive processes have both challenged orthodox positions in, and raised new questions for the disciplines of economics, philosophy, politics, and beyond.

F. Niker (✉)
University of Stirling, Stirling, UK
e-mail: fay.niker@stir.ac.uk

G. Felsen
University of Colorado, Boulder, CO, USA
e-mail: gidon.felsen@cuanschutz.edu

S. K. Nagel
RWTH Aachen University, Aachen, Germany
e-mail: saskia.nagel@humtec.rwth-aachen.de

M. J. Blitz et al. (eds.), *The Law and Ethics of Freedom of Thought, Volume 1*, Palgrave Studies in Law, Neuroscience, and Human Behavior,
https://doi.org/10.1007/978-3-030-84494-3_6

To give a brief example, findings relating to the automaticity and context-dependency of our rational processing and the dual-process theories of cognition that purport to explain them (Kahneman, 2011; Stanovich & West, 2000) have challenged our traditional views on rationality and suggest that *situated* conceptions of reason may be more appropriate (Hurley, 2011). In economics, this body of empirical research has served to establish behavioral economics as a distinct way of modeling human behavior (Simon, 1972). In psychology, these findings were instrumental in directing attention toward the emotions and their role in practical and moral reasoning (Bagnoli, 2011; May & Kumar, 2018), precipitating debates over the viability of virtue ethics as a metaethical enterprise (Doris, 2002) and setting the contours for revisionary theories of moral responsibility (Doris, 2015). In public policy and political theory, empirical research (Tversky & Kahneman, 1981) informed the shift toward the use of "nudges" as a public policy lever (Thaler & Sunstein, 2009) and reopened philosophical debates about the nature and wrongness of both paternalism and manipulation (Coons & Weber, 2013, 2014).

Within this broad tradition of inquiry, there are questions that can be raised about the relationship between empirical work in the cognitive sciences and the concept of autonomy. Specifically, one might ask whether empirical insights from fields such as neuroscience, psychology, and experimental philosophy can enrich our understanding of the nature of personal autonomy.[1] This is the focus of the current work.

P. B. Reiner
University of British Columbia, Vancouver, Canada
e-mail: peter.reiner@ubc.ca

[1] This is distinct from the metaphysical question asking whether neuroscientific experiments have shown that free will is an illusion. For those who understand "autonomy" as within the family of metaphysical freedom terms (e.g., Mele, 1995, 2012), this metaphysical question is the same as asking whether neuroscience has shown that there are no autonomous human beings. There has been a lively debate over this issue (see Lavazza, 2016). Yet, it is more common to make a distinction between personal autonomy and freedom, and we take this route. Freedom concerns the ability to act (and on some conceptions, "having sufficient resources and power to make one's desires effective"); whereas autonomy concerns "the independence and authenticity of the desires (values, emotions, etc.) that move one to act in the first place" (Christman, 2015).

Broadly understood, to be autonomous "is to be one's own person, to be directed by considerations, desires, conditions, and characteristics that are not simply imposed externally upon one, but are part of what can somehow be considered one's authentic self" (Christman, 2005). An agent who exercises this capacity to direct herself thus is said to be self-governing. The ability of individuals to exercise this self-government over their lives is a central value of many (though certainly not all) cultures and political systems, and it plays a weighty role in moral and political theorizing. As fundamental as this concept has been in the development of liberal thought, the task of specifying more precisely the conditions for autonomy has proven controversial.

A central distinction within this debate is between *internalist* and *externalist* conditions for autonomy. Some accounts are purely internalist insofar as they hold that whether an agent, or one of her decisions, can be described as autonomous or not depends entirely on features concerning her mental states. On Frankfurt's view (Frankfurt, 1971), for instance, a decision is autonomous if the first-order desire that motivates it coheres with the person's higher-order attitude on the matter. This is deemed to be the central factor relevant to autonomy ascription, regardless of how this higher-order attitude came about. This makes such coherentist accounts doubly internalist: Autonomy ascription depends *neither* on how we came to have the relevant higher-order attitude or make the decision at hand—"a fact that is prior to (and in this sense external to) the action itself"—*nor* on how our beliefs and attitudes relate to reality—"a fact that is independent of (and in this sense external to) the beliefs and attitudes themselves" (Buss & Westlund, 2018). This reveals two ways in which external factors may be relevant to autonomy ascription. First, we might be concerned with various ways in which the internal conditions of autonomy, such as the quality of our rational deliberation, are affected by external factors such as socialization or manipulation. For this reason, many autonomy theorists place a procedural constraint on such internal conditions: What matters is that an individual's preferences and values have (or could have) survived the right kind of critical reflection (Christman, 2010; Dworkin, 1988; Friedman, 2003). Second, we might move beyond the effects of external factors on internal, subjective criteria, and instead hold that there are external conditions for autonomy concerning, for example, how our beliefs and attitudes relate to reality.

We aim to show in this chapter that empirical research can provide some insights into the nature of autonomy, in particular, the internalist and externalist character of two broadly consensus conditions for autonomy. We'll be assuming an account that requires: (1) critical reflection on one's pro-attitudes and (2) that one's decisions are not subject to undue external influence. We explore some ways in which empirical work interacts with this philosophical view, so as to bring additional nuance to the way in which the internalist–externalist distinction plays out with respect to each of these two conditions. More specifically, we explore an overlooked aspect of the relation between critical reflection and autonomy, which adds to the existing externalist concern about the historical formation of a person's higher-order attitudes another concern about how her beliefs and values relate to reality (Section "Critical Reflection and Evidence-Responsiveness"). We then explore a novel, internalist dimension of the way in which a person's decision making is influenced by a range of external factors and actors, and consider the complex relationship between what we have termed "pre-authorization" and autonomy (Section "External Influence and Pre-authorization"). We do so with particular reference to research that we have conducted on this topic in recent years (Felsen & Reiner, 2011, 2015; Nagel & Reiner, 2013; Niker et al., 2016, 2018a, 2018b), and with the aim of integrating this with other relevant work in philosophy and neuroscience. We then apply our analysis to practical situations in which infringement of autonomy is a concern—specifically, with respect to public policy nudges and the design of persuasive technologies—in order to draw out some of the implications of our theoretical discussion (Section "Implications for the Ethics of Influence").

Critical Reflection and Evidence-Responsiveness

Making autonomous decisions requires certain competencies, such as capacities for internal self-reflection and for forming and revising one's beliefs and values. Moreover, these autonomy competencies must be exercised in ways that ensure that the resulting decision is authentic to the person in question—that it is her own decision in the relevant sense (Christman & Anderson, 2005). Classically, models of authenticity ensure this by claiming that autonomy requires critically reflecting upon and endorsing (or rejecting) one's motives. Critical reflection is generally considered to be the principal *internalist* condition for autonomy, so-called because the process occurs entirely within the bounds of the mind. Through this process, a person shapes the attitudes that guide her decisions and actions. It is therefore both an important competency for autonomous decision making, as well as a key part of the story about how the authenticity that is required for autonomy is achieved.

As noted above, this process of critical reflection is often thought to aim at bringing our first-order desires into coherence with our more reflective, higher-order desires (Frankfurt, 1971), thereby ensuring that a person identifies with or endorses her motives (Dworkin, 1988).[2] Other autonomy theorists maintain that there is more to the capacity for self-reflection than the capacity to hold higher-order attitudes. For instance, when we endorse our motives, we also implicitly make claims about which motives have the support of our practical reasoning (Buss & Westlund, 2018). This understanding of critical reflection has two important implications, both of which take us beyond coherentist accounts. The first is that it captures the intuition that someone who has been unduly influenced with respect to the development of their higher-order attitudes (e.g., indoctrinated or oppressively socialized), or whose practical reasoning has been manipulated in some other sense, would not be properly self-governing. We discuss the idea of undue external influence in more detail in the next section. The second is that, when we take account of practical reasoning, we see that autonomy requires that someone can change her mind when she discovers good reason to do so. We consider this feature of critical reflection in this section. We present a philosophical innovation, and then assess whether this garners empirical support from a neurobiological perspective.

A person's set of *pro-attitudes*—a term we use as shorthand for higher-order desires, preferences, values, beliefs, etc.[3]—underlies her autonomy in important ways. Debate on pro-attitudes in this context has focused either on coherence (i.e., between these attitudes and lower-order desires,

[2] In our earliest studies of the relationship between the cognitive sciences and the concept of autonomy (Felsen & Reiner, 2011), we found that this philosophically defined hierarchical schema broadly aligns with our understanding of the fundamental neurobiology of the brain—in particular with executive control theory in which the prefrontal cortex exerts a top-down influence over other brain regions (Miller & Cohen, 2001).

[3] Elsewhere within the philosophical debate over the nature of autonomy, what we are here labeling as a person's set of pro-attitudes are referred to variously as her "motivational set" (Weimer, 2013), "psychological core" (Noggle, 2005), or "collection of values" (Mele, 1995).

as referenced above) or on history (i.e., how pro-attitudes were initially formed). But, a complete account of autonomy requires deeper consideration of the fact that we exercise and maintain our autonomy competencies *over time*. As experience of the world continues throughout life, our pro-attitudes may need to change in order to accommodate relevant new information—a process we can call ***pro-attitude revision***. A thought experiment developed by Blöser et al. illustrates the matter:

"Pat is a 70-year-old man and a loving father and grandfather. He nevertheless finds it difficult to accept that his children and grandchildren live their lives in ways different from those that he himself pursued at their age. For example, his son has had his children out of wedlock, and Pat is convinced that children can only flourish within a stable family, which he believes to be one in which the children's parents are married. In accordance with the procedural account of autonomy, Pat is able to critically reflect on these issues in light of his existing pro-attitudes. But he holds the same pro-attitudes that he (that is, "younger Pat") authentically acquired half a century ago. What 'old Pat' struggles with is questioning his pro-attitudes in light of new experiences. Although his son's family provides a stable environment in which his grandchildren are flourishing, Pat is unable to reconsider whether marriage really is a basic requirement of good parenthood." (Blöser et al., 2010)

This thought experiment has been constructed to show that something important remains for a complete account of autonomy, even when the standard internalist requirements (i.e., those relating to Pat's capacity to reflect upon and endorse his pro-attitudes) and historical externalist requirements (i.e., that Pat's pro-attitude did not come about via any problematic interference) are met. This remainder relates to Pat's ability, or rather his lack thereof, to reconsider his pro-attitudes in light of new experiences or evidence—or, as we put it above, in light of the reality of the situation.

It would appear, then, that we can draw a distinction between two kinds of critical reflection that are relevant to autonomy: (i) critically reflecting on a pro-attitude in light of our other pro-attitudes and (ii) critically reflecting on a pro-attitude in light of new experiences or evidence (Niker et al., 2018b). The problem with respect to old Pat's autonomy does not have to do with (i), because there is no inconsistency between his various pro-attitudes. Rather, the problem arises from the fact that his value-based childrearing belief is "encrusted" (Blöser et al., 2010).

He does not reflect upon this pro-attitude in light of his new experience of and evidence about childrearing as it applies to his grandchildren; it is his failure to exercise the critical reflection as in (ii) which undermines his autonomy with respect to this pro-attitude. In other words, the intransigence of Pat's previously acquired pro-attitude prevents him from (skillfully) adapting to new situations that merit re-evaluation of his existing pro-attitudes.[4]

If this is correct, then a robust view of autonomy requires that we have the ability to critically reflect upon and to modify our existing pro-attitudes when our experiences or evidence call them into question. Elsewhere, we have described this in terms of the process of "updating ourselves," by appropriately revising our pro-attitudes over time (Niker et al., 2018b). Blöser et al. (2010) label this capability "experience-responsiveness", while Weimer (Weimer, 2013, 2017) refers to the same condition as "evidence-responsiveness".[5] Here, we use the latter term, as this captures both the information acquired via a person's own direct perception as well as information garnered through the testimony of others. If we accept that such *evidence-responsive critical reflection* has a place within a complete account of autonomy, we can see the externalist character of critical reflection *itself*, which goes beyond the weaker externalist character of protecting a person's internal critical reflection process from being unduly shaped by external influences.

To what degree does neurobiological data align with this philosophical innovation? This is a complex issue; here, we outline a set of observations relating to how pro-attitudes might be represented in the brain which suggest the beginnings of a neurobiological framework for evidence-responsive critical reflection. We begin from the claim that, given that pro-attitudes represent a distributed set of desires, beliefs, values, and so on, they are less likely to be instantiated as discrete memories than as

[4] Similar views can be found, more implicitly, in earlier accounts of autonomy. One example is Richard Arneson's view, demonstrated by his claim that, "To live an autonomous life an agent must decide on a plan of life through critical reflection and in the process of carrying it out, remain disposed to subject the plan to critical review if [...] unanticipated evidence indicates the need for such review" (Arneson, 1994).

[5] There is a strand of autonomy theory which defines autonomous decision-making in terms of reasons-responsiveness. Without endorsing this theory, here we simply point out that evidence-responsiveness might plausibly be understood as a specific way of responding to reasons, namely responding to reasons-to-review or reasons-to-revise a pro-attitude that one currently holds (Niker et al., 2018b).

widely dispersed networks of information, consistent with modern theories of information storage in the brain (Dehaene et al., 1998; Squire, 2004). These distributed networks then represent the neurobiological correlates of our pro-attitudes.

Arguably the best candidate for a plausible mechanistic explanation of the process of critical reflection is the phenomenon of Bayesian inference (Knill & Pouget, 2004). In this schema, decisions are represented probabilistically and result from combining two sources of information: internally generated "priors"—our pro-attitudes—and the new information that is associated with a particular decision. The two are provisionally integrated in the brain, generating a new statistical inference. The relative value of this new inference, as well as a measure of confidence in this evaluation, is then determined (Meyniel & Dehaene, 2017). Such evaluation is the essence of critical reflection—appraising the likelihood that the new inference provides a more or less useful strategy for moving forward. When this process makes space for incorporation of new information, it qualifies as evidence-responsive. In addition to influencing specific decisions, new information—if it provides sufficiently compelling evidence—can also be used to update the priors themselves, which will then be applied to subsequent decisions. To return to our example of Pat: If he were capable of revising his pro-attitude that family stability requires marriage, based on the strong evidence provided by his flourishing grandchildren, he would be able to autonomously accept his son's (and even others') decision to have children out of wedlock.

The diffusion-to-bound model, a formal model of perceptual decision making (Ratcliff & Rouder, 2016), helps to illuminate how this might work (Bitzer et al., 2014). The model proposes that one's options are represented as bounds, and a "decision variable" evolves in a multidimensional bounded space as we integrate information relevant to the decision with our priors. When the decision variable reaches one of the bounds, a decision is made which corresponds to selecting that option. This model can explain a range of behavioral phenomena and is consistent with extant neurophysiological data (Gold & Shadlen, 2007; Smith & Ratcliff, 2004); it also provides a useful framework for evaluating external influences on decision making (Bode et al., 2014) and how they affect autonomy of choice (Felsen & Reiner, 2015). While this model is consistent with executive control theory (Miller & Cohen, 2001), and represents an explicit, top-down process of evaluation, it is also possible to incorporate new information *below* the level of conscious awareness

(Dehaene & Changeux, 2011). This does not preclude the possibility of top-down reflection, but it is consistent with the idea that non-conscious processing of inputs such as emotions can provide a useful heuristic for efficient decision making (Gigerenzer & Gaissmaier, 2011).

Critically, the distance between the starting point of the decision variable and the decision bound determines the degree of evidence required to select the option represented by the bound: The further the bound, the more evidence is required. Thus, by setting the bound corresponding to the choice consistent with priors closer to the starting point of the decision variable, that option is more likely to be selected, without precluding the selection of alternatives given sufficient countervailing evidence. Bound setting is under top-down cortical control (Mulder et al., 2012), providing a mechanism for the influence of priors on decisions and for updating the priors themselves. To return again to Pat: Given his pro-attitude that marriage is required for family stability, the variable representing his decision about his son's choice to raise children out of wedlock effectively begins at the "reject" bound. To have any chance of the variable reaching the "accept" bound, the reject bound must be shifted away from the decision variable's starting point in response to the new evidence that Pat's grandchildren are flourishing despite their parents being unmarried.

We hope to have provided a philosophical account of evidence-responsiveness and a sketch of how this process might occur in the brain. While much work remains to link our philosophical and neurobiological explanations (Niker et al., 2018b), we hope that our preliminary work can provide a framework for future studies examining pro-attitudes in terms of priors, how the neural representations of priors are updated by new evidence and the extent to which decisions based on these updated priors are autonomous.

A second stream of neurobiological observations, specifically developed to account for long-term memory formation but likely also relevant to the incorporation of the distributed set of desires, beliefs, values, and so on that represent our pro-attitudes, provides a plausible mechanism for this process. The key finding is that memories are not static but subject to a cycle of deconsolidation and reconsolidation (Nader, 2015). To best understand how this works, think for a moment of a teacher that you had when you were in elementary school. The first salient observation is that you have been able to maintain a memory of that teacher for all these years—for some readers that would be several decades. This is the way we normally think of memory—as a stable feature of normal brain physiology. But while memories are indeed stable for years, the very act of recalling them transforms them from stable to labile. At this very moment, your memory of your elementary school teacher is not protected in the same way that it has been during the years that it lay dormant, but rather is available to develop a new set of associations. These associations, which likely arise via the process of Bayesian inference discussed above, are then stabilized by a process known as reconsolidation, most likely during a subsequent night's sleep (Klinzing et al., 2019; Tononi & Cirelli, 2014). Critically, when the existing memories and the new information are reconsolidated, they are linked; in our example the array of memories about your years in elementary school would be linked to this particular discourse on memory consolidation. Weeks from now, perhaps at a dinner party, you may share this thought experiment with a group of friends. When you do so, you will be drawing upon the association of these two memories to recall how the experiment works. As you delight your friends with your new insight, these memories will once again be labile in our brain, slated for reconsolidation when you return home for a good night's sleep.

Together, this set of observations provides a neurobiological framework for evidence-responsive critical reflection. Bayesian inference draws together extant and new information, providing a mechanism for critical reflection, and then the iterative process of deconsolidation and reconsolidation provides a mechanism for incorporating external information into our existing pro-attitudes—the essence of evidence-responsiveness. This process repeats itself throughout our lives, and we suggest that the ability to engage in evidence-responsive critical reflection represents an important part of this key condition for autonomy.

There is evidence to suggest that older brains are less agile in this regard. While substantial plasticity occurs in the aging brain (Gutchess, 2014), a wealth of data supports the view that fluid cognitive abilities such as working memory, attention, and executive control decline with age, while crystallized cognitive abilities are preserved (Samanez-Larkin & Knutson, 2015). Because fluid cognitive abilities are precisely those that are required to nimbly manage new information, those who are best endowed with these traits will naturally be in the strongest position to utilize them in a process of evidence-responsive critical reflection. It is for this reason that Blöser et al.'s choice of an elderly person in the example of "old Pat" is so plausible: It is certainly not the case that *all* elderly people have strongly fixed pro-attitudes, but it is common to encounter older people who cling to their previously acquired pro-attitudes, and it is this that impairs his ability to engage in evidence-responsive critical reflection.

External Influence and Pre-authorization

The second condition for autonomy that we're assuming in our inquiry is that for a person's decision to be autonomous it must not be the result of undue external influence. It must be "hers" in the appropriate sense. It is relatively simple to agree that certain forms of influence are undue, insofar as they present an obvious threat to a person's autonomy. This is especially the case when it comes to heavy-handed forms of external influence such as brainwashing or coercion (Chen-Wishart, 2006). But in the course of our day-to-day lives, we continuously encounter a range of external influences that run the spectrum from overt to subtle and on to imperceptible to those whom they affect. Determining which of these various influences are to be considered "undue" is a complicated matter.

As mentioned in the introduction, research in the cognitive sciences has shed light upon the extent to which our decisions are influenced by seemingly irrelevant situational factors, and has sought to explain how and why this often happens below the level of our conscious awareness. The robustness of this empirical research has laid the foundations for important shifts in philosophy of mind, including moves toward understanding cognition as embedded in and extended into our external environments. On such situated conceptions, decisions result from an interaction between mind and environment; decisions are, as a matter of fact, always influenced by external factors to some extent. We might worry about this from the perspective of autonomy, perhaps because it makes it more difficult to discern which influences are permissible (insofar as they respect autonomy's authenticity conditions) and which are not; but we might also think that this situated conception of cognition provides some insight into the concept of autonomy itself.

Such an insight, we think, would be related to a second philosophical innovation in the debate over autonomy in recent years. This has centered not on empirical work on cognition, but rather on theoretical work on conceptions of the self. Both kinds of work, though, are connected by the fundamental role that they give to social embeddedness. Constructive critiques from feminist philosophers have led to a reconceptualization of autonomy in light of appropriate appreciation being given to the fact that we are relational beings—beings who are not only continually subject to external influences, but who require them in order to develop and exercise our autonomy competencies (Meyers, 1989). Often collectively termed *relational* (Mackenzie & Stoljar, 2000), the twofold motivation of such accounts is to show, on the one hand, that "rational autonomous capacities are made possible by the support of numerous surrounding agents who enable careful reflection and judgment" and on the other, that "individuals' autonomous capacities can be disabled or oppressed by the withholding of this contextual support" (Specker Sullivan & Niker, 2018). This reconceptualization offers rich opportunities to delve deeper into the question of when and why an influence is considered undue.

Much philosophical attention has been given to determining which types of influences are morally problematic—*how* a decision is influenced, and the ethical character of these various types has been debated in detail. There are, for instance, distinct and in some cases burgeoning philosophical literatures on the nature and (political) morality of coercion (Anderson, 2010; Wertheimer, 2014), manipulation (Coons & Weber, 2013), persuasion (McKenna, 2020), upbringing and socialization (Clayton, 2006), and nudging (Niker, 2018; Sunstein, 2016). Interestingly, though, there has been much less discussion of a different feature that may be relevant to the "dueness" of external influence, namely, *who* is exerting the influence and how the person who is subject to it understands their relationship to this influencing actor. We intuitively allow some people, institutions, and so on to have a greater influence upon our decision making than others. To put it another way, information from certain actors is viewed as a welcome input into our decision making, but this is not so when the very same information comes from other actors. In recent work, we have sought to offer a conceptual tool for better understanding this selective process regarding the source of external influences and to examine how this relates to (relational) autonomy.

It is plausible to think that whether information is regarded as welcomed or not by a given person depends not only upon its relevance to the decision at hand, but also upon that person's perception of the reliability of the source of that information. We have termed the latter sort of consideration *pre-authorization* (Niker et al., 2016). We operationally define pre-authorization as an evaluative stance by which an individual gives certain agents preferential access to influencing her decision-making processes (Niker et al., 2018a). Several reasons can be put forward for pre-authorizing an agent. One prominent example occurs when we perceive that the agent has values, commitments, and goals that are similar to ours—that is, that in some meaningful way they share our worldview. Another common situation is one in which the agent has some relevant knowledge or expertise that we do not have and which we can trust, for example, when we consult with a physician or a lawyer. The result, in both cases, is that we feel comfortable incorporating information from these agents into our decision making. More specifically, the evaluative stance taken by an individual toward some agents means that an influence from a pre-authorized agent is incorporated in relevant future interactions without necessarily needing to be consciously evaluated, and without impacting the individual's perception of the control that she has over, and the authenticity of, her resultant decision (Niker et al., 2018a). We have suggested that the extent to which the source of an influence is pre-authorized contributes to our perception that we are making an autonomous decision. A person's actual autonomy and her perceived autonomy can be distinct—for example, while in practice an intervention does not impact on a person's decision-making capacity, she might perceive that it does, or vice versa. Yet, if pre-authorization can be shown to play a role in what we might call the "folk" conception of autonomy, this would justify consideration of its relation to the autonomy competencies, as understood on a relational account of autonomy.

To further explore whether the concept of pre-authorization has some basis in the way that people view influences upon their decision making, we carried out a set of empirical studies. We particularly examined how people perceive of everyday socio-relational influences on their decisions, such as a news clip on a social media platform, a friend's comment or suggestion, a notification from an app, and so on. The data, derived from carefully balanced contrastive vignettes, demonstrated that the influence of pre-authorized agents with whom we share a worldview—be they individuals or institutions—was judged to be significantly less undue than when that same influence derived from non-pre-authorized agents. One might imagine that this was secondary to our familiarity with the agent, because in the normal course of events we are usually better acquainted with those with whom we share a worldview than those with whom we do not. Yet these effects persisted even after controlling for the familiarity of the agent. Thus, we found that the public's conception of when an influence is welcome or not is indeed dependent upon the source of the influence, providing initial support for the validity of the concept of pre-authorization (Niker et al., 2018a).

Another way of saying this is that we evaluate not just the content of information that arrives at our doorstep but also its pedigree. Does it come from a trusted source? Is it from someone who shares our world-view? Is it from someone who has expert knowledge on the topic? These questions define our attitude toward the source, and that in and of itself affects the degree to which we allow it to have an influence over our decision-making processes. From an empirical perspective, we have hypothesized that our brains have something akin to a *skeptical filter*, and that our evaluation of the pedigree of the information determines the stringency of the skeptical filter we apply to it. When it comes from a pre-authorized source, the skeptical filter is loosened, making it easier for that information to "get through" and influence the decision at hand. When it comes from a source that is not pre-authorized, our evaluation of the information is more rigorous, calling for further cognitive work. We suggest that an important autonomy competency is the ongoing maintenance of this skeptical filter, using it as a means of authorizing external influences that are consistent with one's goals, values, desires, convictions, and life plan.[6] There is a modicum of evidence in support of the existence of this filter. For example, people use more stringent criteria to evaluate others' arguments than when they produce arguments themselves (Bode et al., 2014; Felsen & Reiner, 2015). Moreover, the concept is consistent with neurobiological descriptions of decision making that account for the incorporation of external influences (Bode et al., 2014; Shadlen & Roskies, 2012). Nonetheless, the precise neural circuitry that undergirds this phenomenon is currently unknown.

How does this relate to autonomy? The answer is not entirely straightforward. As noted above, our studies grounding the concept of pre-authorization test a person's *perceptions* about whether an external influence is welcome or not. But whether an influence is considered welcome by a person for the purposes of her decision making is not the same thing as it being a morally permitted influence; it acts merely as a proxy. While often overlapping, a person's autonomy and her perceived autonomy aren't the same thing. Insofar as they overlap, we might say that the phenomenon of pre-authorization is one particular way

[6] This maintenance may include engaging in evidence-responsive critical reflection in order to update the stringencies of the filters when appropriate, so that they don't become "encrusted" in the way discussed in Section "Critical Reflection and Evidence-Responsiveness".

in which we can capture the role and value of interpersonal relations in supporting autonomy—or more specifically, in supporting a person's ability to make autonomous decisions under real-world conditions, where information is both abundant and costly and where time is often limited. Pre-authorization provides us with a possible mechanism by which we exercise autonomy relationally.

We might think, then, that pre-authorization fits into the framework of *autonomy support* (Nagel, 2015; Nagel & Reiner, 2013), which acknowledges the social and relational ties that bind and support individuals in their making of decisions throughout their life (and is discussed in more detail in the Section "Implications for the Ethics of Influence"). On this view, autonomy is an intersubjective phenomenon that is not only developed socially but is also constantly reflected, maintained, and advanced in relational contexts. What is interesting about the concept of pre-authorization is that it posits an empirically plausible (though unverified) means by which an individual can exert control over the differential impact of external sources of influence on her decision-making processes, as determined by how these sources relate to her own beliefs, values, life plan, etc. This is interesting from the perspective of the framework set up by the chapter because, if accepted, pre-authorization highlights a novel *internalist* feature of this externalist condition (i.e., of not being subject to undue external influence). Together with the conclusion of the previous section, this further problematizes any clear distinction between internalist and externalist conditions for personal autonomy.

But, as insinuated above, there is much more to say about the relationship between pre-authorization and autonomy. Our notion of the skeptical filter provides some insight into one of the pitfalls of pre-authorization. As Onora O'Neill has pointed out, trusting others to provide us with information is only valuable if the individual or institution is in fact trustworthy (O'Neill, 2018). Thus, if we pre-authorize an agent and they lead us astray by convincing us of incorrect information that they sincerely believe, or worse, by using our confidence in them to manipulate us, we are in a very bad situation indeed, as the loosening of the skeptical filter causes us to less rigorously assess the veracity of their claims. In this way, we see that the heuristic nature of pre-authorization—a quick and efficient but nonetheless imperfect solution to evaluating external information—can lead to situations in which our autonomy may be subverted.

The ideal version of pre-authorization is one in which a person has reflected upon the issue and intentionally decides to pre-authorize another agent (these days it is probably wise to include algorithmic agents in the mix). But in practice, this is not what normally happens. The canonical example is a friendship that develops over time. Initially, both parties might be open to each others' ideas but still a bit skeptical. Over time, as they get to know and trust each other, they begin to pre-authorize each other to influence their thinking on certain matters. But they are unlikely to stop and say something like, "Wow, my friend Judy seems like a really good person to take advice from. I think I will do so from here on in." Rather, the pre-authorized relationship develops in an implicit manner (Niker & Specker Sullivan, 2018). Indeed, one may not even explicitly realize it has happened, unless prompted to reflect on the issue. What we don't know is how, from a mechanistic point of view, this process plays out. What we do know is that over time, we come to rely upon some individuals more than others, and past experience is one factor that plays into the process. All of this is to say that our vision of the concept of pre-authorization holds less in common with a legally binding grant of power than the sort of power exchange that occurs informally among parties with everyday social interaction.

Another interesting dimension of the relationship between pre-authorization and autonomy comes from the inverse of pre-authorization. Although we have not specifically tested the hypothesis, it seems plausible that actors may not only pre-authorize but also *anti-pre-authorize* other agents. This has become a common trope in modern life in which the partisan nature of political positions and the structure of our informational landscape allows us to ignore information that derives, e.g., from news sources that do not align with our worldview, irrespective of the comparative factual quality of the different outlets (Bessi et al., 2016; Del Vicario et al., 2016). Thus, while pre-authorization may be a useful heuristic insofar as it allows us to more easily integrate information from trusted kith and kin, its inverse—anti-pre-authorization—may be a factor that negatively affects our capacity to make informed decisions and to engage in the evidence-responsive critical reflection discussed in the previous section.

Implications for the Ethics of Influence

We have considered some of the issues involved in two consensus conditions of autonomous decision making—critical reflection and not being subject to undue external influence—from the perspective of both philosophy and neurobiology. We turn our attention now to exploring the practical relevance and potential implications of our theoretical discussion for real-world scenarios about which there is concern over autonomy. We focus in particular on the phenomenon of *nudging*, both as it functions as a public policy lever and the role it plays in the design of persuasive technologies (Thaler & Sunstein, 2009). It makes an illustrative case because: (i) the ethical debate over nudging has centered on autonomy and (ii) we think that both the issues of critical reflection and evidence-responsiveness (Section "Critical Reflection and Evidence-Responsiveness") and pre-authorizing selected sources of external influence (Section "External Influence and Pre-authorization") have interesting implications for the debate over the ethics of nudging. Indeed, our analysis shows that these two aspects of our theoretical discussion are heavily interrelated in the practical case of nudging.

Nudging involves intentionally modifying a person's choice environment in order to predictably, yet non-coercively influence her decision making toward a specified end. Introduced as a public policy lever aimed at promoting individual and social welfare (Thaler & Sunstein, 2009), this form of influence was provocatively termed *libertarian paternalism*.[7] When motivated in this way, a nudge is paternalistic because the "choice architect" intervenes with the best interests of the nudged person in mind. But this welfare-promoting aim is reined in by liberal values, it is thought, because the nudged person is not forced to decide in accordance with the nudge; for it to count as a nudge, she needs to be free to opt out with relative ease. Such interventions find their rationale and operational mechanisms in the empirical research grounding situated conceptions of rational agency. This research has shown that environmental settings have a deep impact on the decisions people make, such that "seemingly trivial changes in the way information is conveyed, choices are arranged, or

[7] Despite the initial equation of nudges with a form of paternalism, it is now well-established that nudging is a type of influence that can be used in service of different ends. While we may be motivated to nudge for paternalistic reasons, we might also use nudges for the purpose of promoting justice, utility, commercial profit, or so on.

default rules are set" can affect the decisions they make (Moles, 2015). For instance, whether an in-work pension scheme or organ donation registration scheme has an opt-in or an opt-out default makes a considerable difference to the uptake of both. Knowledge of the various ways in which cognitive heuristics and biases affect our decisions makes it possible to design choice architecture in a way that steers, or "nudges," people in a particular direction. In recent years, several governments have changed the default of these two schemes with the explicit aim of producing higher rates of savings and cadaveric organ donation; but this move towards nudging in policymaking has not been without its critics (Goodwin, 2012; Waldron, 2014; Yeung, 2012).

Much of this critical engagement has examined nudging's relationship to autonomy (Engelen & Nys, 2020; Felsen et al., 2013; Grune-Yanoff, 2012; Wilkinson, 2013). One under-theorized critique of nudging from this direction is that it may infringe upon the *development* of autonomy competencies. Blöser et al. (2010) emphasize that one must recognize an experience as being new and relevant in some manner as a pre-condition for evidence-responsive critical reflection. Nudges may diminish the opportunity to engage in such reflection. Consider an adolescent who, rather than finding his own way in the world by "learning from their mistakes," has parents who remove obstacles from his path—a situation commonly known as "snowplow parenting." In essence, this adolescent lives in an environment that is designed by choice architects (his parents, in this case) to make the best decisions most likely. He may end up with decisions that are welfare-promoting, or even ideal in some sense, but there is less opportunity for him to develop the fundamental skills involved in decision making. The worry is that a similar sort of diminishment of human decision-making competencies is going on in a world structured by public policy nudges. This is especially so if we agree with critics that nudges work by bypassing our deliberative capacities (e.g., Grüne-Yanoff, 2012); operating in this way would threaten the development and exercise of several autonomy competencies, not only evidence-responsive critical reflection.

But, as Neil Levy has recently argued, there is at least a certain kind of nudge—which he calls *nudges to reason*—which might have an important role to play in helping us to become *more responsive* to genuine evidence (Levy, 2017). In recent years, much attention has been directed toward issues relating to evidence-responsiveness in a so-called post-truth world. This has been bolstered by findings such as the "backfire effect," which describes the phenomenon that occurs when those who are motivated to resist and reject (some kinds of) evidence become more entrenched in their false beliefs after being presented with arguments citing such evidence (Nyhan & Reifler, 2013). This related set of issues clearly pose a threat to the flourishing of democratic systems (e.g., the possibility of having a well-informed electorate), to public health (e.g., the case of anti-vaxxers), and to climate justice (e.g., the case of climate change deniers). Levy suggests that nudges to reason may offer an effective and ethically permissible means of addressing such false beliefs by increasing responsiveness (or, at least, reducing *perverse* responsiveness) to evidence. He accepts the critic's claim that interventions into decision making and belief formation threaten a person's autonomy when they bypass her capacities for deliberation; but nudges to reason, he argues, address themselves to capacities that are partially constitutive of a person's reasoning (Levy, 2017). In so doing, these interventions do not offend against autonomous decision making, and, in fact, they may support autonomy by enabling people to engage in evidence-responsive critical reflection. How might they do this?

One of the ways in which psychologists have found we can become more responsive to evidence relates to recent insights into how we respond to testimony. As Levy explains,

"Children and adults must learn from others: there is a great deal that we cannot check for ourselves, and a great deal more that it would be too time-consuming or otherwise costly to check. In the contemporary world, we rely on medical specialists to diagnose our ills, technology specialists to fix our computers, accountants to manage funds for our retirement and meteorologists to advise us when to hold a picnic. But this reliance on specialists [...] is a feature of traditional societies too. Canoe making, for instance, is a specialised skill, and not everyone has the time to acquire it. Moreover, skill acquisition is itself dependent on the acceptance of testimony: children often cannot discover essential techniques for survival themselves, and must be taught them. [...] For all these reasons, we are often forced to learn from others in the absence of a capacity directly to gauge how reliable they are. We are therefore forced to use cues to reliability; cues which reliably enough correlate with being a good source of testimony." (Levy, 2017)

This relates directly to the concept of pre-authorization discussed above. In essence, a person uses cues of reliability and benevolence to help her to determine which information to take account of in their belief formation and decision-making processes. In the case of correcting false beliefs, it has been shown that a person's sensitivity to these cues plays a role in explaining why some corrections are successful, while others are not. For instance, Nyhan and Reifler (2013) found that the source of the information made a significant difference to whether corrections of myths about President Obama's policies were successful for conservatives or not. In fact, there were two source-based considerations that produced this effect: the perceived ideological leanings of the media outlet that reported the debunking claim, and the source of the claim (i.e., whether it was attributed to a liberal, non-partisan, or conservative think tank) (Levy, 2017; Nyhan & Reifler, 2013).

This evidence opens up the possibility of counteracting public ignorance and misconceptions by designing interventions that present evidence in certain ways. The most relevant case for our purposes concerns intentionally selecting the source(s) of the evidence so as to increase the likelihood that (a certain set of) people will respond to it as they rationally ought to. But there are also other techniques such as "moral reframing," which works by framing a position that an individual would normally not support in a way that is consistent with her values and so positively affects the credence she gives to it (Feinberg & Willer, 2019), in line with the rational significance of genuine evidence. Should these nudge interventions—and, in particular, the testimonial version that is of particular interest to us—be regarded with the same sort of suspicion as other nudges? And if not, why not?

According to Levy, these testimonial nudges count as nudges to reason because rather than modify a person's behavior directly, they do so by seeking to alter her beliefs through the process of making her more responsive to evidence. Nevertheless, critics may accept this while remaining worried about how these nudges affect this change of mind, where the concern is just a variant of the standard worry that such interventions operate by bypassing our deliberative faculties. The real reason explaining why we changed our mind, it might be thought, has to do with the selection of a source that has been intentionally chosen to avoid the backfire effect; and so, "by bypassing our deliberative capacities, [such nudges] may threaten the substantive freedom of our choices even if they succeed in making us more responsive to the evidence" (Levy, 2017). There are different responses available; but the more interesting, from our perspective, is to deny that nudges to reason do in fact bypass an individual's deliberative faculties. Instead, such interventions are "designed to be processed by filters that are partially *constitutive* of reasoning in normal functioning agents" (ibid.). In Levy's terms,

"[a] process is a proper part of reasoning [...] when it regularly and reliably supports better deliberation (either in a domain-general or a domain-specific manner)... Appeals to the mechanisms that weigh testimony by reference to their source are very plausibly appeals to mechanisms that are partially constitutive of rationality, because we likely have such mechanisms in virtue of the role they played in enabling better decision-making... [T]hese mechanisms are sensitive to the previous track record of the source. That is, very obviously, sensitivity to a property that is truth-conducive. We should put less weight on the testimony of those who are frequently wrong than those who have better records. Similarly, sensitivity to the ideological orientation of the source is also truth-conducive. We should be wary of the claims of people who lack benevolence towards us, because they may be motivated to exploit us. We also should put more stock in testimony from agents who have an incentive to reject the claim they affirm... Sensitivity to these properties is sensitivity to considerations that are relevant to the credence we should place on testimony. Appealing to them is appealing to capacities that have as their proper function the assessment of reasons for belief—a function that is obviously partially constitutive of reasoning—in their role as reasoning mechanisms." (Levy, 2017)

If this argument is correct, nudges to reason may permissibly be used to counteract false beliefs held by the public. By presenting evidence via a source that is more likely to be pre-authorized, and hence more likely to make it through the skeptical filter, these nudges support a person's capacity for evidence-responsiveness and for evidence-responsive critical reflection (see also Adams & Niker, 2021).[8] Given our analysis, then, it is plausible that nudges to reason support the exercise of autonomy competencies, especially when autonomy is conceptualized in relational terms. Of course, not all nudges are nudges to reason; indeed, most would not be categorized as such, so our conclusion applies only to a subset of nudges.

[8] Neurobiologically, this could be represented as shifting the starting point of the drift-diffusion process closer to one of the bounds (Felsen & Reiner, 2015). Often, as with encrusted values, bounds are set by internal biases. By changing the relative distances to bounds, nudges can be seen to counteract such internal biases in ways that are (more) consistent with the agent's pro-attitudes.

In a sense, Levy's nudges to reason can be viewed as an example of ***autonomy support***—a strategy introduced in the previous section that aims to help individuals arrive at decisions that are aligned with their values, needs, preferences, and desires. Originally developed as a means of supporting individuals in developing autonomy competencies, particularly in the domains of education and the workplace (Reeve, 1998; Ryan & Deci, 2000), the concept of autonomy support can be thought of as a set of strategies that assist people in developing and executing autonomy competencies throughout their life course (Nagel, 2015). Unlike classic nudges that are designed to make it more likely that an individual arrives at a decision that the choice architect has deemed to be in their best interests, the external influences that comprise autonomy support give extra weight to respect for the person, devoting effort to consider how one might enable individuals to arrive at decisions that are in their own best interests.

But there is a further sense in which pre-authorization seems to be a useful concept for understanding another phenomenon associated with public policy nudging. Namely, pre-authorization may be one of the factors that explain why certain nudges are perceived as more or less welcome. There is empirical data showing that certain contextual factors make a difference to whether any given nudge is perceived by the public as infringing upon or respecting their autonomous decision making (Castelo et al., 2012; Felsen et al., 2013; Jung & Mellers, 2016). In an era in which trust in institutions is weakening, this has substantial implications for public policy initiatives which employ nudges to alter citizens' behavior. Indeed, these data may go some way toward explaining the phenomenon of ***partisan nudge bias***, whereby attitudes toward particular policy goals or policymakers—i.e., whether they align with the actor's goals and commitments—affect attitudes about the moral permissibility of the nudge policy itself (Tannenbaum et al., 2017).

We move now to another example within the ethics of influence that draws together the concepts of nudging, pre-authorization, and autonomy support, namely, the ethical dimensions of persuasive technologies. In the modern world, influence over our decision making is increasingly exerted not by other humans but rather via software on our algorithmic devices, colloquially known as "apps." It is well established that by monitoring our digital footprints, software can predict a great deal about us, from Big Five personality traits to our political views and more (Kosinski et al., 2013; Matz et al., 2017). This information can then be used to micro-target individuals in an effort to persuade—or nudge—them to follow one or another course of action (Calo, 2014; Frischmann & Selinger, 2018; Susser et al., 2019). Karen Yeung calls this "hypernudging," because these Big Data analytic nudges are much more potent than their standard public policy counterparts on account of "their networked, continuously updated, dynamic and pervasive nature" (Yeung, 2016).

The potency and personalization of persuasive technologies make them novel; but so does the fact that, through repeated use, we accept our algorithmic devices—exemplified most obviously by the smartphone—as extensions of our minds (Clark, 2008). As we do so, we increasingly rely upon them as a trusted source of information, social interaction and approval, and a means of offloading cognitive work (Fitz & Reiner, 2016; Reiner & Nagel, 2017). If, as seems to be the case, we treat apps as pre-authorized agents (Niker et al., 2018a), we allow them to have an outsized influence upon our decision making. Although there have already been several substantial efforts to explore these issues (Susser et al., 2019; Williams, 2018; Yeung, 2016), there is much work still to be done in this area of applied ethics.

But rather than simply critiquing persuasive technologies, it is perhaps apropos to highlight how our relationship with persuasive technologies might be constituted such that it is supportive of our autonomy competencies. Consider the app *Moment* which helps people manage their smartphone usage. It resides on the device and, after you grant it sufficient privileges, it monitors most of what you do on your phone during the day. It doesn't prevent you from using your phone (unless you ask it to), but from time to time it gives you feedback on how much you have used your phone, and even includes a reminder of what your goal for phone usage is. In this way, the *Moment* app causes you to critically reflect upon your phone usage by presenting you with evidence of your current usage. This, we suggest, is an existing example of an algorithmic nudge to reason (Levy, 2017). By regularly prompting you to reflect on your choices of phone usage, the app helps you to make an autonomous decision to keep your phone usage at a level that you wish it to be (Specker Sullivan & Reiner, 2019). This represents a plausible example in which nudges can be harnessed to support autonomy, at once helping humans make better decisions and become better decision makers.[9]

Conclusion

Autonomy, with its implications for moral, political, and philosophical thought, is a well-studied concept in Western intellectual thought. Nonetheless, there remain opportunities to advance our knowledge in this realm, and this chapter represents our attempt to explore recent progress in our understanding of two consensus conditions of autonomy—critical reflection and not being subject to undue influence. Our consideration of these matters has attempted to integrate conceptual work with empirical research in the cognitive sciences. In both cases, our analysis has put pressure on the idea that we can draw any clear distinction between internalist and externalist conditions for personal autonomy.

Critical reflection upon one's pro-attitudes is a fundamental internalist condition of autonomy. We have suggested that the critical reflection required for autonomy includes critically reflecting in direct response to new experiences and genuine evidence, in order to assess how our beliefs and values relate to reality. This evidence-responsive critical reflection requires that we consider and revise our pro-attitudes wherever these are found to be called into question by relevant external factors, such as reliable evidence garnered by first personal experience or from trustworthy third-party experts. By exploring what is known about relevant neurobiology, we have been able to suggest a neurobiological framework for evidence-responsive critical reflection. We have also deepened our understanding of the concept of undue influence, in particular in the realm of the sorts of everyday influences that we experience in virtue of being socially embedded. As part of this exploration, we have developed the concept of pre-authorization, which suggests that the pedigree of information that might influence us has some bearing upon how we view such information—admitting it with relatively little skepticism or examining it more carefully. Not being subject to undue external influence on our

[9] We do recognize, though, that most of the worries about nudges to reason are diminished in the case of *Momentum* (vis-à-vis public policy nudges to reason) by the fact that a person has intentionally granted permission to the app to influence her decision-making in this way.

decision-making processes tends to be viewed as an externalist condition for autonomy, but pre-authorization, with its role in determining who counts as the external actors whose influence is welcomed in our decision making, represents a novel internalist aspect that is relevant to understanding when this condition has and has not met.

We brought both sets of insights together to analyze the ethics of influence, with a particular focus on nudging carried out by governments and by our increasingly technologically enriched environment. Taken together, these investigations add to the existing body of knowledge about autonomy and its discontents, recognizing our desire for control over our own decisions as well as helping us to better understand how we might preserve autonomy as socially embedded beings.

References

Adams, M., & Niker, F. (2021). Harnessing the epistemic value of crises for just ends. In *Political Philosophy in a Pandemic: Routes to a More Just Future*, F. Niker & A. Bhattacharya (Eds.), pp. 219-232.

Anderson, S. (2010). The enforcement approach to coercion. *Journal of Ethics and Social Philosophy, 5*, 1–31.

Arneson, R. (1994). Autonomy and preference formation. In *In harm's way: Essays in honor of Joel Feinberg*, J. Feinberg, J. L. Coleman, & A. E. Buchanan (Eds). Cambridge University Press, pp. 42–75.

Bagnoli, C. (2011). *Morality and the emotions*. Oxford University Press.

Bessi, A., Zollo, F., Del Vicario, M., Puliga, M., Scala, A., Caldarelli, G., Uzzi, B., & Quattrociocchi, W. (2016). Users polarization on Facebook and Youtube, T. Preis (Ed.) *PLoS ONE, 11*: e0159641.

Blöser, C., Schöpf, A., & Willaschek, M. (2010). Autonomy, experience, and reflection. On a neglected aspect of personal autonomy. *Ethical Theory & Moral Practice, 13*, 239–253.

Bitzer, S., Hame, P., Felix, B., & Stefan, K. (2014). Perceptual decision making: Drift-diffusion model is equivalent to a Bayesian model. *Frontiers in Human Neuroscience, 8*, 1-17.

Bode, S., Murawski, C., Soon, C. S., Bode, P., Stahl, J., & Smith, P. L. (2014). Demystifying "free will": The role of contextual information and evidence accumulation for predictive brain activity. *Neuroscience and Biobehavioral Reviews, 47*, 636–645.

Buss, S., & Westlund, A. C. (2018). Personal autonomy. *The Stanford Encyclopedia of Philosophy.*

Calo, R. (2014). Digital market manipulation. *Stanford Technology Law Review, 82*, 995–1051.

Castelo, N., Reiner, P. B., & Felsen, G. (2012). Balancing autonomy and decisional enhancement: An evidence-based approach. *American Journal of Bioethics, 12*, 30–31.

Chen-Wishart, M. (2006). Undue influence: Vindicating relationships of influence. *Current Legal Problems, 59*, 231–266.

Christman, J. (2005). Autonomy, self-knowledge, and liberal legitimacy. In *Autonomy and the challenges to liberalism: New essays*, J. Christman & J. Anderson (Eds.). Cambridge University Press, pp. 330–358.

Christman, J. (2010). *The politics of persons: Individual autonomy and socio-historical selves.* Cambridge University Press.

Christman, J. (2015). Autonomy in moral and political philosophy. *The Stanford Encyclopedia of Philosophy.*

Christman, J., & Anderson, J. (Eds.) (2005). *Autonomy and the challenges to liberalism: New essays.* Cambridge University Press.

Clark, A. (2008). *Supersizing the mind: Embodiment, action, and cognitive extension.* Oxford University Press.

Clayton, M. (2006). *Justice and legitimacy in upbringing.* Oxford University Press.

Coons, C., & Weber, M. (2013). *Paternalism: Theory and practice.* Cambridge University Press.

Coons, C., & Weber, M. (2014). *Manipulation: Theory and practice.* Oxford University Press.

Dehaene, S., & Changeux, J.-P. (2011). Experimental and theoretical approaches to conscious processing. *Neuron, 70*, 200–227.

Dehaene, S., Kerszberg, M., & Changeux, J. P. (1998). A neuronal model of a global workspace in effortful cognitive tasks. *Proceedings of the National Academy of Sciences, 95*, 14529–14534.

Del Vicario, M., Bessi, A., Zollo, F., Petroni, F., Scala, A., Caldarelli, G., Stanley, H. E., & Quattrociocchi, W. (2016). The spreading of misinformation online. *Proceedings of the National Academy of Sciences of the United States of America, 113*, 554–559.

Doris, J. M. (2002). *Lack of character.* Cambridge University Press.

Doris, J. M. (2015). *Talking to our selves.* Oxford University Press.

Dworkin, G. (1988). *The theory and practice of autonomy.* Cambridge University Press.

Engelen, B., & Nys, T. (2020). Nudging and autonomy: Analyzing and alleviating the worries. *Review of Philosophy Psychology, 11*, 137–156.

Feinberg, M., & Willer, R. (2019). Moral reframing: A technique for effective and persuasive communication across political divides. *Social and Personality Psychology Compass, 13*, e12501.

Felsen, G., Castelo, N., & Reiner, P. B. (2013). Decisional enhancement and autonomy: Public attitudes towards overt and covert nudges. *Judgment and Decision Making, 8*, 202–213.

Felsen, G., & Reiner, P. B. (2011). How the neuroscience of decision making informs our conception of autonomy. *AJOB Neuroscience, 2*, 3–14.

Felsen, G., & Reiner, P. B. (2015). What can neuroscience contribute to the debate over nudging? *Review of Philosophy Psychology, 6*, 469–479.

Fitz, N. S., & Reiner, P. B. (2016). Perspective: Time to expand the mind. *Nature, 531*, S9–S9.

Frankfurt, H. G. (1971). Freedom of the will and the concept of a person. *The Journal of Philosophy, 68*, 5–20.

Friedman, M. (2003). Autonomy and social relationships: Rethinking the feminist critique. In *Autonomy, gender, politics*, M. Friedman (Ed.). Oxford University Press pp. 81–97.

Frischmann, B. M., & Selinger, E. (2018). *Re-engineering humanity*. Cambridge University Press.

Gigerenzer, G., & Gaissmaier, W. (2011). Heuristic decision making. *Annual Review of Psychology, 62*, 451–482.

Gold, J. I., & Shadlen, M. N. (2007). The neural basis of decision making. *Annual Review of Neuroscience, 30*, 535–574.

Goodwin, T. (2012). Why We Should Reject "Nudge." *Politics, 32*, 85–92.

Grune-Yanoff, T. (2012). Old wine in new casks: Libertarian paternalism still violates liberal principles. *Social Choice and Welfare, 38*, 635–645.

Gutchess, A. (2014). Plasticity of the aging brain: New directions in cognitive neuroscience. *Science, 346*, 579–582.

Hurley, S. (2011). The public ecology of responsibility. In *Responsibility and distributive justice*, C. Knight, & Z. Stemplowska (Eds.). Oxford University Press, pp. 187–217.

Jung, J. Y., & Mellers, B. A. (2016). American attitudes toward nudges. *Judgment and Decision Making, 11*, 62–74.

Kahneman, D. (2011). *Thinking, fast and slow*. Penguin.

Klinzing, J. G., Niethard, N., & Born, J. (2019). Mechanisms of systems memory consolidation during sleep. *Nature Neuroscience, 35*, 1–13.

Knill, D. C., & Pouget, A. (2004). The Bayesian brain: The role of uncertainty in neural coding and computation. *Trends in Neurosciences, 27*, 712–719.

Kosinski, M., Stillwell, D., & Graepel, T. (2013). Private traits and attributes are predictable from digital records of human behavior. *Proceedings of the National Academy of Sciences of the United States of America, 110*, 5802–5805.

Lavazza, A. (2016). Free will and neuroscience: From explaining freedom away to new ways of operationalizing and measuring it. *Frontiers in Human Neuroscience, 10*, 1-17.

Levy, N. (2017). Nudges in a post-truth world. *Journal of Medical Ethics, 43*, 495–500.

Mackenzie, C., & Stoljar, N. (2000). *Relational autonomy: Feminist perspectives on automony, agency, and the social self*. Oxford University Press.

Matz, S. C., Kosinski, M., Nave, G., & Stillwell, D. J. (2017). Psychological targeting as an effective approach to digital mass persuasion. *Proceedings of the National Academy of Sciences, 114*, 12714–12719.

May, J., & Kumar, V. (2018). Moral reasoning and emotion. In *The Routledge Handbook of Moral Epistemology*, pp. 139–156.

McKenna, R. (2020). Persuasion and epistemic paternalism. In Guy Axtell & Amiel Bernal (Eds.), *Epistemic paternalism: Conceptions, justifications, and implications*. Rowman & Littlefield, pp. 91-106.

Mele, A. R. (1995). *Autonomous agents*. Oxford University Press.

Mele, A. R. (2012). Another scientific threat to free will? *The Monist, 95*, 422–440.

Meyers, D. T. (1989). *Self, society, and personal choice*. Columbia University Press.

Meyniel, F., & Dehaene, S. (2017). Brain networks for confidence weighting and hierarchical inference during probabilistic learning. *Proceedings of the National Academy of Sciences of the United States of America, 71*, 201615773.

Miller, E. K., & Cohen, J. (2001). An integrative theory of prefrontal cortex function. *Annual Review of Neuroscience, 24*, 167–202.

Moles, A. (2015). Nudging for liberals. *Social Theory and Practice, 41*, 644–667.

Mulder, M. J., Wagenmakers, E. J., Ratcliff, R., Beokel, W., & Forstmann, B. U. (2012). Bias in the brain: A diffusion model analysis of prior probability and potential payoff. *Journal of Neuroscience, 32*, 2335–2343.

Nader, K. (2015). Reconsolidation and the dynamic nature of memory. *Cold Spring Harbor Perspectives in Biology, 7*, 1–16.

Nagel, S. K. (2015). When aid is a good thing: Trusting relationships as autonomy support in health care settings. *The American Journal of Bioethics, 15*, 49–51.

Nagel, S. K., & Reiner, P. B. (2013). Autonomy support to foster individuals' flourishing. *American Journal of Bioethics, 13*, 36–37.

Niker, F. (2018). Policy-led virtue cultivation: Can we nudge citizens towards developing virtues? In *The theory and practice of virtue education*, T. Harrison & D. Walker (Eds). Routledge, pp. 153–167.

Niker, F., Reiner, P. B., & Felsen, G. (2016). Pre-authorization: A novel decision-making heuristic that may promote autonomy. *American Journal of Bioethics, 16*, 27–29.

Niker, F., & Specker Sullivan, L. (2018). Trusting relationships and the ethics of interpersonal action. *International Journal of Philosophical Studies*, 26, 173-186.

Niker, F., Reiner, P. B., & Felsen, G. (2018a). Perceptions of undue influence shed light on the folk conception of autonomy. *Frontiers in Psychology, 9*, 1–11.

Niker, F., Reiner, P. B., & Felsen, G. (2018b). Updating our selves: Synthesizing philosophical and neurobiological perspectives on incorporating new information into our worldview. *Neuroethics, 11*, 273–282.

Noggle, R. (2005). Autonomy and the paradox of self-creation: Infinite regresses, finite selves, and the limits of authenticity. In *New essays on personal autonomy and its role in contemporary moral philosophy*, J. S. Taylor (Ed.). Cambridge University Press.

Nyhan, B., & Reifler, J. (2013). Which corrections work? Research results and practice recommendations. *New America Foundation.*

O'Neill, O. (2018). Linking trust to trustworthiness. *International Journal of Philosophical Studies, 26*, 1–8.

Ratcliff, R., & Rouder, J. N. (2016). Modeling response times for two-choice decisions. *Psychological Science : A Journal of the American Psychological Society, 9*, 347–356.

Reeve, J. (1998). Autonomy support as an interpersonal motivating style: Is it teachable? *Contemporary Educational Psychology, 23*, 312–330.

Reiner, P. B., & Nagel, S. K. (2017). Technologies of the extended mind: Defining the issues. In *Neuroethics: Anticipating the future*, J. Illes (Ed.). Oxford University Press, pp. 111–126.

Ryan, R. M., & Deci, E. L. (2000). Intrinsic and extrinsic motivations: Classic definitions and new directions. *Contemporary Educational Psychology, 25*, 54–67.

Samanez-Larkin, G. R., & Knutson, B. (2015). Decision making in the ageing brain: Changes in affective and motivational circuits. *Nature Reviews Neuroscience, 16*, 278–289.

Shadlen, M. N., & Roskies, A. L. (2012). The neurobiology of decision-making and responsibility: Reconciling mechanism and mindedness. *Frontiers in Neuroscience, 6*, 1-12.

Simon, H. (1972). Theories of bounded rationality. In *Decision and Organization*, C. B. McGuire, & R. Radner (Eds.). North-Holland, pp. 161-176.

Smith, P. L., & Ratcliff, R. (2004). Psychology and neurobiology of simple decisions. *Trends in Neurosciences, 27*, 161–168.

Specker Sullivan, L., & Niker, F. (2018). Relational autonomy, paternalism, and maternalism. *Ethical Theory & Moral Practice, 21*, 649–667.

Specker Sullivan, L., & Reiner, P. B. (2019). Digital wellness and persuasive technologies. *Philosophy & Technology, 34*, 413–424.

Squire, L. R. (2004). Memory systems of the brain: A brief history and current perspective. *Neurobiology of Learning and Memory, 82*, 171–177.

Stanovich, K. E., & West, R. F. (2000). Individual differences in reasoning: Implications for the rationality debate? *The Behavioral and Brain Sciences, 23*, 645–665.

Sunstein, C. R. (2016). *The ethics of influence: Government in the age of behavioral science*. Cambridge University Press.

Susser, D., Roessler, B., & Nissenbaum, H. F. (2019). Online manipulation: Hidden influences in a digital world. *Georgetown Law Technology Review, 4*, 1-45.

Tannenbaum, D., Fox, C. R., & Rogers, T. (2017). On the misplaced politics of behavioural policy interventions. *Nature Human Behaviour, 1*, 0130.

Thaler, R. H., & Sunstein, C. R. (2009). *Nudge*. Penguin.

Tononi, G., & Cirelli, C. (2014). Sleep and the price of plasticity: From synaptic and cellular homeostasis to memory consolidation and integration. *Neuron, 81*, 12–34.

Tversky, A., & Kahneman, D. (1981). The framing of decisions and the psychology of choice. *Science, 211*, 453–458.

Waldron, J. (2014). It's all for your own good. *The New York Review of Books*.

Weimer, S. (2013). Evidence-responsiveness and autonomy. *Ethical Theory & Moral Practice, 16*, 621–642.

Weimer, S. (2017). Evidence-responsiveness and the ongoing autonomy of treatment preferences. *HEC Forum, 13*, 1–23.

Wertheimer, A. (2014). *Coercion*. Princeton University Press.

Wilkinson, T. M. (2013). Nudging and manipulation. *Political Studies, 61*, 341–355.

Williams, J. (2018). *Stand out of our light: Freedom and resistance in the attention economy*. Cambridge University Press.

Yeung, K. (2012). Nudge as fudge. *The Modern Law Review, 75*, 122–148.

Yeung, K. (2016). "Hypernudge": Big data as a mode of regulation by design. *The Information Society, 20*, 118–136.

Niker, F., & Specker Sullivan, L. (2018). Trusting relationships and the ethics of interpersonal action. *International Journal of Philosophical Studies, 26*, 173-186.

Adams, M., & Niker, F. (2021). Harnessing the epistemic value of crises for just ends. In *Political Philosophy in a Pandemic: Routes to a More Just Future*, F. Niker & A. Bhattacharya (Eds.), pp. 219-232.

CHAPTER 7

The Ethics of Memory Dampening

Adam J. Kolber

Suppose we could erase memories we no longer wish to keep. In such a world, the victim of a terrifying assault could wipe away memories of the incident and be free of the nightmares that such memories often cause. Some memories, however, even quite unpleasant ones, are extremely valuable to society and ought not be eliminated without due consideration. An assault victim who hastily erases memory of a crime may thereby impede the investigation and prosecution of the perpetrator. In a world

---

This chapter is adapted, with permission, from Adam J. Kolber, Therapeutic Forgetting: The Legal and Ethical Implications of Memory Dampening, 59 Vanderbilt Law Review 1561 (2006). The original article is considerably longer with more detailed citations and acknowledgments.

A. J. Kolber (✉)
Law at Brooklyn Law School, New York City, NY, USA
e-mail: adam.kolber@brooklaw.edu

M. J. Blitz et al. (eds.), *The Law and Ethics of Freedom of Thought, Volume 1*, Palgrave Studies in Law, Neuroscience, and Human Behavior,
https://doi.org/10.1007/978-3-030-84494-3_7

with memory erasure, our individual interest in controlling our memories may conflict with society's interest in maintaining access to those memories.

While true memory erasure is still the domain of science fiction,[1] (Eternal Sunshine of the Spotless Mind, 2004; Men in Black, 1997), less dramatic means of dampening the strength of a memory may have already been developed. Some experiments suggest that propranolol, an FDA-approved drug, can dull the emotional pain associated with the memory of an event when taken within six hours *after* the event occurs (Pitman et al., 2002; Vaiva et al., 2003). The effects have been hard to replicate, however, and researchers have turned to a variety of other approaches to alter the factual and emotional components of memory.[2] I will address such efforts generally in ways that aren't tied to propranolol or any currently existing technology so that we can look at the underlying ethical issues that might someday be presented.

The President's Council on Bioethics (the "Council")[3] engaged in a similar exploration in a series of hearings in 2002 and 2003 and in a report that came out of those hearings, *Beyond Therapy: Biotechnology and the Pursuit of Happiness* (President's Council on Bioethics, Beyond Therapy: Biotechnology and the Pursuit of Happiness, 2003 [Beyond Therapy]). By and large, the Council was skeptical of the merits of memory dampening, raising concerns that memory dampening may: (1) prevent us from truly coming to terms with trauma, (2) tamper with our identities, leading us to a false sense of happiness, (3) demean the genuineness of human life and experience, (4) encourage us to forget memories that we are obligated to keep, and (5) inure us to the pain of others. While the Council

[1] In *Freedom of Memory Today*, I describe legal and ethical issues raised by what purports to be a real-life case of memory erasure (Kolber, 2008).

[2] In recent years, other studies have both provided additional findings about propranolol's capacity to dampen memories about other drugs' effects on memory formation. (*see, e.g.*, Kindt & Soeter, 2018, reporting successful use of propranolol and sleep to dampen fearful memories in humans, Vallejo, et al., 2019, using propofol and sleep to impair reconsolidation of human episodic memories, and Kaser, et al., 2017, finding that subjects with remitted depression given modafinil scored higher on tests of episodic memory).

[3] In 2001, George W. Bush created the Council by executive order. Exec. Order No. 13237, 66 Fed. Reg. 59851 (Nov. 28, 2001). In recent decades, all US presidents have had some sort of bioethics commission of their own with the exception of Donald Trump (Appel, 2019).

did not make policy recommendations concerning memory-dampening drugs, one might ask whether the kinds of concerns raised by the Council could justify prohibiting or broadly restricting their use. I argue that many of these concerns are rooted in controversial premises about whether it is prudent to modify our natural abilities to remember and, as such, they do not offer widely-shared reasons to broadly restrict memory dampening. Other concerns expressed by the Council can be addressed with only modest regulation. In this chapter, I analyze the novel ethical issues that could be presented by memory-dampening technology and argue that the Council's concerns do not provide grounds for broad legal restrictions on its use.

Prudential Concerns

One series of concerns set forth by the Council suggests that memory dampening will in some way damage the psychological well-being of patients or otherwise degrade or dehumanize the quality of their lives. The Council claims, for example, that the old-fashioned process of dealing with negative memories has adaptive effects on the individual and that pharmaceutical solutions may sever our connection with real world experiences and weaken or otherwise damage our sense of identity. I call these the Council's "prudential concerns," because, though they are presented as ethical concerns, they focus on ways in which memory dampening may prevent a particular individual from leading a meaningful, flourishing life. They are not quintessentially ethical concerns because the Council does not argue that we have ethical obligations *to other people* to lead our lives in the ways that the Council finds meaningful and fulfilling.

I will argue that this set of concerns serves principally to offer guidance to individuals and medical professionals about when to dampen memories. Taken as advisory comments, the Council's prudential concerns may prove helpful to those who accept the widely disputed premises on which they are based. More importantly, however, because they are founded on widely disputed premises, they fail to carry sufficient force or to be of sufficient generality to justify broad-brushed restrictions on memory dampening.

A. The Tough Love Concern

The Council claims that memory dampening, by offering us a solution in a bottle, allows us to avoid the difficult but important process of coming to terms with emotional pain. There are two ways to understand the concern. The first is that there is something false or undeserved about the manner in which memory dampening eases distress. Gilbert Meilaender makes this point in his essay on memory dampening where he claims that, rather than erasing traumatic experiences, "it might still be better to struggle—with the help of others—to fit them into a coherent story that is the narrative of our life" (Meilaender, 2003, 21–22). "Our task," according to Meilaender, "is not so much to erase embarrassing, troubling, or painful moments, but, as best we can and with whatever help we are given, to attempt to redeem those moments by drawing them into a life whose whole transforms and transfigures them" (Id., 22).

People have divergent views, however, about what it means to transform and transfigure our experiences into "a coherent story" (Id., 21). It seems quite plausible that one could craft a coherent life narrative punctuated by periods of dampened memories. Moreover, it is open to debate how important it is that one's life story be coherent or otherwise neatly packaged. Some research suggests that those with narcissistic, self-enhancing personalities tend to be particularly resilient after traumatic experiences (Bonanno, 2004, 25–26; Bonanno, 2005, 984–6, 994). Yet, while such personality traits may make it easier to cope with traumatic events, they do not necessarily serve us well in other aspects of our lives[4] (Bonanno, 2005, 985). Thus, it is at least a complicated matter whether we should seek to develop those aspects of our personalities that help us rebound after trauma.

Furthermore, even if one shares Meilaender's preference to redeem and transform our experiences without memory dampeners, two additional

[4] Bonanno writes: "[B]ehaviors or dispositions that help people to cope with unusual and extremely aversive events might also carry with them a serious cost" (Bonnano, 2005, 985). Those with a self-enhancing bias, although they appear to be particularly resilient to trauma, "score highly on measures of narcissism... and with repeated contacts, tend to evoke negative impressions in unfamiliar peers" (Id. (citations omitted)).

responses are suggested. First, many experiences are simply tragic and terrifying, offering virtually no opportunity for redemption or transformation. For example, after a 1978 plane crash in San Diego, desk clerks and baggage handlers were assigned to retrieve dead bodies and clean up the crash site[5] (Butcher & Hatcher, 1988, 728). Emotionally unprepared for this task, many of them were so distraught that they were unable to return to work. In such cases, it seems unlikely that the traumatized employees should, in Meilaender's words, "redeem those moments by drawing them into a life whose whole transforms and transfigures them" (Meilaender, 2003, 22). Most would agree that such employees should not have participated in the cleanup in the first place, and, hence, they should not be required or expected to bear the emotional burden of having done so.[6]

Second, even if it is better to weave traumatic events into positive, life-affirming narratives, many people are never able to do so. Memory-dampening drugs may enable such people to make life transformations that they would be *incapable* of making in the absence of the drugs. For others, pharmaceuticals may drastically shorten the time it takes to recover from a traumatic experience. Suppose a person spends ten years coming to terms with a traumatic event that could have been surmounted in two years with pharmaceutical assistance. While he might be viewed as heroic by Meilaender, others might view him as extremely obstinate. Therefore, even in those instances when positive human transformation should accompany traumatic experience, there may well be a role for memory dampening to facilitate the process.

The more modest version of the "tough love" concern merely states that "[p]eople who take pills to block from memory the painful or hateful aspects of a new experience will not learn how to deal with suffering or sorrow"[7] (Beyond Therapy, 2003, 291). This concern, however, merely fights the hypothetical existence of effective memory-dampening drugs.

[5] This example was also raised by James McGaugh at the Council's hearing.

[6] The Council acknowledges that if "bitter memories are so painful and intrusive as to ruin the possibility for normal experience of much of life and the world," the "impulse" to dampen those memories is "fully understandable." The Council quickly retreats, however, adding: "And yet, there may be a great cost to acting compassionately for those who suffer bad memories, if we do so by compromising the truthfulness of how they remember" (Beyond Therapy, 2003, 230).

[7] The Council asks: "What qualities of character may become less necessary and, with diminished use, atrophy or become extinct, as we increasingly depend on drugs to cope with misfortune?") (Beyond Therapy, 2003, 208).

If a memory-dampening drug increases the overall psychological distress of patients by being addictive or by otherwise leading them to make poor choices, it will be unappealing to doctors and patients, not as a matter of ethics, but as a matter of science. Such drugs would not be deemed effective psychiatric tools. To even launch the interesting policy questions related to memory dampening, we must assume the existence of a drug that is not highly addictive and that satisfies basic requirements of medical efficacy and safety.

Assuming that we identify such a drug, legitimate but manageable concerns may arise about overuse. If the drug is used principally for victims of motor vehicle accidents and violent crimes, the drug is not likely to be used often by the same people. Furthermore, many of those with good coping skills have never had a motor vehicle accident nor been the victim of a violent crime; thus, working through these experiences cannot be critical to the development of these skills. If, however, a person frequently dampens memories for comparatively insignificant events, then the Council's fear seems more plausible. Yet, virtually every medication runs a risk of overuse, and barring evidence that a medication is addictive, we usually manage that risk with our ordinary restrictions on prescription medications.

B. The Personal Identity Concern

Memory and identity are closely linked.[8] We feel a special connection to our past selves largely because we remember having our past experiences. For example, when I get out of bed in the morning, I consider myself the same person who went to sleep there the night before, in part, because I remember doing so. Those with extreme memory disorders, like advanced Alzheimer's disease, may lack such memories and may lose a stable sense of self[9] (Cf. Jaworska, 1999, 105). While memory is not the sole constituent of personal identity, it creates much of the psychological continuity that makes us aware of our continuing existence over time (Parfit, 1984, 208).

[8] On the relationship between memory and identity, see Parfit (1984, 208), Perry (1975) collecting essays. Persons 199–345 (1984); Personal Identity (John Perry ed., 1975) (collecting essays).

[9] Jaworska argues that we should respect the autonomy interests of those Alzheimer's patients who retain a capacity to value even after they have lost a coherent life narrative.

John Locke deemed memory and identity to be so closely connected that he claimed that we should not punish a person for a crime he no longer remembers committing[10] (Locke, 1975, 48). According to Locke, the person who cannot recall the crime is a different person than the perpetrator because the two lack an essential connection through memory, and the former should not be punished for the crime of the latter.

While courts have not accepted Locke's overstated conclusion, some courts have held that a genuine inability to recall participation in a crime (even if one had full mental faculties at the time of the crime) can help support a finding of incompetence to stand trial[11] (Wilson v. United States, 1968, 463–64; State v. McIntosh, 1988, *23–4). Rather than absolving a defendant of responsibility, however, courts considering a defendant's competence may simply deem it procedurally unfair to require a defendant to stand trial if his memory loss makes him unable to "assist properly in his defense."[12]

Nevertheless, a glimmer of the Lockean view may be found in various places in the law of insanity where we are disinclined to hold people responsible for actions taken by their psychologically discontinuous alter egos. For example, in a case of dissociative identity disorder (formerly known as multiple personality disorder), the court held that the defendant—more specifically, the dominant personality of the defendant—could not be held responsible for the crimes of an alternate personality when the dominant personality was unaware of those crimes at the time they were committed, even if the alternate personality was legally sane[13]

[10] Locke wrote:"[I]n the great day, wherein the secrets of all hearts shall be laid open, it may be reasonable to think, no one shall be made to answer for what he knows nothing of....") (Locke, 1975, 48). As Parfit writes, "Locke claimed that someone cannot have committed some crime unless he now remembers doing so" (Parfit, 1984, 208).

[11] In Wilson v. United States, the D.C. Circuit Court of appeals remanded to the district court for further fact-finding as to whether defendant's permanent retrograde amnesia for the events surrounding his alleged participation in a robbery interfered with his due process right to present an adequate defense. In State v. McIntosh, the Wisconsin Court of Appeals relied on *Wilson* to find that defendant did not receive a fair trial where there was a "real possibility that the amnesia may be 'locking in' exculpatory information".

[12] Such claims are usually unsuccessful, however, as the consensus view is that "loss of memory due to amnesia is not alone an adequate ground upon which to base a finding" of incompetence (LaFave, 2003, §8.01(a)).

[13] In United States v. Denny-Shaffer, the 10th Circuit Court of Appeals ordered retrial with an insanity instruction where the defendant presented sufficient evidence that her

(United States v. Denny-Shaffer, 1993, 1016). In addition, the Supreme Court has held it unconstitutional to execute an insane death row inmate, even if the inmate was sane at the time of the murder (Ford v. Wainwright, 1986, 399, 410). Our unwillingness to execute the insane may recognize, in some measure, the psychological discontinuity between an insane inmate and his sane counterpart who committed the crime[14] (Beyond Therapy, 2003, 211–212).

Recognizing the important connection between memory and identity, the Council suggests that memory dampening may weaken our sense of identity by dissociating memories of our lives from those lives as they were actually lived. Selectively altering our memories, according to the Council, can distort our identity, "subtly reshap[ing] who we are, at least to ourselves" (Id., 212). "[W]ith altered memories," the Council writes, "we might feel better about ourselves, but it is not clear that the better-feeling 'we' remains the same as before" (Id., 212).

Yet, even in the absence of memory dampeners, we cannot help but selectively remember. Memories have a natural rate of decay and are far more a synthesis and reconstruction of our past than a verbatim transcript[15] (Gazzaniga, 2005, 120–142). Just to process the tremendous amount of information that is presented to our senses, we must constantly abstract away from the "real" world. As the Council acknowledges, "individuals 'naturally' edit their memory of traumatic or significant events—both giving new meaning to the past in light of new experiences and in some cases distorting the past to make it more bearable" (Beyond Therapy, 2003, 217, n*). In fact, such selective reconstruction of our lives seems to be at the very heart of the creation of a coherent life story that Gilbert Meilaender advocates. Nevertheless, we do not worry whether our better-feeling naturally reconstructed selves remain the same as before.

It is, thus, not at all clear why we ought to revere the selective rewriting of our lives that we do without pharmaceuticals, yet be so skeptical of pharmaceutically-assisted rewriting. In fact, memory dampening

dominant personality was not in control during the offense and was not aware that another personality was controlling her physical actions.

[14] Such a view is far from explicit, however, in the Court's decision in *Ford v. Wainwright*, which notes that there is no "[u]nanimity of rationale" behind the rule. *Id.* at 408 (Ford v. Wainwright, 1986, 408).

[15] Gazzaniga describes myriad ways in which memory can fail to accurately represent past experience.

may strengthen our sense of identity. By preventing traumatic memories from consuming us, memory dampeners may allow us to pursue our own life projects, rather than those dictated by bad luck or past mistakes. As David Wasserman has noted, "pharmacologically-assisted authorship may strengthen rather than reduce narrative identity," by allowing one to "edit his autobiography, instead of having it altered only by the vagaries of neurobiology" (Wasserman, 2004, 14). Thus, to the extent that people voluntarily make changes to their mental processes, such changes may be perceived as bolstering self-identity. In fact, many people who begin taking antidepressants report feeling like themselves for the first time.[16] This suggests that some deliberate shifts in identity may not seem alienating at all.

C. Genuine Experiences Concern

The Council also worries that a memory-dampened life, chemically-altered as it is, is somehow a less genuine life[17] (Beyond Therapy, 2003, 213). According to the Council, "we might often be tempted to sacrifice the accuracy of our memories for the sake of easing our pain or expanding our control over our own psychic lives. But doing so means, ultimately, severing ourselves from reality and leaving our own identity behind" (Id., 233–34). This, according to the Council, "risks making us false, small, or capable of great illusions" (Id., 234). It also risks making us "capable of great decadence or great evil" (Id.)

Unfortunately, the Council never explains what makes a life genuine and truthful (nor how leading a life that is otherwise makes us capable of great evil). Is a memory-dampened life thought less genuine simply because some of the memories associated with it decay at a faster rate than they otherwise would have? Given that memories never precisely replicate our past experiences, do undampened memories provide a standard of

[16] Peter Kramer quotes a patient who, after starting the SSRI antidepressant Prozac, said she felt "as if I had been in a drugged state all those years and now I am clearheaded." Eight months after beginning Prozac, the same patient stopped the treatment and said she felt like "I am not myself" (Kramer, 1993, 18).

[17] The Council writes: "[B]y disconnecting our mood and memory from what we do and experience, the new drugs could jeopardize the fitness and truthfulness of how we live and what we feel..." (Beyond Therapy, 2003, 213).

genuineness? How important is it to lead a "genuine" life, whatever that means?[18]

In the case of those who are emotionally traumatized, traumatic memories can be overwhelming and trigger exaggerated responses to harmless stimuli. Such overreactions are themselves divorced from reality. Memory dampeners, by preventing people from being overtaken by trauma, may actually make them more genuine, more true to what they take their lives to be, than they would be if they were gripped by upsetting memories.

Furthermore, we are not always troubled by discrepancies between our perceptions and the world as it "genuinely" is. It has been widely observed that in many areas of life, people systematically overestimate their abilities and prospects relative to others (Brown, 1986, 353: Elga, 2005, 117).

Suppose there were a pill that eliminated these systematic self-enhancing biases. On the one hand, one could argue, those who took such pills would lead less genuine lives, as they would no longer understand the world in the way that they would in the absence of the pill. Their lives would be less genuine in the sense that they would lack a characteristically human understanding of the world. On the other hand, those who took the pill might lead more genuine lives, freed from the ruby-colored lenses that nature has given us.

No doubt, as a general life strategy, we do well to firmly commit ourselves to reality and to discovering the truth about ourselves and the world around us. Yet such a strategy might, at times, be worse for us all things considered; or, at least, the Council has not shown otherwise. To make the case that memory-dampening drugs will harmfully affect our lives, the Council must be much clearer about what makes a life genuine, how these drugs make lives less genuine, and why that should matter so much to us that we ought to suffer in distress to preserve our unadulterated memories.

[18] Robert Nozick's famous "experience machine" thought experiment is often taken to show that we want our lives to be closely connected to reality (Nozick, 1974, 42–5). For criticism, see Kolber (1994/95).

General Response to the Prudential Concerns

I have argued that many of the Council's concerns about memory dampening are founded on controversial premises. Not all of us will agree with the Council about how we ought to cope with emotional pain, what changes to our memory will damage our sense of self, and what makes one set of experiences more genuine and, therefore, better than another. While the concerns expressed by the Council and some of its members may prove insightful to likeminded patients or medical professionals, they are insufficiently developed to provide a basis for broad restrictions on memory dampening.

Each of the concerns presented reflects a bias for our natural, pharmaceutical-free mechanisms of responding to trauma. The Council implicitly or explicitly defended: (1) our natural ability to surmount difficult life obstacles, (2) our natural memories as the desirable basis for our sense of identity, and (3) our natural memories as more genuine and more desirable than those that are pharmaceutically altered.

There are two reasons commonly given for this preference for the status quo. The first is that we doubt that human intervention can improve upon our natural endowments when it comes to responding to difficult memories. We generally do an astonishingly good job of remembering what we need to remember and forgetting what we can do without. This delicate balance, some claim, has been optimized by evolution, such that "[w]hat looks to be an improvement could have hidden downsides" (Douglas et al., 2005, 28–9). The Council reflected a similar sentiment, stating that "[t]he human body and mind, highly complex and delicately balanced as a result of eons of gradual and exacting evolution, are almost certainly at risk from any ill-considered attempt at 'improvement'" (Beyond Therapy, 287). If millions of years of evolution have tended to select for brains that optimally balance retained and deleted memories, then we may find it very difficult indeed to improve upon our natural endowment.

However, while evolution has made the human brain remarkably adept at balancing our needs to retain and to forget memories, it surely did not lead each of us to an optimal balance. The conditions and needs of modern society differ substantially from those during most of our evolution. Furthermore, some people have better memories than others, and some are more susceptible to PTSD than others. It is very unlikely that we each have a brain optimized for our individual needs, especially because our needs can change during the course of a lifetime. And as a general matter, pharmaceutical tinkering with memory is not always counterproductive, as witnessed by the millions of people being treated for Alzheimer's disease.

The Council is surely correct that it is difficult to improve upon our natural endowments, and for this reason, we are justifiably skeptical that any particular drug will constitute an improvement. It is certainly possible, however, to improve on our endowments and to suggest otherwise, rather than resolving the interesting policy issues raised by memory dampening, merely avoids or postpones them.

A second reason to defend our natural balance of retention and forgetting is that, with such a balance, we lead distinctively human lives and perhaps doing so is itself valuable. The Council expresses such a sentiment, acknowledging that its concerns with memory dampening and certain other new technologies "may have something to do with challenges to what is naturally human, what is humanly dignified, or to attitudes that show proper respect for what is naturally and dignifiedly human" (Beyond Therapy, 2003, 286–87).

A running theme in the Council's report is that memory dampening dehumanizes us by giving us too much control over our life experiences. According to the Council, "We are not free to decide everything that happens to us; some experiences, both great joys and terrible misfortunes, simply befall us. These experiences become part of who we are," part of our lives "as truthfully lived" (Id., 233). The Council stated:

Acknowledging the giftedness of life means recognizing that our talents and powers are not wholly our own doing, nor even fully ours, despite the efforts we expend to develop and to exercise them. It also means recognizing that not everything in the world is open to any use we may desire or devise. Such an appreciation of the giftedness of life would constrain the Promethean project and conduce to a much-needed humility (Id., 288).

Yet the Council acknowledges exactly what makes this view so unpersuasive: "The 'giftedness of nature' also includes smallpox and malaria, cancer and Alzheimer [sic] disease, decline and decay" (Id., 289). Surely we are not expected to accept everything in the world that is "given." The Council, however, offers no principled basis for deciding when to intervene, insisting that a "respectful attitude toward the 'given'" is "both necessary and desirable as a restraint," (Id.) even though "[r]espect for the 'giftedness' of things cannot tell us which gifts are to be accepted as is, which are to be improved through use or training, which are to be housebroken through self-command or medication, and which opposed like the plague" (Id.). At some point, one must wonder whether this distinction actually serves to distinguish. Indeed, what is "given" may itself be dynamic, for our "given" nature might be to transcend our boundaries and constantly improve ourselves. At one point, the Council makes exactly that suggestion[19] (Id., 291.n*). It is, therefore, very difficult to understand why human enhancement should be restrained by our "given" nature.

The weaknesses of a status quo preference can be illustrated by imagining a world called Dearth, where the inhabitants are very much like us except that, on average, they are less likely than we are to suffer from traumatic memories. Perhaps Dearthlings are less emotionally aroused by traumatic experiences than humans typically are. One day, the government of Dearth establishes a commission that holds hearings on an emerging technology, called traumatic memory *enhancement*. Using memory-enhancing drugs, Dearthlings can make their traumatic memories more vivid, more persistent, and otherwise more like those of typical humans.[20] Ought Dearthlings enhance their responses to trauma to make them more like the responses of typical humans?

With limited facts, it is difficult to say. Without the drug, Dearthlings suffer less; on the other hand, they might, in some sense, experience a richer, more meaningful life with the drug. Most would agree, however, that a Dearthling should not be forced to take a drug that will create a significant risk that he will develop upsetting memories from a recent traumatic experience. Similarly, a human being with a significant risk of developing upsetting memories from a recent traumatic experience should be permitted to use memory-dampening drugs to prevent those memories from forming. The only difference between a Dearthling at risk from traumatic memory-enhancement and a human at risk from refraining from memory dampening is whether the risk comes from taking a pill or from not taking it. If the Dearthling is permitted to avoid a bad state of affairs

[19] *The Council writes:* "By his very nature, man is the animal constantly looking for ways to better his life through artful means and devices; man is the animal with what Rousseau called 'perfectibility.'" (Id., 291.n*).

[20] In our world, David Wasserman has observed that such affect-enhancing memory drugs could someday be used to punish criminals by forcing them to reflect more intensely on their criminal behavior (Wasserman, 2004, 14–15).

by not taking a pill, the human should be able to avoid that same bad state of affairs by taking one. Otherwise, the preference for the status quo begins to seem like an unprincipled taboo on pill taking.[21]

Some Council members might respond by saying that there is a very important difference between these two individuals—namely, one is a human and one is a Dearthling—and the human ought to deal with traumatic memories in characteristically human rather than Dearthling ways. In response, I must present the chilling news that there are Dearthlings among us, for some humans are quite resilient in the face of traumatic experiences while others are prone to PTSD. In fact, one sibling may be quite sensitive to trauma while another is the human equivalent of a Dearthling. Given the amount of variation among humans, appeals to human nature tell us little about whether we must respond to trauma like a Dearthling or like a statistically-typical human.

At this point, the Council might reiterate that our human nature may require each of us to accept his own personal "given" response to trauma whatever it might be. Yet the Council encourages us to change our "given" response to traumatic memories so long as we do so the old-fashioned way. It is difficult, however, to see why the method of change matters if it leads to the same end point. Perhaps the Council doubts that a pharmaceutical intervention will get us to the same end point as a non-pharmaceutical intervention. That, however, would merely serve as a critique of some particular imperfect form of memory dampening rather than a critique of memory dampening in general.

To recap, we considered two potential reasons to prefer our status quo methods of dealing with trauma over memory dampening. The first was that our status quo methods are simply the best methods possible. I argued that this is highly implausible as an empirical matter. The second was that our status quo methods are best because they are, in some sense, given to us as part of our human nature. I argued that there is little reason to prefer some state of affairs simply because it is the status quo, and it is virtually impossible to determine when human nature dictates that we leave some state of affairs alone and when it dictates that we do whatever we can to change it.

One reason the Council's concerns about memory dampening do not translate well into legal restrictions on memory dampening is that the

[21] Nick Bostrom and Toby Ord have offered a more generalizable version of the Dearthling thought experiment (Bostrom & Ord, 2006).

concerns discussed so far are not quintessentially ethical in nature. For example, the Council advises each of us to lead a genuine life because such a life is valuable to the person living it. To the extent that there is an ethical obligation to lead such a life, it is an obligation one has to one's self. Yet the notion of having an obligation to one's self is controversial. If A has an obligation to B, then, ordinarily, B can choose to release A from that obligation. Now suppose that A has an obligation to himself. Can A release himself from an obligation to himself? If so, it is not clear that A is obligated in any meaningful way[22] (Singer, 1959, 202–203).

While it may be possible to resurrect the notion of having an obligation to one's self, as a matter of legal regulation, we are more reluctant to restrict an individual's liberty to interfere with his own well-being than with another's. Thus, even if we were uniformly convinced of the strength of the three prudential concerns presented here, for the purposes of our inquiry, some additional argument would be needed to justify broad restrictions on memory dampening.[23]

Restrictions based on what I call the Council's prudential concerns are paternalistic in nature. Paternalistic limitations on our freedom may "serve[] the reflective values of the actor," or "impose[] values that the actor rejects" (Greenawalt, 1995, 718). The "soft" paternalism that is consistent with our own values is usually thought less invasive and more respectful of individual autonomy than the "hard" paternalism that imposes values foreign to the actor. To the extent that I have shown that the Council's concerns in the last Section are founded on controversial premises and do not reflect quintessentially ethical obligations, I have thereby suggested that interventions based on those concerns are of the more suspect variety.

The Council's prudential concerns provide little ground for doubting the ability of individual patients and their doctors to collectively decide when to use memory-dampening drugs, much as they would collectively decide to use any other physical or psychiatric medical treatment. The

[22] Singer writes: "[A] duty to oneself, then, would be a duty from which one could release oneself at will, and this is self-contradictory. A 'duty' from which one could release oneself at will is not, in any literal sense, a duty at all." Daniel Kading raises some objections to Singer's position (Kading, 1960).

[23] Such arguments typically suggest that individuals are incapable of making appropriate decisions, perhaps because the behavior at issue is addictive or people lack information needed to decide appropriately. I discuss the latter issue in more detail in the context of informed consent in Kolber (2006, 1586–89).

possibility remains, however, that the concerns described here could be reconfigured in terms of the effects that they would have on others. In that case, perhaps one could formulate non-paternalistic reasons for restrictions. Indeed, in the next two sections, I describe concerns of the Council that I take to be somewhat stronger because thy do identify more widespread societal effects of memory dampening.

A. *Obligations to Remember*

In the Supreme Court's most influential "right to die" case, *Cruzan v. Director, Missouri Department of Health*, Nancy Cruzan's family failed in its effort to obtain a court order to disconnect Nancy from the artificial feeding and hydration equipment that kept her alive in a persistent vegetative state (Cruzan v. Director, Missouri Department of Health, 1990). Writing in dissent, Justice John Paul Stevens emphasized that "[e]ach of us has an interest in the kind of memories that will survive [us] after death"[24] (Id., 356). Stevens dissented, in part, because Nancy Cruzan may have had "an interest in being remembered for how she lived rather than how she died," and he feared that "the damage done to those memories by the prolongation of her death is irreversible"[25] (Id., 353).

Stevens suggests that people have strong interests in being remembered in certain ways for who they are and what they do. If Stevens is correct, then we may have obligations to satisfy these interests by appropriately remembering people and events. Because memory dampeners may facilitate violations of these obligations, we arguably have grounds to heavily restrict their use.

I will suggest otherwise. First, I will describe the concerns of Council members that memory dampening may violate obligations to remember. Then, I will argue that even if we sometimes have ethical obligations to

[24] Stevens states in his dissent that the most famous declarations of Nathan Hale and Patrick Henry "bespeak a passion for life that forever preserves their own lives in the memories of their countrymen" (Cruzan v. Director, Missouri Department of Health, 1990, 344).

[25] Stevens also noted that her surviving family members have "an interest in having their memories of her filled predominantly with thoughts about her past vitality rather than her current condition" (Cruzan v. Director, Missouri Department of Health, 1990, 356).

others to remember, these obligations cannot, by themselves justify broad restrictions on memory dampening.

Council member Gilbert Meilaender suggests, albeit meekly, that we may have ethical obligations to remember those "treated unjustly... to remember the evil done them," which "might be necessary not just for the sake of the victims themselves but for our common humanity" (Meilaender, 2003, 22). While Meilaender merely "suspect[s] we can imagine circumstances in which we might think that there is indeed an obligation not to forget," (Id.) I think that prima facie obligations to remember are commonly recognized, stemming from interests in respect, honor, and justice (see generally Margalit, 2002).

In a world without memory dampening, it may seem that one cannot possibly be responsible for failing to remember, as we have limited control over our memories,[26] and voluntary control is often thought to be a prerequisite to responsibility.[27] On further examination, however, we clearly hold people responsible for failing to remember. For example, we blame those who forget an important birthday or anniversary, and we penalize those who forget to file a timely tax return. Some of the most tragic instances of failed memory occur when parents unintentionally cause the death of their young children by leaving them stranded in the backseats of automobiles on hot days, sometimes leading to criminal punishment.

The nature of our obligations to remember are radically underexplored, however, partly because, prior to the realistic possibility of memory dampening, there was relatively little one could do to consciously alter one's memories, and there was correspondingly little one could do to consciously fulfill or escape obligations to remember. One explanation for the observation that we do, in fact, hold people responsible for forgetting is that, in the examples given above—failing to commemorate a special occasion, to file tax returns, and to care for one's children—we are actually faulting people, not for their involuntary forgetfulness, but

[26] On whether and how we may be responsible for states of affairs beyond our control, see Statman ed. (1993). For an argument against the existence of genuine moral luck, see Kolber (1996) (unpublished senior thesis, Princeton University) (on file with author).

[27] . In criminal law, we require that every offense contain either a voluntary act or an omission to act when there is a duty to do so. This requirement prevents us from punishing people based merely on thoughts beyond their control (see, e.g., Proctor v. State, 1918: Packer, 1968, 73–79).

for some intentional failure at an earlier point in time[28] (Kelman, 1981, 593–94, 600–16). For example, perhaps the neglectful taxpayer intentionally decided not to record his filing deadline on his calendar or made other deliberate choices not to develop those attributes that would have prevented his memory failure. In a world with memory-altering drugs (either enhancing or dampening), we would have more opportunities to consciously alter our inclinations to remember or forget, leading perhaps to more responsibility for whatever memories we keep or discard.

Even if we can have obligations to remember, however, it is easy to overestimate the strength of these obligations. Perhaps the Council does so when it states that it may have been inappropriate for those with firsthand experiences of the Holocaust to dampen their traumatic memories:

Consider the case of a person who has suffered or witnessed atrocities that occasion unbearable memories: for example, those with firsthand experience of the Holocaust. The life of that individual might well be served by dulling such bitter memories, but such a humanitarian intervention, if widely practiced, would seem deeply troubling: Would the community as a whole—would the human race—be served by such a mass numbing of this terrible but indispensable memory? Do those who suffer evil have a duty to remember and bear witness, lest we all forget the very horrors that haunt them? (Beyond Therapy, 2003, 291).

There is something harsh about expecting trauma sufferers to bear the additional burden of carrying forward their traumatic memories for the benefit of others. The Council, recognizing this, goes on to soften its perspective somewhat, stating that "we cannot and should not force those who live through great trauma to endure its painful memory *for the benefit of the rest of us*" (Beyond Therapy, 2003, 230–231).

Yet, even for those who suffer from the most tragic of memories, the Council is ambivalent about the ethics of pharmaceutical dampening:

[A]s a community, there are certain events that we have an obligation to remember—an obligation that falls disproportionately, one might even say unfairly, on those who experience such events most directly. What kind of people would we be if we did not "want" to remember the Holocaust, if we sought to make the anguish it caused simply go away? And yet, what

[28] Kelman describes the "arational choice between narrow and broad time frames" in the criminal law (Kelman, 1981, 593–94, 600–16).

kind of people are we, especially those who face such horrors firsthand, that we can endure such awful memories? (Id., 231).

According to the Council, we are sometimes obligated to remember some person or set of events because doing so pays respect to that person or set of events. (Id.) For example, we may have obligations to remember great sacrifices that others make on our behalf, not because these memories will guide our actions, but rather because retaining the memory demonstrates a kind of respect or concern for these others.

The case for legally restricting memory dampening is particularly weak when it comes to such "homage" memories. What makes the retention of a traumatic homage memory significant is that the person who bears the traumatic memory has chosen to identify with it in some way. In fact, memory-dampening drugs, by giving us the opportunity to consciously choose to keep a memory intact, may actually facilitate our identification with it. On the other hand, if an individual retains an homage memory simply because he has no choice—because the tragic memory was indelibly imprinted into his brain by stress hormones or because memory dampening has been prohibited—the holding of the homage memory loses much of its significance. Such memories are not truly homages at all.[29]

Nevertheless, we can easily imagine situations where our obligations to remember are much stronger. For example, suppose a bystander is the only person to see the face of a serial rapist fleeing the home of his latest victim. Though the bystander may find the memory of the perpetrator's appearance quite upsetting, virtually everyone would agree that the bystander ought to retain the memory if doing so will ultimately help prosecute the perpetrator and protect potential future victims. Such a conclusion would be much less likely, however, if we consider instead the point of view, not of a mere bystander-witness, but of the traumatized victim who, let us now suppose, is the only one to see the perpetrator's face. In that case, we might still expect the victim to experience even this more intense trauma for, say, an hour until a police sketch artist can preserve the memory. It is much less clear, however, if the victim should be obligated to wait more than six hours to begin memory dampening in a world (like ours today, perhaps) where memory dampening would no longer be effective. At a minimum, however, it is clear that some people

[29] Admittedly, the analysis is complicated, however, by the inability to recover a previously dampened or erased memory.

have obligations to remember because there are strong societal interests in preserving certain memories.

Translating ethical obligations to remember into legal restrictions on memory dampening is no simple matter.[268] Memory dampening is a kind of medical treatment, and we do not ordinarily limit a person's access to medical resources simply to further police investigations.[30] On the other hand, memory dampening can destroy evidence, and we have plenty of laws prohibiting that (Kolber, 2006, 1579–92). It, therefore, seems plausible that some balancing of interests should occur when a person wishes to dampen memories that hold substantial instrumental value to society.

Yet even if we sometimes have ethical obligations to retain memories that ought sometimes be backed by legal sanctions, there is little reason to think that broad restrictions on memory dampening are needed. So, for example, an expansion of obstruction of justice statutes could further limit the use of memory-dampening drugs when patients have memories that are needed to protect societal interests in justice and safety. Alternatively, physicians could be required to make certain inquiries before prescribing memory-dampening drugs and could perhaps be obliged to notify authorities if a patient seeks to dampen or erase memories, where doing so may endanger someone else's life.[31] (Cf. Tarasoff v. Regents of Univ. of Cal., 1976, 340). Limited restrictions like these derive from concerns about memory dampening that, unlike those previously discussed, are based on ethical obligations we have to others and do not rely on much disputed conceptions of human nature or controversial preferences for what is deemed natural.

B. *Coarsening to Horror*

The Council also expressed concern that memory dampening will coarsen our reactions to horror and tragedy. If we see the world from

[30] According to psychiatrist Roger Pitman, if a crime victim has severe physical pain requiring the administration of morphine, we do not restrict it even though morphine can interfere with the victim's memory (Dupree, 2004, 9–10) (stating a claim made by Pitman).

[31] The Court in Tarasoff stated: "When a therapist determines, or pursuant to the standards of his profession should determine, that his patient presents a serious danger of violence to another, he incurs an obligation to use reasonable care to protect the intended victim against such danger" (Tarasoff v. Regents of Univ. of Cal., 1976, 340).

a chemically-softened, affect-dulled perspective, we may grow inured to trauma and its associated distress, "making shameful acts seem less shameful, or terrible acts less terrible, than they really are" (Beyond Therapy, 2003, 228).

As an example, the Council describes a hypothetical witness to a murder who dampens his memory and eventually perceives the crime as less severe than he would have without pharmaceutical assistance:

Thanks to [a memory-dampening] drug, [the memory of the murder] gets encoded as a garden-variety, emotionally neutral experience. But in manipulating his memory in this way, he risks coming to think about the murder as more tolerable than it really is, as an event that should not sting those who witness it. For our opinions about the meaning of our experiences are shaped partly by the feelings evoked when we remember them. If, psychologically, the murder is transformed into an event our witness can recall without pain—or without any particular emotion—perhaps its moral significance will also fade from consciousness. (Id.)

One concern suggested by this example is that memory dampening will make it more difficult to accurately convey evidence and other kinds of information to each other. According to the Council, the person described above "would in a sense have ceased to be a genuine witness of the murder," and when later asked about the event, "he might say, 'Yes, I was there. But it wasn't so terrible.'" Though the Council asks whether this person was a "genuine witness of the murder," the implicit reference to the natural is more appropriate here than it was with respect to the Council's prudential concerns. If this person were to appear before a jury, his description of the events surrounding the murder will be interpreted by listeners against a backdrop of *natural* linguistic conventions that help connect a speaker's affect to the events he describes. Similarly, in the military context, some worry that memory-dampened soldiers will come back from battle with unnatural affect-reduced descriptions of their experiences, making combat seem less horrific than it would otherwise[32]

[32] The Council writes: "Even if they existed, and even in times of great peril, we might resist drugs that eliminate completely the fear or inhibition of our soldiers, turning them into 'killing machines' (or 'dying machines'), without trembling or remorse" (Beyond Therapy, 2003, 154–5); Wasserman discusses how our willingness to engage in actions, like combat, may be affected by expectations that one can engage in "emotional amnesia" (Wasserman, 2004, 17–18).

(Id., 154–155). Against a standard backdrop of communicative conventions, we would understandably be puzzled by a flat, lifeless description of human tragedy.

Indeed, if memory dampening has a tendency to alter our perceptions and our understanding of events in the world, then, as the Council's example suggests, it may affect more than just the ways we communicate. A deeper concern is that memory dampening will coarsen our feelings and make us less willing to respond to tragic situations. Along these lines, one can imagine a would-be-famous civil rights leader in the 1960s who, in order to combat the memory of childhood injustices, would have gone on to revolutionize our social institutions but, due to his use of memory dampeners, instead pursues a more mundane life plan and is never so much as mentioned in the history books.

Not only might our coarsened emotions disincline us to take positive action, it has been suggested that memory dampeners could reduce our inhibitions to engage in socially destructive action. Thus, violent criminals could use memory dampeners to ease feelings of guilt, making them more likely to recidivate (Id., 224). In addition, it has been claimed, memory-dampened soldiers, freed from burdens of conscience, may be more effective at killing (Id., 154). Council member Paul McHugh asks, "If soldiers did something that ended up with children getting killed, do you want to give them beta blockers so that they can do it again?" (Mundell, 2005). The question is lacking in some important details but, more importantly, these examples suggest that fear and remorse or expectations of fear and remorse inhibit certain antisocial behaviors and that memory dampening may interfere with this desirable control mechanism. While this concern is far from universal, it may warrant studying whether any proposed memory-dampening agent actually has such effects.

Even if there is some empirical basis for these concerns, however, it is important not to overstate their importance. For even if memory dampening does make some trauma *seem* less horrible, this happens in part because memory dampening can *actually make* trauma less horrible. That is, much of what is bad about traumatic experience is that it traumatizes those who survive it. So, for example, to the extent that we can ease the traumatic memories of those involved in military conflict (without leading to a significant increase in total military conflict), then memory dampening makes combat somewhat better than it would otherwise be. Furthermore, when soldiers are injured in battle, we heal their physical wounds using advanced technology, even if doing so makes war seem less horrible; so it is unclear why their emotional wounds should be treated any differently.

While the coarsening concern is far from overwhelming, it at least shows how the widespread use of memory dampeners can potentially affect the lives of those who do not use them. Nevertheless, this concern cannot alone justify broad restrictions on memory dampening, at least not if such restrictions are consistent with our typical policies of drug regulation. For example, people consume alcohol to relieve themselves of the pain of traumatic events. Whether or not this leads to some general inurement to tragedy in society (which seems doubtful), most would not address the problem with a comprehensive prohibition of alcohol. Similarly, even if antidepressants are used for relief from the pain of traumatic experiences, we would not generally prohibit them for fear that society will be less compassionate. Likewise, the world may benefit from the inspired artwork of a Vincent van Gogh, yet few would deprive a tortured soul of antidepressants in order to foster artistic creation.

We likely permit the use of such drugs, despite whatever minimal effects they may have on our reactions to tragedy, because their costs are outweighed by other benefits. So even if data someday support the Council's concern that memory-dampening drugs can have negative effects on soldiers' battlefield reactions or on societal reactions more generally, we can surely tailor limits on their use in particular contexts. And if the testimony of memory-dampened witnesses has a different emotional tone than that of ordinary witnesses, experts can explain the differences to jurors.

While memory dampening has its drawbacks, such may be the price we pay in order to heal intense emotional suffering. In some contexts, there may be steps that ought to be taken to preserve valuable factual or emotional information contained in a memory, even when we must delay or otherwise impose limits on access to memory dampening. None of these concerns, however, even if they find empirical support, are strong enough to justify broad-brushed restrictions on memory dampening.

Freedom of Memory

I have argued that concerns over memory dampening are insufficient to justify broad restrictions on the therapy. Furthermore, having the choice to dampen memories supports our interests in self-determination and in avoiding mental illness and upset, and, as noted, enables us to identify more strongly with memories that we decide to keep. Given the potential that memory dampening has to ease the pain of so many people, and that, at a minimum, memory dampening ought not be entirely prohibited, it follows that we should have some right to dampen our memories.

Such a right can be thought of as just a piece of a much larger, as-yet-poorly-defined bundle of rights to control what happens to our memories. For example, we may have some right to be free from forced memory dampening were the government to try to make us forget a trade secret or a voyeuristic memory.[33] Neuroscientists are also hard at work developing drugs to enhance memory retention to treat Alzheimer's disease, as well as less severe age-related memory problems (see McGaugh (2003), 68–79). In the context of memory enhancement, we might have rights to enhance the emotions we attach to our memories (perhaps to increase affect attached to positive memories) as well as rights to enhance the factual content of the memories we store (to avert memory disorders or, more controversially, to perform better in school). We may also have rights to prevent forced enhancement of the factual richness of our memories by those who would make us better spies, soldiers, students, or employees or to prevent forced enhancement of our memory-related affect by those who think doing so would make us more responsive to conscience and less likely to violate social norms (see Wasserman, 2004).[34]

In addition to enhancing and dampening memories, we may have rights to keep memories private. Such a right is already circumscribed by the government's subpoena power—the power to demand that we answer (or at least try to answer) certain questions, under oath, about the content of our memories (see Slobogin, 2005). Advances in neuroscience, however, have led to the creation of neuroimaging technologies, like functional magnetic resonance imaging ("fMRI"), that will make questions about the privacy of memory even more important. For example, neuroscientists are trying to develop brain imaging techniques to determine if

[33] Such autonomy interests are frequently noted in important constitutional law cases. See, e.g., Cruzan v. Director, Mo. Dep't of Health, 1990, 278 ("The principle that a competent person has a constitutionally protected liberty interest in refusing unwanted medical treatment may be inferred from our prior decisions."); Washington v. Harper, 1990, 229 (1990) ("The forcible injection of medication into a nonconsenting person's body represents a substantial interference with that person's liberty."); Riggins v. Nevada, 1992, 134.

[34] Wasserman notes: "Some might suggest that for particularly heinous crimes, enhancement of guilt-ridden memory could serve as a form of punishment, a kind of forced internalization".

an experimental subject recognizes a person in a photograph (i.e., has a memory of that person) using brain imaging alone, without relying on the subject's own (possibly deceptive) report (Thompson, 2005, 1602; see generally Keckler, 2006; Wade, 2005, A19). The emergence of such technologies led one group of researchers to make the controversial claim that "[f]or the first time, using modern neuroscience techniques, a third party can, in principle, bypass the peripheral nervous system–the usual way in which we communicate–and gain direct access to the seat of a person's thoughts, feelings, intention, or knowledge" (Wolpe, 2005, 39; Kamitani & Tong, 2005, 679).[35]

Related to the right to keep memories private is the right to make memories public. One such "publicity right," if it may be called such, concerns the means by which we can voluntarily demonstrate the content of our memories in court. In Harrington v. State, convicted murderer Terry Harrington[36] sought to offer unconventional evidence of his memories in the form of so-called brain fingerprinting, a kind of electroencephalography[37] (Harrington v. State, 515). The brain fingerprinting results purportedly showed that Harrington did not have memories of the crime scene that the actual perpetrator would have had and that Harrington did have memories that supported his alibi (Harrington v. State, 516, n.6). The Iowa District Court, ruling for the first time on the admissibility of such evidence, found some of the brain fingerprinting results to be admissible, but, for a variety of reasons, dismissed Harrington's petition for a new trial (Harrington v. State, 216). When Harrington appealed to the Supreme Court of Iowa, his conviction was vacated on due process grounds unrelated to his evidentiary claim, and the court never ruled on the admissibility of his brain fingerprinting evidence

[35] The reason the claim in the text is controversial is that it is not clear that one can ever, even in principle, have direct access to these features of another's mind.

[36] Harrington was convicted of first degree murder in the late 1970s, State v. Harrington, and was then sentenced to life imprisonment without possibility of parole.

[37] Electroencephalograms measure brain signals known as "event related potentials" that can be detected "on the scalp 300–500 ms after the subject is exposed to a stimulus" (Wolpe, 2005, 41). Farwell's brain fingerprinting technique is supposed to use electroencephalography to determine whether a subject is exposed to a familiar or unfamiliar stimulus by measuring event related potentials that are "associated with novelty and salience of incoming stimuli" (Id).

(Harrington v. States, 512, 516; Slaughter v. State, 1054).[38] In the lower court, however, Harrington did win a narrow right to admit unconventional evidence related to his memory, setting the stage for future battles in this arena.

Before these new neuroscience imaging techniques and pharmaceuticals appeared on the horizon (distant as it may still be), it made little sense to speak of a "freedom of memory." There was simply too little we could do as human beings to affect our own memories to warrant clarifying our rights. In light of these developing technologies, however, we can begin to envision a bundle of rights associated with memory, including perhaps: rights to dampen memories; rights to enhance memories or memory-retention skills; rights to keep memories private (or to allow us to publicize them in court); and rights to be free of certain invasions of our memories by forced enhancement, forced dampening, or even the secret implantation of false memories.[39]

References

Articles and Books

Appel, J. M. (2019). Where's Trump's Bioethics Commission? *Baltimore Sun* May 31, 2019.

Bonanno, G. A. (2004). Loss, trauma, and human resilience. *American Psychologist*, 59, 20, 25–26

Bonanno, G. A. (2005). Self-enhancement among high-exposure survivors of the September 11th terrorist attack: Resilience or social maladjustment, *Journal of Personality and Social Psychology*, 88.

Bostrom, N., & Ord, T. (2006). The reversal test: Eliminating status quo bias in applied ethics. *Ethics*, 116, 656.

Brown, J. D. (1986). Evaluations of self and others: Self-enhancement biases in social judgments. *Social Cognition*, 4, 353.

Butcher, J. N., & Hatcher, C. (1988). The neglected entity in air disaster planning. *American Psychologist*, 43.

[38] In Slaughter, the Oklahoma Court of Criminal Appeals found that the issue of brain fingerprinting "could have been previously raised in the direct appeal" and that there was "insufficient evidence to support a conclusion that brain fingerprinting, based solely upon the MERMER effect, would survive a Daubert analysis".

[39] Elizabeth Loftus and her research team have implanted so-called false memories into experimental subjects under a variety of conditions (see Loftus, 2003-A; Loftus 2003-B).

Douglas, K. et al. (2005). 11 Steps to a better brain. *New Scientist*, 186, 28, 28–29.

Dupree, C. (2004). Cushioning hard memories. *Harvard Magazine*, 106.

Elga, A. (2005). On Overrating Oneself ... and Knowing It, 123 Phil. *Studies, 115*, 117.

Gazzaniga, M. (2005). The ethical brain, 120–142.

Greenawalt, K. (1995). Legal enforcement of morality. *Journal of Criminal Law & Criminology, 85*, 710–718.

Kading, D. (1960). Are there really "No Duties to Oneself"? *Ethics*, 70, 155.

Kamitani, Y., & Tong, F. (2005). Decoding the visual and subjective contents of the human brain. *Nature Neuroscience*, 8, 679.

Kaser, M., et al. (2017). Modafinil improves episodic memory and working memory cognition in patients with remitted depression: A double-blind, randomized, placebo-controlled study. *Biological Psychiatry: Cognitive and Neuroscience Neuroimaging, 2*, 115–122.

Keckler, C. N. W. (2006). Cross-examining the brain: A legal analysis of neural imaging for credibility impeachment. Hastings L.J., 57, 509.

Kelman, M. (1981). Interpretive construction in the substantive criminal law. *Stanford Law Review*, 33, 591, 593–94, 600–16.

Kindt, M., & Soeter, M. (2018). Pharmacologically induced amnesia for learned fear is time and sleep dependent. *Nature Communications, 9*, 1316.

Kolber, A. J. (2008). Freedom of memory today. *Neuroethics*, 1, 145.

Kolber, A. J., Mental statism and the experience machine. *Bard Journal of Social Science*, 3, 10.

Kolber, A. J. (Winter 1994/1995). The moral of moral luck (Apr. 29, 1996) (unpublished senior thesis, Princeton University) (on file with author).

Kolber, A. J. (2006) Therapeutic forgetting: The legal and ethical implications of memory dampening. *Vanderbilt Law Review*, 59, 1561.

Kramer, P. (1993). Listening to Prozac.

Jaworska, A. (1999). Respecting the margins of agency: Alzheimer's patients and the capacity to value. *Philosophy & Public Affairs*, 28, 105.

LaFave, W. R. (2003). Criminal Law §. 8.01(a) (4th ed.)

Locke, J. (1975). Of identity and diversity. In Perry, John (Ed.). *Personal identity*.

Loftus, E. (2003). Our changable memories: Legal and practical implications. *Nature Reviews Neuroscience*, 4, 231

Loftus, E. (2003). Make-believe memories. *American Psychologist*, 58, 867.

Margalit, A. (2002). The ethics of memory.

McGaugh, J. (2003). Memory and emotion.

Meilaender, G. (2003). Why remember? *First Things*, 135.

Mundell, E. J. (2005). Heart drugs could ease trauma memories. *Health Day News*, July 29, 2005.

Nozick, R. (1974). Anarchy, state & utopia.
Packer, H. L. (1968). The limits of the criminal sanction, 73–79.
Parfit, D. (1984). Reasons and persons.
Perry, J., ed. Personal identity (1975).
Pitman, R. K. et al., (2002). Pilot Study Of Secondary Prevention Of Posttraumatic Stress Disorder With Propranolol. *Biological Psychiatry*, 51, 189
President's Council on Bioethics, Beyond Therapy: Biotechnology and the Pursuit of Happiness (2003).
Singer, G. M. (1959). On duties to oneself. *Ethics*, 69, 202, 202–03.
Statmen, D., ed. Moral luck (1993).
Thompson, S. K. (2005) The legality of the use of psychiatric neuroimaging in intelligence interrogation. *Cornell Law Review*, 90, 1601.
Vaiva, Guillaume, et al. (2003). Immediate treatment with propranolol decreases posttraumatic stress disorder two months after trauma. Biological Psychiatry, 54, 947.
Vallejo, A. G., et al. (2019). Propofol-induced deep sedation reduces emotional episodic memory reconsolidation in humans. *Science Advances,* 5, 3.
Wade, Nicholas, Improved Scanning Technique Uses Brain as Portal to Thought, N.Y. Times, Apr. 25, 2005, at A19.
Wasserman, David, Making Memory Lose Its Sting, 24 Phil. & Pub. Pol'y Q. 12 (Fall 2004)
Wolpe, Paul Root, Emerging Neurotechnologies for Lie-Detection: Promises and Perils, 5 Am. J. of Bioethics 39, 39 (2005);

Films

Eternal Sunshine of the Spotless Mind, Focus Features, 2004.
Paycheck, Paramount, 2004.
Men in Black, Sony Pictures, 1997

Cases and Regulations

Cruzan v. Director, Missouri Department of Health, 497 U.S. 261 (1990).
Harrington v. State, 659 N.W.2d 509 (Iowa 2003).
Proctor v. State, 176 P. 771 (Okla. Crim. App. 1918).
Riggins v. Nevada, 504 U.S. 127, 134 (1992)
Slaughter v. State, 108 P.3d 1052, 1054 (Okla. Crim. App. 2005).
Tarasoff v. Regents of Univ. of Cal. , 551 P. 2d 334, 340 (Cal. 1976).
Washington v. Harper, 494 U.S. 210, 229 (1990).
Order No. 13237, 66 Fed. Reg. 59851 (Nov. 28, 2001).

CHAPTER 8

Cognitive Liberty of the Person with a Psychotic Disorder

Mari Stenlund

Introduction

In many, if not all countries, the law enables people to be subject to involuntary psychiatric hospital treatment when they are suffering from a psychotic disorder and they are considered to be a danger to themselves or to others (see, e.g., Mielenterveyslaki 1990/116, 8§). International ethical guidelines for psychiatric treatment direct, as well as commit, individuals with psychotic disorders to involuntary psychiatric treatment in such cases (see. e.g., Council of Europe, 2004, Articles 17–19; MI Principles, 1991, Principle 16). Involuntary treatment often utilizes antipsychotic medication with the goal of reducing or removing psychotic symptoms. The involuntary use of mind-altering medication is accepted in laws and ethical principles guiding psychiatric treatment (Council of Europe, 2004, article 28:1; MI Principles, 1991, Principle 11:6; Mielenterveyslaki 2001/1423, 22b).

M. Stenlund (✉)
Mikkeli, Finland

M. J. Blitz et al. (eds.), *The Law and Ethics of Freedom of Thought, Volume 1*, Palgrave Studies in Law, Neuroscience, and Human Behavior,
https://doi.org/10.1007/978-3-030-84494-3_8

This chapter examines how cognitive liberty is affected when a person is diagnosed as psychotic. I explore what cognitive liberty ultimately protects in this situation, taking into account that from the perspective of psychiatry, delusions and hallucinations are considered to be symptoms of psychosis and are viewed as something that the sufferer has the right to be treated for. According to diagnostic manual DSM-V (2013, 819), delusions are false beliefs "based on incorrect inference about external reality that is firmly held despite what almost everyone else believes." Hallucinations are defined as "perception-like experiences that occur without an external stimulus" (DSM-V, 2013, 87). Like all other mental disorders, psychotic disorders are also, according to DSM-V (2013, 20–21), usually associated with significant distress or disability.

The experiences of patients undergoing involuntary treatments vary. A large proportion of people subjected to involuntary treatment have, in hindsight, concluded that they have benefited from the treatment (Lönnqvist et al., 2014, 741). On the other hand, there are patients who feel that the involuntary treatment has infringed their cognitive liberty (see, e.g. Stenlund, 2018, 1; Uudempaa maailmaa toivoo Joni, 2018). Also, the antipsychiatric movement has argued that the patient's internal freedom of thought is being restricted by compulsory psychiatry (see Gosden, 1997; Szasz, 1990).

Cognitive Liberty as a Human and Fundamental Right

In this chapter, cognitive liberty is understood as a bundle of different rights for believing, thinking, and expressing opinions, and the focus is on the internal dimension of these rights. The human and fundamental rights concerning believing, thinking, and expressing opinions are numerous. We can talk of freedom of religion, freedom of belief, freedom of conscience, freedom of thought, freedom of opinion, and freedom of expression (see ICCPR, 1966, articles 18–19; Rainey et al., 2014, 411–413, 435). When the bundle of these rights is examined both from the perspective of internal and external dimensions, a broader term, "freedom of belief and opinion" is used in this chapter. When I focus on the internal dimension of these rights, I discuss the *forum internum* dimension or cognitive liberty.

When cognitive liberty is understood as a bundle of these freedom rights, we can say that it is a human right inscribed in the international human rights conventions, that protects the rights of people to search for truth, the meaning of life, and for connectedness with other people (see ICCPR, 1966, article 18–19). These rights belong to all people based on their humanity. The starting point is that a person has, and should have, freedom of belief and of opinion even when they are experiencing mental health challenges or have received a psychiatric diagnosis (see MI Principles, 1991, Principle 5:1).

However, the standing challenge is that in discussions about cognitive liberty, and in the definition of different rights of belief and opinion, the background assumption has been that the subjects of cognitive liberty are mentally healthy adults. Due to this, conceptions about the contents and limits of cognitive liberty contained in human rights theory seem to conflict with laws on psychiatry and the praxis of mental health work (see Stenlund, 2014, 89–91; Stenlund & Slotte, 2018).

Forum Externum and Forum Internum

Human rights concerning freedom of belief and opinion have both an external and an internal dimension. Cognitive liberty in essence refers to the internal aspect of these rights.

In human rights theory, the external dimension of freedom of belief and opinion is called the *forum externum*. *Forum externum* literally means "an external forum": the exercising of one's freedom of beliefs and opinions among other people. In other words, it means acting upon and expressing beliefs and opinions. When someone is a churchgoer, or stops another person on the street to tell them about "the good message," uses religious symbols, or reads the Book of Mormon in the commuter train or at home, they exercise the *forum externum* dimension of the freedom of beliefs and opinions. They also act within the *forum externum* dimension when taking part in a demonstration, voting, or expressing their views on social media (see Partsch, 1981, 214, 217; Tahzib, 1996, 26–27, 87. Further reading Stenlund, 2013, 2014).

The internal dimension of freedom of belief and opinion, or *forum internum*, refers to events taking place in the person's "internal forum." When a person ponders whether to believe in God, or when they pray a silent prayer in their minds they act within the *forum internum* dimension. Similar actions are when they ponder about the meaning of their lives or about the nature of the world. In some discussions, membership in religious communities has been considered to belong to the *forum internum*, but at its narrowest, it has to do with the dimension of freedom of belief and opinion that protects the internal workings of the human mind: how and what a person thinks, believes, and ponders (see Evans, 2001, 68, 72–74; Nowak, 1993, 314–315; Partsch, 1981, 214, 217; Rainey et al., 2014, 412; Tahzib, 1996, 25–26. Further reading Stenlund, 2013, 2014; Stenlund & Slotte, 2018).

According to human rights conventions, a person's *forum externum* dimension may be restricted if a person exercises these freedoms in a way that poses a threat to other people's rights. Thus you cannot do and say anything you please in the name of the freedom of belief and opinion. The situation regarding the *forum internum* dimension is different. It is defined in human rights conventions and the human rights theory examining these conventions as an absolute human right, that cannot be restricted in any situation or for any reason. It has been suggested that the right to free thinking, to opinion formation, and the right to any content of one's mind is absolute. It has been claimed that manipulating a person's mind or affecting the mind with involuntary medication is in breach of this absolute right (Evans, 2001, 68, 72–74; ICCPR, 1966, Article 19:1; Nowak, 1993, 314–315; Partsch, 1981, 214, 217; Tahzib, 1996, 87–88. Further reading Stenlund, 2013, 2014; Stenlund & Slotte, 2018).

Forum Internum and Involuntary Psychiatric Medication

How should we perceive the *forum internum* dimension of freedom of belief and opinion that is cognitive liberty in individuals with mental illness? If the human rights conventions and human rights theory were interpreted literally, we should think that a person should have the right even to so-called sick thoughts or psychotic delusions. In human rights theory, it has been claimed that even delusions are a kind of thought or opinion that people have a right to hold in their minds (Stenlund, 2013).

However, in practice most people who are guided by legislation and by ethical principles don't think this way. In psychiatric care people can be forcibly medicated against their expressed will, in order to reduce or remove psychotic symptoms, which are at the same time inner beliefs, thoughts, and experiences. The laws guiding mental health work allow these kinds of restrictive measures. The tensions and contradictions between human rights theory and the laws and praxis of psychiatry reveal that people whose mental health is shaken have not been taken properly into account when developing the human rights theory regarding freedom of beliefs and opinions (Stenlund, 2013, 2014, 89–91; Stenlund & Slotte, 2018).

This tension concerning the rights of people with psychotic disorders reveals that it is unclear what the *forum internum* dimension, i.e., cognitive liberty, fundamentally protects. Legal cases decided in Europe and the United States have not been able to solve these deep problems (see Stenlund, 2013; Stenlund & Slotte, 2018). Moreover, Jan-Christoph Bublitz (2013) observes that "not even the outspoken and critical legal commentaries define [*forum internum's*] contours in more detail."

In this article I present three different ways of understanding the freedom of belief and opinion, and the cognitive liberty contained in these rights. How freedom of belief and opinion is understood substantially affects what we try protect in the case of a person with a psychotic disorder, and how the freedom of belief and opinion is valued in relation to other human rights (see more about different conceptions of freedom, Stenlund, 2014, 92–320).

Cognitive Liberty as a Negative Liberty

When posing the question "When is a person free?" most people perhaps first suggest a scenario where the person is not restricted, that they can do whatever they please at that moment. This way of understanding liberty is the so-called classical way of understanding the rights to freedom that encompass freedom of belief and opinion. Freedom of belief and opinion is realized, according to this viewpoint, when other people do not interfere with an individual's beliefs, thoughts, and opinions using concrete, biological, or legal means, but leave the person free to think and do as they please. This right to liberty is like a shield that protects the thinker from attacks originating from other people. This kind of concept of liberty is often called "negative," since its essence consists of the lack of obstacles and lack of boundaries, i.e., that a person is permitted to be and to act without outsiders concretely interfering with or restricting their being and action (see Berlin, 2005, 169–170; Feinberg, 1973, 7–15). When freedom of belief and opinion is understood as a negative right, the *forum internum* or cognitive liberty primarily protects the contents of thought and belief that the person already has in their mind (Stenlund, 2014, 103, 326; Stenlund & Slotte, 2018). In human rights theory, the *forum internum* is classically defined according to this understanding of freedom. The tensions in relation to the use of involuntary antipsychotic medication arise especially from the viewpoint of negative liberty.

Involuntary Treatment as a Limitation to Cognitive Liberty

In the case of a person with a psychotic disorder, negative freedom of belief and opinion is actualized when the person is not obstructed from acting based on their beliefs and opinions and they are free to think whatever they choose (Stenlund, 2014, 101). The negative understanding of liberty is central in our legal system and sense of justice. This can be seen in the way that involuntary psychiatric treatment is commonly considered as a restriction on the person's freedom rights. If a person is forced into treatment it is considered problematic per se (Stenlund, 2014, 106–117).

During involuntary treatment, freedom of belief and opinion may be restricted in a number of ways. The patient's movements may be restricted, so that they cannot go to the places that would be essential for their practice of religion or opinion. Their communication with people may be restricted. Similarly, their belongings can be confiscated in the event that the mental health staff considers that these restrictions are to protect their health and the well-being of other people. As to the *forum internum* dimension of the freedom of belief and opinion, i.e., cognitive liberty, the most interesting restriction is as presented above, that a patient receiving involuntary treatment may be forced to use psychiatric medication, and when deemed necessary these medications may be administered as injections regardless of the patient's opposition to it (Stenlund, 2014, 117–122; see Council of Europe, 2004, article 28:1; MI Principles, 1991, Principle 11:6; Mielenterveyslaki 2001/1423, 22a–22j§).

Forced medication is often justified by the claim that medication is in the patient's best interests, but from the viewpoint of the negative understanding of liberty this is meddling in the *forum internum* dimension of freedom of belief and opinion, i.e., to the dimension of right, which should never and under no justification be restricted (Stenlund, 2014, 121–129; Stenlund & Slotte, 2018).

Sufficient Competence as a Requirement

Even as involuntary treatment is generally considered to be restricting the freedom of the person, this restriction is often considered as justified from the viewpoint of the negative understanding of freedom. The reason is that negative freedom is not considered to be the only important value, and other values and rights are prioritized above it in situations where a person is not considered to be competent enough to decide on their own affairs (Stenlund, 2014, 129–153; Stenlund & Slotte, 2018).

Thus, in law, the negative sense of the freedom of belief and opinion requires sufficient competency. A person is considered competent if they adequately understand the consequences of their actions and the nature of reality in order to make decisions about themselves. Only an adequately competent person can decline treatment or give their approval for the treatment. So in essence involuntary (or more precisely, non-voluntary treatment) means that when lacking competency the person is treated regardless of what their opinion on the issue is. On the same basis, their religious or ideological practices can be constrained if it is determined that it is harmful to them. It is thought that these restrictions are justified paternalism (Stenlund, 2014, 129–153; see Beauchamp & Childress, 1989, 69, 79).

According to the so-called antipsychiatric point of view, negative freedom should be valued more than the right of the person to well-being, and it should not be interfered with even when it is determined that the person is a danger to themselves. If a person poses a danger to other people, then according to the antipsychiatric standpoint the situation should be dealt with in the same way as in other situations, where a person's threatening behavior or violence is forcibly curbed by the police and juridical sanctions. The antipsychiatric view states that society should not commit anyone to treatment on the grounds that they do not understand their own best interests and is "messed up in the head," for that is underestimating the person's own responsibility for their behavior and choices (see, e.g., Szasz, 2008, 112–117).

The opposition between involuntary treatment practices based on the paternalistic use of power and the antipsychiatric way of leaving people to their own devices is clear, though the former clearly reflects the mainstream in Western societies. The paternalistic use of power is accepted in law. Many people who have been subjected to involuntary treatment have, a posteriori, been grateful that paternalistic use of power was applied to them. Nevertheless, a proportion of patients are very much against involuntary treatment, not only during treatment but after treatment as well, i.e., even when they are in an adequately competent state (see Kaltiala-Heino, 1995, 84, 112–113; Lönnqvist et al., 2014, 741).

The Relation of Competence to the Forum Internum

When viewing freedom of belief and opinion from the negative perspective, the challenge seems to be the conflict between the praxis of paternalistic use of power and the *forum internum* dimension. In involuntary psychiatric treatment the executed involuntary medication aims for the patient's best interests, but it is accomplished by attempting to affect the person's thoughts and beliefs through biological means. The fact that those thoughts and beliefs have been defined as symptoms of an illness is not significant, because in the human rights theory the *forum internum* dimension is viewed as protecting absolutely all kinds of beliefs and thoughts. The idea is that a human being must have an absolute right to any mental content (Stenlund, 2013, 2014, 82–89, 121–129; Stenlund & Slotte, 2018).

If the *forum internum* dimension must not be restricted in any situation, why is forced medication being practised? Usually the justification is that it is necessary in order to protect the person, and that the person should not be left abandoned. However, this good reason is not, in principle, justified because restricting absolute rights such as the forum internum is not allowed for any situation or for any reason (Stenlund, 2013, 2014, 82–89, 121–129). The question arises as to whether sufficient competence can be a requirement for the *forum internum* dimension. However, as Mari Stenlund and Pamela Slotte (2018) ask, is it possible that some human beings could fall completely outside of the realm of rights which should belong to everyone? Does it follow from a person's incompetence that they do not hold rights that are generally considered absolute? If so, are these rights genuinely absolute? Two options remain: first, it is possible that forced medication must be stopped because it is a breach of human rights. Given how the *forum internum* dimension is defined, and what it is thought to protect, prohibiting forced medication would be logical. The second option is to specify more precisely what is meant by freedom of belief and opinion and the *forum internum* dimension or cognitive liberty contained therein. If full abolition of involuntary medication seems unethical and careless toward people with mental health disorders, it is worth pondering if the freedom of belief and opinion can be understood from different viewpoints apart from the negative one.

Cognitive Liberty as Authenticity

Particularly in philosophical discourse, the concept of liberty is sometimes understood from the viewpoint of authenticity. Freedom of belief and opinion understood as authenticity protects people's right to beliefs and opinions that are genuinely their own and formed by themselves. (On the concept of authenticity, see Brison, 1996; Dworkin, 1985, 353–359; Guignon, 2004; Oshana, 2007; Scanlon, 1972). When freedom of belief and opinion and cognitive liberty are understood from the viewpoint of authenticity, the primary targets of protection are the thinking and believing processes which are intended to be authentic (Stenlund, 2014, 186, 326; Stenlund & Slotte, 2018). When the right to freedom of belief and opinion understood in this way is actualized, the beliefs and opinions can be granted "a certificate of authenticity."

While in the negative understanding of freedom of belief and opinion, the restrictions on these rights are understood as concrete biological or legal restrictions, in the understanding of freedom emphasizing authenticity, psychological means and reasons might also restrict the freedom of belief and opinion. For example, manipulating other people psychologically or with so-called religious brainwashing is understood to distort a person's authentic beliefs and thoughts and therefore infringe on their rights to freedom of belief and opinion, especially in the *forum internum* dimension, or cognitive liberty (see Beltran, 2005). Different mental health problems, especially disturbances of a psychotic level, can also be seen as factors restricting a person's cognitive liberty.

Psychosis as a Threat to Cognitive Liberty

Freedom of belief and opinion is understood from the viewpoint of authenticity in many discourses on the philosophy and ethics of psychiatry, though in these discourses attention is usually given to the patient's freedom, autonomy, and agency in general. What is being evaluated in assessing the effects of different mental health problems is the question of to what degree the beliefs and thoughts are really a person's own beliefs and thoughts, and to what degree they are distorted by the mental health disorder, and therefore in essence foreign or inauthentic to that person (see, e.g., Erler & Hope, 2014).

Psychosis, particularly, is often considered as a foreign or outside force that makes the person inauthentic and distorts their beliefs so that they become delusional. So psychotic delusions are understood as products of "a psychotic self" that has been distorted into inauthenticity, and not as authentic views of the genuine self (see, e.g., Gutheil, 1980). Jonathan Glover (2003, 537–538) suggests that serious mental health disorders can even change the core of a human being. Under part of the freedom of belief and opinion. we can reason that from the viewpoint of authenticity, psychosis (or when understood more broadly also other mental health disorders) are like external forces that violate the person's *forum internum*, i.e., their cognitive liberty.

Some people who have experienced psychosis perceive it to be like a foreign entity that has seized them in its grip. Luciane Wagner and Michael King (2005) noticed in their study that many people who have experienced psychosis regarded their disorder as something distinct from their being, and that they had difficulty understanding their psychotic thoughts. Alexandre Erler and Tony Hope (2014) reported a patient suffering from bipolar disorder, who saw the darkness as a stranger who "lodged within my mind" and as an "outside force that was at war with my natural self." In a study by Eeva Iso-Koivisto (2004, 11, 98) it similarly was discovered that some people who have experienced a psychosis try to differentiate the psychosis from themselves.

When psychosis is considered to be this kind of external force imposing itself to the *forum internum* dimension then psychiatry seems to try to liberate the person. Even involuntary treatment and use of involuntary antipsychotic medication are seen only as an effort to free the person from the power of the psychosis (see Gutheil, 1980, 327; Kaltiala-Heino et al., 2000, 213). From this point of view, the conflict between involuntary medication and protecting the *forum internum* dimension subsides, since the goal of the medication and other involuntary treatment is not to restrict, but instead to return, the patient's cognitive freedom.

An Idealistic Understanding of Humanity?

Even if the authenticity point of view for cognitive liberty seems to be sensible to some people who have experienced psychosis and to some parties offering psychiatric treatment, there exist various problems in this approach.

First of all, not everyone who has experienced psychosis has perceived it as a foreign threat. Some persons consider it as a genuine part of their life, and internal life, or as authentic suffering (see Stenlund 2014, 215–218). Second, there is danger in the conception that some actions that appears restrictive on the surface, such as involuntary treatment and the involuntary use of medication therein, would be freedom-increasing in the end. When expanded, such a conception might justify even totalitarian use of power (see Berlin, 2005, 180). Third, the view of humanity underlying the authenticity point of view seems to be very idealistic. Therein it is assumed that humans form their beliefs independently of each other—or at least that this kind of independence is presented as a criterion of genuine humanity. Evil, suffering, dependence, susceptibility to influences, and senselessness on the other hand, are presented as qualities or experiences that are not part of genuine human experience. We can ask whether this kind of understanding of humanity is realistic.

The nature of cognitive liberty appears quite differently, depending on whether it is examined from the point of view of a negative understanding of freedom, or from the point of view of authenticity; to the questions of involuntary treatment and the use of involuntary medication the answers may be opposite depending on the point of view chosen. However, when the focus is on antipsychiatric medication, many questions regarding the psychiatric praxis and societal structures are left without attention (see Stenlund, 2017a). The key question isn't necessarily whether to medicate or not, but that of how to support a person's ability to think, believe, and live according to their values, while living with others.

Cognitive Liberty as Capabilities

Freedom of belief and opinion can also be understood from the perspective of the capabilities approach. In this view, freedom of belief and opinion is meant to protect the person's capability of making choices concerning the beliefs they follow, as well as the ways of life they consider valuable and which are worthy of human dignity (see Stenlund, 2017a). The right to freedom, in a way means the right to the tools with which persons can act, and to good opportunities or "working spaces" in which people can use those tools in a meaningful way.

This kind of approach to the freedom of belief and opinion is in line with current human rights discussion where different civil, political, economic, social, and cultural rights are considered to be interdependent and interrelated, and are understood as giving rise to positive and negative obligations on the part of other actors (see Stenlund & Slotte, 2018).

It is emphasized that freedom is both negative and positive in its nature. To be free, a human being must not only be free from interference by other people. They must also have various resources with which they can lead the kind of life they desire (see Nussbaum, 2006, 287; Sen, 1999, 3–11).

The capabilities point of view has been developed by Amartya Sen (1999, 2009) and Martha Nussbaum (2006, 2011), among others. Nussbaum has listed central capabilities which should be secured for all people. Among these capabilities are several that are pivotally connected to the freedom of belief and opinion. First of all, the freedom of belief and opinion is connected to the capabilities to use the senses, imagination, and thought. Second, it is connected to practical reason, which means the capability of forming conceptions about a good life and how to pursue it. Third, a key to the freedom of belief and opinion is the capability of associating with others. Also, the capability of controlling one's environment and the expression of one's emotions are significant capabilities linked to the freedom or belief and opinion (Nussbaum, 2011, x, 18–19, 33–34). When freedom of belief and opinion is understood according to the capabilities approach, cognitive liberty protection focuses on the abilities of the human mind, instead of the contents of the mind or the belief and thought processes (Stenlund, 2014, 326; Stenlund & Slotte, 2018).

Psychosis and Treatment from the Capabilities Point of View

From the capabilities point of view, several questions are central for people recovering from a psychotic disorder, questions which are left in the sidelines by the negative liberty and authenticity approaches. First, the capabilities approach emphasizes that persons whose actions are based on delusions can find it hard to reach their goals, because the world does not seem to work in the way they assume. Also, forming social relationships may prove to be difficult if the person understands reality very differently from the people around them. There can be difficulties in understanding and being understood, and an atmosphere of chaos may arise. Therefore the person may have difficulties in living the kind of life that they wish for themselves (see Bolton & Banner, 2012, 94; Gillet, 2012, 242).

Second, in many cases, psychosis includes the deterioration of cognitive abilities, such as difficulty in concentrating and lowered motivation. In the context of the capabilities theory, these can be factors interfering with the fulfillment of human rights and which might be alleviated with suitable psychiatric treatment (see Kuosmanen, 2009, 11). It must be mentioned that it is unclear to what extent such difficulties are caused by the psychosis, as opposed to life crises and stigmatization, or to the undesired effects of psychiatric medication. Antipsychotic medication can lower the person's motivation and the ability to feel longing and pleasure (see Gøtzsche, 2015; Kapur, 2003; Whitaker, 2016).

From the point of view of the capabilities theory, a tendency to psychotic delusions and hallucinations can also be seen as a problem, one that will require significant personal struggle, drain energy, and narrows possibilities for choice. Even if the persons themselves consider, for example, the voices they hear as symptoms of a mental health disorder to which they should pay no attention, their "voices" may occasionally become so strong, and communicate about themes that seem so significant, that they put extraordinary strain on the person (see Gillett, 2012, 242; Romme & Escher, 2010, 22–24). Some people with a tendency to have delusions can also refrain from developing new ideas in order to avoid delusional thinking. For example, John Nash is said to have avoided politically oriented thinking after learning to identify and to be aware of his tendency toward paranoid thinking (Nasar, 1998, 353, 356; Radden, 2011, 127–128;). If someone's delusions and hallucinations have been related to religion they may feel the need to put themselves at a distance from anything religious in order to stay sane (see, e.g., cases presented by Iso-Koivisto, 2004, 85, 91). In these ways, mental health difficulties can become an obstacle to a person continuing to live as an adherent to a persuasion of a religious or political nature, and as a person who develops new thoughts.

It must be noted that psychotic experiences are not unequivocally and solely negative and capabilities-reducing experiences. Some people perceive that during periods of psychosis they become more aware of their life and its meaning. Sometimes psychotic experiences are life-enriching. They can also be positive crises that direct the person to see the meaninglessness of his or her earlier life, and to make choices that lead in new directions (Fulford & Radoilska, 2012; Iso-Koivisto, 2004, 84; Kapur, 2003, 13, 18; Roberts, 1991;).

Experiences of psychoses can therefore, in some cases, also add to cognitive liberty understood as a capability or set of capabilities. This does not necessarily mean that people should be encouraged to go through psychoses. Understanding the plurality of the psychosis experiences nevertheless helps us to see that psychoses can be something else, besides just experiences that are solely bad, to be avoided, and immediately treated to eliminate them (Stenlund, 2014, 277–279; 2017a). A wider approach to psychoses can also shed light on which kinds of treatments and support are seen as sensible and possible. The perspective in the treatment can focus on the quality and the meaningfulness of the person's life, instead of only observing symptoms and trying to control them.

What Abilities Should the Forum Internum *Protect?*

It seems that different conceptions of freedom of belief and opinion protect different things, especially when it comes to the *forum internum* dimension of these freedom rights or in other words, cognitive liberty. Whereas negative liberty primarily protects the contents of a person's mind, the viewpoint emphasizing authenticity is interested especially in whether a belief and thought process that led to it has originated from the self. When viewed from the perspective of the capabilities approach, the focus is on the abilities of the person.

When the *forum internum* dimension, i.e., cognitive liberty, is examined, especially from the perspective of the rights of psychotic people, the capabilities approach seems the most reasonable. When the focus is on the abilities of the person we can see that the *forum internum* dimension protects something crucial, simultaneously avoiding the carelessness of the negative understanding of freedom and the looming threat of totalitarianism from the authenticity point of view, where freedom is restricted in the name of freedom (see Stenlund, 2017a).

However, what kinds of abilities the *forum internum* dimension protects requires clarification. It would seem that the protection includes, at least, those cognitive abilities that are connected to competency. It would violate the *forum internum* if such abilities of the person would be destroyed in psychiatric treatment or in other settings. Also, emotional life could, at least for some parts, be included within the sphere of protection of the *forum internum*. From the capabilities perspective it can be argued that actions that irreversibly destroy the person's ability to believe and to think, and their ability to a rich emotional life, are absolutely forbidden and against human rights. For example, some brain surgical "treatments" (the so-called lobotomy procedure, for example), fortunately are no longer among the treatments used in modern psychiatry and can be considered as contrary to absolute human rights (Stenlund, 2014, 305–310; 2017b).

Additionally, the capabilities approach makes it possible to assess the risks of other psychiatric treatments and the side-effects they pose to capabilities. For example, the undesired effects on thought and affective life are an important point of consideration, and psychiatric treatment should not be pursued at all possible cost. The labeling of patients as mentally ill, and the relatively few opportunities for such patients to impact their society are, in the capabilities approach, key topics for discussion regarding the freedom of belief and opinion and its core area—cognitive liberty (see Stenlund, 2017a).

Acknowledgements This article was written based on a Finnish-language article published in the book '*Vapaa mieli: Uskonnon- ja mielipiteenvapaus mielenterveyden järkkyessä*' (Free Mind: Freedom of Religion and Opinion when Mental Health is Shaken) published by the Asiantuntijaosuuskunta Mielekäs cooperative. I would like to thank Mr. Juho Kunsola, who translated into English the article used as the base for this chapter.

References

Beauchamp, T. L., & Childress, J. F. (1989). *Principles of biomedical ethics*. Oxford University Press.

Berlin, B. (2005). *Liberty* (pp. 169–170). Hardy, H. (ed). Oxford University Press.

Bolton, D., & Banner, N. (2012). Does mental disorder involve loss of personal autonomy? In L. Radoilska (Ed.), *Autonomy and mental disorder* (pp. 77–99). Oxford University Press.

Brison, S. J. (1996). The autonomy defence of free speech. *Ethics, 108*, 312–339.

Bublitz, J. C. (2013). My mind is mine!? Cognitive liberty as a legal concept. In E. Hildt & A. G. Franke (Eds.), *Cognitive enhancement: An interdisciplinary perspective* (pp. 233–264). Springer.

Council of Europe. (2004). Recommendation No. Rec (2004)10 of the Committee of Ministers to members States concerning the protection of the human rights and dignity of persons with mental disorder and its Explanatory Memorandum. https://www.coe.int/t/dg3/healthbioethic/Activities/08_Psychiatry_and_human_rights_en/Rec(2004)10%20EM%20E.pdf. Accessed May 2, 2018.

DSM-V, Diagnostic and Statistical Manual of Mental Disorders. (2013). Fifth edition. American Psychiatric Publishing.

Dworkin, R. A. (1985). *A matter of principle*. Harvard University Press.

Evans, C. (2001). *Freedom of religion under the ECHR*. Oxford University Press.

Feinberg, J. (1973). *Social philosophy*. Prentice Hall.
Fulford, K. W. M., & Radoilska, L. (2012). Three challenges from delusion for theories of autonomy. In L. Radoilska (Ed.), *Autonomy and mental disorder* (pp. 44–74). Oxford University Press.
Gillett, G. (2012). How do I learn to be me again? Autonomy, life skills, and identity. In L. Radoilska (Ed.), *Autonomy and mental disorder* (pp. 233–251). Oxford University Press.
Glover, J. (2003). *Towards humanism in psychiatry, the tanner lectures on human values*. Princeton University, February 12–14, 2003. http://tannerlectures.utah.edu/_documents/a-to-z/g/glover_2003.pdf. Accessed May 2, 2018.
Gosden, R. (1997). Shrinking the freedom of thought: How involuntary psychiatric treatment violates basic human rights. *Monitors: Journal of Human Rights and Technology*, 1(Feb). http://web.archive.org/web/20030603222242.html.http://www.hri.ca/doccentre/docs/gosden.html Accessed May 2, 2018.
Guignon, C. (2004). *On being authentic*. Routledge.
Gutheil, T. G. (1980). (1980) In search of true freedom: Drug refusal, involuntary medication, and "Rotting with Your Rights On." *American Journal of Psychiatry, 137*(3), 327–328.
Gøtzsche, P. C. (2015). *Deadly psychiatry and organised denial*. People's Press.
ICCPR (International Covenant on Civil and Political Rights). (1966). United Nations. https://www.ohchr.org/en/professionalinterest/pages/ccpr.aspx.
Iso-Koivisto, E. (2004). "Pois sieltä, ylös, takaisin" – ensimmäinen psykoosi kokemuksena. Diss, Turun yliopisto.
Kaltiala-Heino, R. K., Korkeila, J., Tuohimäki, C., & Tuori, T. (2000). Lehtinen V (2000) Coercion and restrictions in psychiatric inpatient treatment. *European Psychiatry, 15*(3), 213–219.
Kapur, S. (2003). (2003) Psychosis as a state of aberrant salience: A framework linking biology, phenomenology, and pharmacology in Schizophrenia. *American Journal of Psychiatry, 160*(1), 13–23.
Kuosmanen, J. (2009). Personal liberty in psychiatric care: Towards service user involvement. Turun yliopiston julkaisuja, Sarja D: 841. Diss, Turun yliopisto.
Lönnqvist, J., Moring, J., Henriksson, M. (2014). Hoitoon ohjaaminen. In Lönnqvist, J. et al. (eds.), *Psykiatria*. Duodecim.
MI Principles (Principles for the protection of persons with mental illness and the improvement of mental health care), A/RES/46/119, 75th plenary meeting, December 17, 1991.
Mielenterveyslaki (Mental Health Act) 1990/116, 2001/1423. Unofficial translation. www.finlex.fi/en/laki/kaannokset/1990/en19901116.pdf. Accessed May 2, 2018.
Nasar, S. (1998). *A beautiful mind*. Faber and Faber Limited.

Nowak, M. (1993). *U.N. Covenant on civil and political rights, CCPR commentary*. N.P Engel Publisher.

Nussbaum, M. C. (2006). *Frontiers of justice. Disability, nationality, species membership*. The Belknap Press of Harvard University Press.

Nussbaum, M. C. (2011). *Creating capabilities. The human development approach*. The Belknap Press of Harvard University Press.

Oshana, M. (2007). (2007) Autonomy and the question of authenticity. *Social Theory & Practice, 33*(33), 411–429.

Partch, K. J. (1981). Freedom of conscience and expression, and political freedoms. In L. Henkin (Ed.), *The international bill of rights* (pp. 209–245). New York, Columbia University Press.

Radden, J. (2011). *On delusion*. Routledge.

Rainey B, Wicks E, Ovey C (2014) *Jacobs, White & Ovey. The European convention on human rights* (6th ed.). Oxford University Press.

Roberts, G. (1991). Delusional belief systems and meaning in life: A preferred reality? *British Journal of Psychiatry, 159*(suppl.14), 19–28.

Romme, M., & Escher, S. (2010). *Making sense of voices. A guide for mental health professionals working with voice-hearers*. Mind Publications.

Scanlon, T. (1972). (1972) A theory of freedom of expression. *Philosophy and Public Affairs, 1*(2), 204–226.

Sen, A. (1999). *Development as freedom*. Oxford University Press.

Sen, A. (2009). *The idea of justice*. Allen Lane.

Stenlund, M. (2013). Is there a right to hold a delusion? Delusions as a challenge for human rights discussion. *Ethical Theory and Moral Practice, 16*(4), 829–843.

Stenlund, M. (2014). Freedom of delusion. Interdisciplinary views of freedom of belief and opinion meet the individual with psychosis. Diss, The University of Helsinki. http://urn.fi/URN:ISBN:978-952-10-9747-8. Accessed May 2, 2018.

Stenlund, M. (2017a). The freedom of belief and opinion of people with psychosis: The viewpoint of the capabilities approach. *International Journal of Mental Health, 46*(1), 18–37.

Stenlund, M. (2017b). Promoting the freedom of thought of mental health service users: Nussbaum's capabilities approach meets values-based practice. *Journal of Medical Ethics*. Online first, August 9, 2017.

Stenlund, M. (2018). Oikeuksia, ei vastakkainasettelua. – Vapaa mieli: Uskonnon- ja mielipiteenvapaus mielenterveyden järkkyessä. *Asiantuntijaosuuskunta Mielekkään julkaisuja*. BoD. 9–17.

Stenlund, M., & Slotte, P. (2018). *Forum Internum* revisited: Considering the absolute core of freedom of belief and opinion in terms of negative liberty, authenticity, and capability. *Human Rights Review*. Online first, June 12, 2018.

Szazs, T. (1990). Law and psychiatry: The problems that will not go away. *Journal of Mind and Behavior, 11*(3–4), 557–563.

Szasz, T. (2008). *Psychiatry: The science of lies*. Syracuse University Press.

Tahzib, B. G. (1996). *Freedom of religion or belief. Ensuring effective international legal protection*. Martinus Nijhoff Publishers.

Uudempaa maailmaa toivoo Joni (2018). Silmänreikiä tynnyrissä. – Vapaa mieli: Uskonnon- ja mielipiteenvapaus mielenterveyden järkkyessä. Asiantuntijaosuuskunta Mielekkään julkaisuja. BoD. 87–95.

Whitaker, R. (2016). The case against antipsychotics. A review of their long-term effects. *Mad in America*. https://www.madinamerica.com/wp-content/uploads/2016/07/The-Case-Against-Antipsychotics.pdf. Accessed September 20, 2017.

CHAPTER 9

Technology Against Technology: A Case for Embedding Limits in Neurodevices to Protect Our Freedom of Thought

Andrea Lavazza

New Threats to Freedom of Thought and Conscience

The defence of freedom of thought and conscience has always involved the possibility of expressing one's ideas without constraints or limitations, as well as the possibility of living according to one's convictions, manifested in certain behaviors like the publication of books or the celebration of religious rites—or, more recently, the publication of content and messages on the Internet. The implicit premise to this conception is that every human being can develop and entertain thoughts and beliefs, modify them, and mature the will to share or express them, or to keep them secret. Consequently, this is one of the most precious abilities we have, to which we attach great value. We all use this capacity differently within our own minds, so to speak. We can be prevented from manifesting

A. Lavazza (✉)
Centro Universitario Internazionale, Arezzo, Italy

M. J. Blitz et al. (eds.), *The Law and Ethics of Freedom of Thought, Volume 1*, Palgrave Studies in Law, Neuroscience, and Human Behavior,
https://doi.org/10.1007/978-3-030-84494-3_9

our thoughts, but no one can force us not to think what we want or to change our minds.

In this chapter, I will focus on defining the key concepts (this section) and the new neuroscientific technologies and devices which are capable, or are taken to be capable in the future, to literally read our mind/brain (Sect. "Neuroscience Crossing the Final Frontier"). In Sect. "New Technologies that Influence Cognitive Processes and Mental Contents", I will explain how digital technologies and devices can influence cognitive processes and mental contents. In particular, the risk exists that, when exposed to similar stimuli, the "common brain" or "collective brain" of the human being may end up conforming and becoming very similar from individual to individual. In this sense, digital technology may not be as neutral as some theories suggest. In Sect. "The Need for and Right to Cognitive Freedom", I will provide a definition of mental integrity and explain why there is a need for and a right to cognitive freedom. In Sect. "Using Technology as a Defence Against Technology Itself", I will argue how we can try to defend mental integrity, namely by stating that functional limitations should be incorporated into any devices capable of interfering with mental integrity, and I will give examples of it and arguments in order to justify that rule.

Some examples taken from literary classics clearly introduce the idea of freedom of thought. In a 1634 work entitled *Comus*, the poet John Milton tells the story of a young noblewoman ("The Lady") who utters these well-known words: "Thou canst not touch the freedom of my mind," signaling that, whatever an individual might suffer, she is able to safeguard her own freedom of thought, which cannot be affected by any external assault. If the body can be subject to the control of other people, the inner self—which has a long history in the Western tradition (but not only), from Socrates through Augustine of Hippo to the present day—can neither be accessed nor bound by others. And Jorge Luis Borges wrote in a short story of his (*The Secret Miracle*, 1943) about a prisoner who is about to be killed by the Nazis and before being shot manages to mentally write the play he was thinking about. Again, our mind is the only place where our enemies cannot enter.

Another example is a famous text by Henry David Thoreau dated 1849, which illustrates the resistance that can be opposed to the influence of the state thanks to one's freedom of thought, even if one is subject to the full physical domination of a stronger authority.

"I have paid no poll-tax for 6 years. I was put into a jail once on this account, for one night; and, as I stood considering the walls of solid stone, two or three feet thick, the door of wood and iron, a foot thick, and the iron grating which strained the light, I could not help being struck with the foolishness of that institution which treated me as if I were mere flesh and blood and bones, to be locked up. I wondered that it should have concluded at length that this was the best use it could put me to, and had never thought to avail itself of my services in some way. I saw that, if there was a wall of stone between me and my townsmen, there was a still more difficult one to climb or break through before they could get to be as free as I was. I did not for a moment feel confined, and the walls seemed a great waste of stone and mortar. I felt as if I alone of all my townsmen had "paid my tax" and "The plainly did not know how to treat me, but behaved like persons who are underbred. In every threat and in every compliment there was a blunder; for they thought that my chief desire was to stand the other side of that stone wall. I could not but smile to see how industriously they locked the door on my meditations, which followed them out again without let or hindrance, and they were really all that was dangerous. As they could not reach me, they had resolved to punish my body; just as boys, if they cannot come at some person against whom they have a spite, will abuse his dog. I saw that the State was half-witted, that it was timid as a lone woman with her silver spoons, and that it did not know its friends from its foes, and I lost all my remaining respect for it, and pitied it. Thus, the State never intentionally confronts a man's sense, intellectual or moral, but only his body, his senses. It is not armed with superior wit or honesty, but with superior physical strength" (Thoreau, *Resistance to Civil Government*).

In the situation described by Thoreau, the state's repressive apparatus could not do much against the ideas that supported civil disobedience: those were rooted in the individual's mind, where the state cannot reach.

This representation of freedom of thought is certainly idealized even compared to the past. In fact, it is known that the way one has been raised, the schools one has attended, external pressures, implicit or explicit social conditioning, restrictions on access to different ideas, up to actual indoctrination, all influence one's freedom of thought and conscience, by restricting its perimeter, so to speak. We cannot always rely on an open, pluralistic and tolerant environment where rationality and well-founded information prevail. The fact remains, however, that even in non-ideal conditions, the mind has always been (and still is in many ways) the place where freedom of thought can generally be exercised without the direct interference of others. As George Orwell famously wrote in his novel *1984*: "Nothing was Your Own Except the Few Cubic Centimeters Inside Your Skull."

But today, at the beginning of the third decade of the twenty-first century, things are changing. There are three elements to consider that are causing the change and make it no longer true that "Thou canst not touch the freedom of my mind," or that the brain is always "our own." The first element is conceptual, although it is linked to scientific progress. The other two are related to technological progress: neurosciences, and algorithms and digital devices that implement them.

Let's start with the first element. The medical-scientific knowledge that has accumulated since the nineteenth century has meant that the distinction between mind and brain has slowly eroded, arriving at the identification of the first with the second. As Nobel Prize winner Francis Crick famously wrote, "'You', your joys and your sorrows, your memories and your ambitions, your sense of personal identity and free will, are in fact no more than the behaviour of a vast assembly of nerve cells and their associated molecules" (Crick, 1984). In this sense, *Descartes' Error*, as the equally famous book by Antonio Damasio (Damasio, 1994) is titled, seems to have finally been corrected, at least in the opinion of most neuroscientists and intellectuals, even though the majority of people still seem to have dualistic intuitions. What Descartes' mistake was is well-known: to consider the mind as an entity distinct from the body (or the brain) from an ontological (dual substances) and, consequently, qualitative point of view. Mind and brain, for him, had different characteristics, some of which were related to the privacy and inaccessibility of the self, which remains outside the reach of both science and other individuals.

With the loss of credibility of this metaphysical conception (which, however, has not been fully falsified: consider the difficulties of providing a scientific account of phenomenal consciousness; Bayne et al., 2009), everything we thought was typical of our "thinking mind" has been attributed to the electrochemical functioning of our brain. From romantic love to mental illness, from the ability to resist temptation to deviant behavior, from the ability to compute to artistic creativity, it all depends on our cerebral makeup (Gazzaniga, 2009). And the brain is primarily accessible and "manipulable" by medicine: this fact has made it possible to offer tremendous relief for many people suffering from anxiety, depression, or schizophrenia. But the brain is also open to other types of stimulation in order to improve or limit its activity (Khan & Aziz, 2019; Lavazza, 2019a).

That is why today we are facing the challenges of human enhancement (Clarke et al., 2016; Lavazza, 2019b) and so-called neuro-interventions, namely interventions which in one way or another operate directly on the brain of a subject as a possible way to prevent offenders from engaging in the future criminal activity (cf Ryberg, 2019). In fact, the conceptual twist at play is that the brain does not need to be protected from prying eyes, so to speak. Conceived as mind/brain, as the physical organ that processes the mental contents, by its very nature, the brain is neither protected nor protectable, precisely because everything passes through it. And in the age of technology not to "open" and connect one's brain means condemning oneself to isolation, disease, lack of job opportunities and relationships, preventive suspicion of increasingly advanced surveillance systems, and so forth (Zuboff, 2018).

The other two elements that must be considered are related to the technology available today, whose progress is accelerating exponentially (at least in its current phase). The first element in this respect is given by the advances in applied neuroscience that go beyond the attempt to modify the brain by acting on its outputs, such as treating a disease with a drug that is effective even if we do not know exactly how it works, or improving creativity with a non-invasive brain stimulation that is an unspecific and still controversial technique. Indeed, what we are beginning to do today is "read people's minds," to use an evocative expression. To put it more clearly, it is now possible to achieve a "thought apprehension" through sophisticated tools that can be external or internal to the brain. And this is a decisive step in relation to the presumed inviolability of our mind as the last refuge of freedom of thought (Meynen, 2019).

The second element related to technology is that of the digital devices that carry most of the information we refer to in our lives. Not only is the power of computers growing, but we are also witnessing two phenomena resulting from the interaction between programmers and commercial companies on the one hand and new possibilities offered by algorithms, increasingly intelligent and capable of handling huge amounts of data, on the other (Schneider, 2019). As we shall see, this type of technology is not neutral, but tends to "capture" the user, both in terms of pleasantness of the activity and in terms of the creation of a cognitive and informative environment that strongly conditions the user, to the point of threatening their "internal" freedom of thought. One might think that these "attention capturing devices" are not different from other older pleasant activities and that we use them voluntarily but, as I'll explain below, they are different in nature as, unbeknownst to us, they exploit specific mechanisms which were out of reach for the older pastimes.

It is therefore time to look closely into the impact of technology on our cognitive freedom. I will try to explain why this impact should not go unnoticed, and why we should grant citizens a new form of freedom of thought and conscience, giving them the tools to fight against possible attacks and unintentional threats to which their freedom of thought may now be subjected.

Neuroscience Crossing the Final Frontier

"Guess what I'm thinking?"—"You cannot understand the pain I'm going through right now"—"I can't tell how she's feeling about that"—"What is going through the mind of a person who commits a massacre?" These are just some of the classic questions and statements that show the opacity of our mental states, even to those closest to us. Today, several philosophers, psychologists, and cognitive scientists, in the wake of a refined behaviorism that we can trace back to Gilbert Ryle (1949), are skeptical about the vividness and precision of our autobiographical experience as we grasp it through introspection. They argue that, in fact, we ourselves have only access to our own "third-person" mental states (cf. Carruthers, 2013). To the purpose of this discussion, however, we can overlook this radical vein of research and focus on some examples of how new neuroimaging technologies are able to cross the final frontier of our mind/brain, namely what has been called "thought apprehension" (Meynen, 2019). I am not interested here in providing a precise chronology of this undertaking. A few examples of very recent experiments will suffice to give a picture of this rapidly evolving situation.

A recent medical achievement is linked to the possibility of better diagnosing patients in a state of altered consciousness, by communicating with them despite the impossibility of manifest forms of expression of thought. When it comes to disorders of consciousness, the rate of misdiagnosis is approximately 40%, and new methods are required, especially if the patient's capacity to show behavioral signs of awareness is diminished. For example, Monti and colleagues (2010) used functional magnetic resonance imaging (fMRI) to assess every patient's ability to generate willful, neuroanatomically specific, blood-oxygenation-level–dependent responses during two established mental-imagery tasks. As the authors of this pioneering study explain, a technique was developed to determine whether such tasks could be used to communicate yes-or-no answers to simple questions. Of the 54 patients enrolled in the study, 5 were able to willfully modulate their brain activity. In three of these patients, additional bedside testing revealed some sign of awareness. One patient was able to use the technique to answer yes or no to questions during functional MRI. To answer yes, he was told to think of playing tennis, a motor activity. To answer no, he was told to think of wandering from room to room in his home, visualizing everything he would expect to see there, creating activity in the part of the brain governing spatial awareness.

These results show that a small proportion of patients in a vegetative or minimally conscious state are capable of brain activation which reflects a degree of awareness and cognition. Careful clinical examination will result in the reclassification of the state of consciousness in some of these patients. This technique may thus be useful in establishing basic communication with patients who appear to be unresponsive. As is evident, this is an extraordinary step forward for the well-being of those patients who are literally trapped in their own bodies, without the possibility of communicating with their surroundings. But at the same time, it is the demonstration that one can literally and with very good precision "read a person's mind." Non-clinical examples indicate that the path is open for even more advanced applications.

To make another example, Mason and Just (2016) used fMRI to assess neural representations of some concepts of physics (momentum, energy, etc.) in students majoring in physics or engineering. The goal, in their words, was to identify the underlying neural dimensions of these representations. Using factor analysis to reduce the number of dimensions of activation, they obtained four physics-related factors that were mapped to sets of voxels. The four factors were interpretable as causal motion visualization, periodicity, algebraic form, and energy flow. The individual concepts were identifiable based on their fMRI signatures with a mean rank accuracy of 0.75 using a machine-learning (multivoxel) classifier. Furthermore, "there was commonality in participants' neural representation of physics; a classifier trained on data from all but one participant identified the concepts in the left-out participant (mean accuracy = 0.71 across all nine participant samples). The findings indicate that abstract scientific concepts acquired in an educational setting evoke activation patterns that are identifiable and common, indicating that science education builds abstract knowledge using inherent, repurposed brain systems" (Mason & Just, 2016).

What's more, the clinical assessment of suicidal risk appears to be substantially complemented by a biologically based measure that assesses alterations in the neural representations of concepts related to death and life in people who engage in suicidal ideation. The study by Just and colleagues (2017) used machine-learning algorithms to identify such individuals (17 suicidal ideators versus 17 controls) with high (91%) accuracy, based on their altered fMRI neural signatures of death-related and life-related concepts. The most discriminating concepts were "death," "cruelty," "trouble," "carefree," "good," and "praise." A similar classification accurately (94%) discriminated nine suicidal ideators who had made a suicide attempt from eight who had not. "Moreover, a major facet of the concept alterations was the evoked emotion, whose neural signature served as an alternative basis for accurate (85%) group classification. The study established a biological, neurocognitive basis for altered concept representations in participants with suicidal ideation, which enables highly accurate group membership classification" (Just et al., 2017).

Is it also possible to predict the freely chosen content of voluntary imagery from prior neural signals? Koenig-Robert and Pearson (2019) have shown that the content and strength of future voluntary imagery can be decoded based on activity patterns in visual and frontal areas well before participants engage in voluntary imagery. In this study, participants chose which of two images to imagine. Using functional magnetic resonance (fMRI) and multi-voxel pattern analysis, the authors decoded the imagery content up to 11s before the voluntary decision was made in the visual, frontal, and subcortical areas. The ability of decoding this imagery in the visual areas, in addition to the generalization of perception-imagery, suggested that predictive patterns correspond to visual representations. "Importantly, activity patterns in the primary visual cortex (V1) from before the decision, predicted future imagery vividness. The results suggest that the contents and strength of mental imagery are influenced by sensory-like neural representations that emerge spontaneously before volition" (Koenig-Robert & Pearson, 2019). Finally, brain imaging may predict individual mental traits and behavioral dispositions from data through machine-learning approaches and this can raise very relevant ethical issues, as such approaches hold the potential to gain substantial influence in fields such as human resource management, education, or criminal law (Eickhoff & Langner, 2019).

On a different note, it is worth underlining that in October 2019, patients with severe opioid addiction were given brain implants to help reduce their cravings, in the first trial of its kind in the US and possibly in the world (Wakefield, 2019). A young man who has struggled with substance abuse for more than a decade, with many relapses and overdoses, underwent the surgery as the first patient to be ever treated in this way. Lead doctor Ali Rezai described the device as a "pacemaker for the brain." "It starts with a series of brain scans. Surgery follows with doctors making a small hole in the skull in order to insert a tiny 1 mm electrode in the specific area of the brain that is supposed to regulate impulses such as addiction and self-control. A battery is inserted under the collarbone, and brain activity will then be remotely monitored by the team of physicians, psychologists and addiction experts to see if the cravings recede" (Wakefield, 2019). Over the next two years, the patients will be closely monitored. So-called deep brain stimulation (DBS) has been approved by the US Food and Drug Administration for treating a range of conditions including Parkinson's disease, epilepsy, and obsessive–compulsive disorder. Some 180,000 people around the world have brain implants, but this is the first time DBS has been approved for drug addiction.

Dr. Rezai told the *BBC News*: "Addiction is complex, there are a range of social dynamics at play and genetic elements and some individuals will have a lack of access to treatments so their brains will slowly change, and they will have more cravings. I think it is very good for science and we need more science to advance the field and learn more about the brain. This is not for augmenting humans and that is very important. This is not a consumer technology." Yet, even if this intervention is specifically clinical, one can easily imagine its possible applications to different scenarios. If the cerebral pacemaker worked, it would mean that complex and multifactorial aspects of our behavior can be directly manipulated in a rather simple technical way by means of an electronic device (once the electrode has been implanted). And once this is established, it would not be so strange for the criminal system to require a person convicted of a violent crime to undergo such an intervention so as to control any new aggressive impulses. And a criminal system commanded by an undemocratic regime or by governments that do not respond to public opinion could use this route to try to control not only inappropriate violence, but a number of other mental states, including explicit and active dissent.

This framework also involves private companies, which try to use new technologies for both clinical uses (often as a justification for research that is barely ethically acceptable) and commercial uses. Neuralink, for example, has applied to start human trials in the US in which to insert electrodes into the brains of patients with paralysis. In this respect, scientists have already created devices capable of both interpreting brain activity and stimulating neurons in the brain (Martin, 2019; Samuel, 2019). A demonstration of the technology was carried out in 2012 when paralyzed patients were able to control a robotic arm. Elon Musk has said that, apart from treating neural conditions such as Parkinson's, he hopes that Neuralink could one day facilitate a "symbiosis" between humans and AI. He also announced that the company had successfully got a monkey to "control a computer with its brain," and that Neuralink hopes to start human testing very soon.

A key feature of Neuralink's system is the sheer number of electrodes it plans to implant via its "sewing machine," in which a stiff "insertion needle" rapidly shoots thin-film polymer probes containing arrays of electrodes into the brain. A brain-computer link could go in both directions, both recording neural activity and stimulating it, even though such devices seem not to be in view for the next few years. In any case, these are advanced technologies capable of making the human–machine interaction increasingly easier, with the aim of enhancing the cognitive and operational capacities of the human being, but with the consequence of making the human being itself transparent to the machine, with all the uncertainties that follow.

Along the same line, Facebook is supporting research into a headset able to transcribe words at a rate of 100 per minute, starting from the user's thoughts. In this case, assuming that such a project is really feasible, the certainly fascinating objective of doing away with manual typing or vocal dictation, which takes much longer the rapid flow of thought, exposes our thinking in words to the possibility of being recorded, stored on the servers of the corporation and eventually spread or used in unauthorized forms. Of course, this already applies to emails written by us, but the headset envisioned by Facebook would put us at risk of recording even things that we would not want to put in writing.

Yet, another example is the new project recently announced by the U.S. Defense Advanced Research Projects Agency (DARPA), which "has officially funded a program to come up with a brain-machine interface — in the form of a headset designed to let military personnel control anything from 'active cyber defense systems' to 'swarms of unmanned aerial vehicles' through brain activity alone" (Tangermann, 2019). This form of merging between soldier and weapon is particularly sensitive, as it implies that the soldier himself becomes part of the weapon system, thereby losing his prerogatives of private thinking during a war action, as everything must be shared and coordinated for the effectiveness of the mission and the safety of all participants.

Finally, recent developments suggest that "a stable, secure, real-time system" may be created that would allow for interfacing the cloud with the human brain. One promising strategy for enabling such a system, denoted as a "human brain/cloud interface" ("B/CI"), would be based on technologies referred to as "neuralnanorobotics." "A specialized application might be the capacity to engage in fully immersive experiential/sensory experiences, including what is referred to here as 'transparent shadowing' (TS). Through TS, individuals might experience episodic segments of the lives of other willing participants (locally or remotely)" (Martins et al., 2019).

Also, artificial neurons that mimic the way our body's nerve cells transfer electrical signals could one day help patients with nerve damage, but this silicon chips, once implanted, could be easily used to directly monitor, record, influence, or even hijack our brain activity (Abu-Hassan et al., 2019). The whole field of brain-computer interfaces raises relevant ethical issues quite overlooked so far (Hendriks et al., 2019; Klein et al., 2016; Lázaro-Muñoz et al., 2018).

This development seems to involve the crossing of yet another threshold, since the fusion of subjective experiences, however, fascinating and enriching it may be, implies by definition the end of the privacy and cognitive freedom that everyone can currently enjoy (cf Martins et al., 2019). As already mentioned, the point I want to defend here is not that one should restrict research or prohibit all discoveries that could potentially threaten our cognitive freedom. Rather, I want to highlight the risks posed by new technologies.

New Technologies that Influence Cognitive Processes and Mental Contents

Neuroscientist Lamberto Maffei, who has studied neuronal plasticity for many years, has underlined the risk that, when exposed to similar stimuli, the "common brain" or "collective brain" of the human being may end up conforming and becoming very similar from individual to individual. Let's have a look at his argument, based on the development mechanisms of the nervous system (Maffei, 2014).

The structural and functional similarities between brains far outweigh the differences: this allows us to speak of such a thing as the "human brain" as opposed to the brain of a particular individual. But from the study of the functional and structural variations of the brain both during fetal development and in adult life, we know that the nervous system and the cortex in particular are very capable of "reprogramming" (neuroplasticity). During a period immediately after birth, the potential for changes in function and structure is very high. In the early years of life, the nervous system can very easily modify, under appropriate conditions, the connections and size of some structures, given also that the number of neurons and synapses is much greater in early childhood than in adulthood. In adult life, however, experience refines and specializes nerve connections and functions, with moderate possibilities for change.

In this sense, the way in which the stimuli—the messages—are received by the human subject is particularly important. Neurophysiology suggests that in order to be effective, such stimuli must be repeated, pass through a neurologically powerful sensory channel such as the visual one, and be connected to other messages that have emotional value or are biologically relevant to survival, such as food or sex. Maffei's thesis (2014) is that the great plasticity of the nervous system, in today's world, could potentially become the instrument of a "mental imprisonment" that could, in turn, lead to the "globalized brain." In fact, equal experiences are likely to produce similar or equal changes in the brain (or at least at a cortical level), as confirmed by a large number of experiments on both animal models and human beings. If all individuals were subjected to exactly the same motor, sensory, and "cultural" stimulation in general, there would be similar brain development and modification. Similar but not identical, of course, because the underlying genetic variability, in terms of alleles, would make it impossible to have the same exact response to the same stimuli.

Something similar concerning science and scientific discoveries has been stated by Geman and Geman (2016), according to whom an excess of communication, and the presence of the Internet itself, has produced group thinking that reduces creative independence. "It may not a coincidence – they write – (...) that two of the most profound developments in mathematics in the current century (...) were the work of iconoclasts with an instinct for solitude and, by all accounts, no particular interest in being "connected."

Now, to some extent, the intersubjective similarity of brain development is useful for the formation of a society and for increasingly large groups of individuals to be able to understand each other easily and to interact effectively. In this sense, globalization in its classical sense is positive, and the spread of digital technologies has also contributed to it. An excess of uniformity, though, might be created as a result of the repeated and massive use of social media. This effect of creating a "globalized" brain is not caused by the sharing of increasingly similar contents—for example, a certain pop culture in its Western synthesis produced by the few large entertainment corporations—but by the functioning mechanisms and communication logics of digital platforms.

In fact, the phenomena related to the digital world—the predominance of the visual mode, the rapidity of interaction, the emotional coloring of the communication, the possibility of easily accessing enormous amounts of content and switching effortlessly from one to another, the focus on specific themes and approaches on the basis of previously expressed preferences—may end up conditioning the basic information at our disposal, our way of processing it cognitively and our capacity of judgment. According to Maffei (2014), this, if taken to its extreme forms, would risk threatening our cognitive freedom, since we would not be aware of being strongly limited by a technical apparatus that does not formally exert any form of coercion on us, but still has an enormous power to direct our thinking—a power that probably no other apparatus has ever possessed in the history of humankind. In the past, critical philosophy, for example, by the Frankfurt School, stated that the culture industry conditioned our mind and deployed new modes of oppression. We don't have to share those criticisms to acknowledge that new technologies might realize what Frankfurt theorists feared.

As has already been pointed out, everyone's cognitive freedom, in a broad sense, is limited by the fact that one grew up in a certain cultural environment, which may sometimes be very restricted. This, however, does not prevent one from dissociating oneself from it, as so many individual cases show: for example, political dissidents born, raised, and educated within totalitarian states; people who reject the religion of their group, even if it is the basis of all rules and accepted behavior; or geniuses in the fields of art and science who spent their first years of life in culturally deprived families. The real risk of conformity therefore comes from something else, something which we are all exposed to, sometimes for many hours every day: digital technology.

Indeed, digital technology may not be as neutral as some theories suggest: it might instead have a special force of attraction that should be made explicit, offering the possibility of countering its most invasive and potentially damaging effects. The most important one is not so much the uniformity of a "globalized" brain, but the "narrowing" of the mind caused by the prevalence of cognitive automatisms favored by the type of technology in use, which can trigger a form of dependence due to the brain rewards it produces. In this sense, it is neuroscience itself that provides the creators and managers of computer platforms and apps with the knowledge they need to "capture" the brains of their users and, as an unintended consequence, to eventually render them severely limited in their ability to adopt different cognitive styles (Horwitz et al., 2021).

This framework also includes the debate on the interpretation of the relationship between technology and human behavior (Fasoli, 2019, 2020). "Technological instrumentalism" suggests that technological artifacts are "simple means" (Heersmink, 2015; Pitt, 2014), which have no influence on us and are therefore not subject to ethical evaluation. At the opposite end of the continuum lies the "deterministic" conception of technology, which instead conceives the latter as capable of determining human behavior. Instrumentalism presupposes a strong decision-making autonomy on the part of the users and a fundamental inability of the artifacts to interfere with these processes. If technological instrumentalism were true, we would have no reason to worry about what technologies we own and use, nor how they are built, because they would be irrelevant and the responsibility of our behavior and perceptions would fall solely on us: in this case, we would be "free to use every object as we please, and it would entirely depend on us." Technological determinism, in its standard formulation, seems instead to attribute a strong power to technological objects and low decision-making autonomy to the human being, who in this perspective passively "undergoes" the influence of technology.

The post-phenomenological theory of technological mediation, instead, places itself in an intermediate position (Ihde, 1990; Latour, 1994; Verbeek, 2015). This position differs from both determinism and instrumentalism and denies some assumptions regarding the presumed neutrality of technology. According to the theory of technological mediation, artifacts constitute mediators in the relationship between human beings and the world. As such, they shape our actions, experiences, and practices and should not be conceived as mere tools. When we look at a tree with an infrared camera, for example, many aspects of the tree that are visible to the naked eye are lost, but at the same time, a new feature of the tree becomes visible: a subject can thus see if the tree is healthy (Verbeek, 2006).

In fact, some uses of digital technology exploit human cognitive mechanisms in such a way as to favor the deterministic thesis (understood as a theory with a limited radius, that is, limited to specific fields and specific situations). In general, instead, a mid-range position, such as that of technological mediation, may seem more plausible. Some examples taken from Fasoli (2019) can help illustrate this point. Some perceptual illusions to which we are subject are exploited for the design of web pages, as in the case of the "infinite newsfeed." Experimental evidence (Wansik et al., 2005) shows that, by altering the perceptual references related to a subject's consumption, it is possible to push them to consume more than they would in a standard condition. Based on this, web designers have created web pages that never end because, when the user scrolls down, new contents are continuously inserted. By concealing the visual signals that usually allow one to monitor the time spent on a page, i.e., by imperceptibly moving the sidebar higher and higher, and by always inserting new content, the user is encouraged to spend more time on the platform.

The suggestions created by algorithms, instead, exploit a different mechanism, linked to the fact that human beings often do not have well-defined preferences when making a choice (Ariely, 2008). In these cases, if the website intervenes by suggesting alternatives, the user's choice can be easily affected. This is the case with the suggestions that are provided to users when formulating queries in search engines. A similar case is the default position. The more complex the technological artifacts are and the more numerous the parameters that can be modified by users, the greater the probability that the default setting will not be changed by them. This is because this operation requires both a cognitive effort and digital skills.

In this sense, the designer's choice of a particular default setting will have a major impact on the way in which their product will be used by most users. For example, if an instant messaging application uses by default a visual clue (such as a double blue tick) to signal when a message has been read, users are implicitly encouraged to keep that setting.

One can therefore describe "technological prescriptiveness" (Fasoli, 2019) as the ability of an artifact to modify users' perceptions and to stimulate or discourage certain behaviors through the provision of affordances and functions, as well as through designs that exploit specific cognitive-behavioral phenomena. Through design, in fact, it is possible to do many things: to create a superstimulation, to make a choice salient, to exploit a bias, to randomize a reward, to deceive the perceptual system, to introduce a cost or a reward, to create new affordances, to make affordances perceptible or not, to exploit default options, etc. Therefore, in most cases, this type of prescriptiveness is not deterministic, since behavior is not caused but stimulated by exploiting specific neurobiological phenomena (such as dopamine circuits that are sensitive to the reward given, e.g., by likes or retweets) and cognitive processes, in a way that the "guided" subject is often unaware of.

In light of what has emerged so far, the prescriptiveness of technological artifacts arises first of all from the various affordances that technological artifacts offer us. Using and, in some cases, just owning (i.e., having at our disposal) different technologies means accepting these artifacts—which are not mere tools—and the ways in which they shape our perception of the world, of others and of ourselves, also by stimulating different behaviors with variable degrees of persuasion. In a second phase, a different prescriptive capacity of technological artifacts arises from the application of behavioral techniques and design choices, which are effective to the extent that they intercept certain innate propensities and reactions of our brain. In this sense, prescriptiveness seems to come about as a consequence of an effective coupling between our basic cognitive architecture and the structure of certain technological artifacts.

All this is further highlighted by the fact that the main and most visited platforms (both in terms of numbers of users and in terms of user time spent on them), from Facebook to Twitter, from Instagram to TikTok to Amazon, have become as centralized as the media infrastructures of the last century. This is due to the scale economies of those who manage them and the possibility they offer of engaging in effective advertising and even surveillance (cf. Lovink, 2019). All in all, therefore, it can be said that the platforms largely induce users to proceed along the beaten track, so to speak. And this confirms the risk pointed out by Maffei of eventually producing an increasingly cognitively uniform "collective" brain, with a consequent and corresponding loss of cognitive freedom.

The Need for and Right to Cognitive Freedom

What has been described so far are actual or potential dangers for our cognitive freedom, understood as the possibility of elaborating one's own thoughts autonomously, without interference, and of revealing them totally, partially, or not at all on the basis of a personal decision. Cognitive freedom—the contemporary version of freedom of thought and conscience—seems to have both intrinsic value and instrumental value (Farina & Lavazza, 2021). In this sense, it deserves special protection, not only for the individual but for society as a whole. I have introduced elsewhere the term "mental integrity" (Lavazza, 2018), which seems to be a suitable way to extend the concept I want to express here ("mental integrity" is also used in Article 3 of the European Charter of Fundamental Rights). In the normative sense, one could adopt the following definition:

> Def 1: Mental Integrity is the individual's mastery of their mental states and brain data so that, without their consent, no one can read, spread, or alter such states and data in order to affect the individual in any way.

This definition broadens the one offered by Ienca and Andorno (2017), by which the right to mental integrity "should provide a specific normative protection from potential neurotechnology-enabled interventions involving the unauthorized alteration of a person's neural computation and potentially resulting in direct harm to the victim." The definition I propose addresses both the issue of privacy and that of cognitive freedom, which Bublitz defined as "the right to alter one's mental states with the help of neurotools as well as to refuse to do so" (Bublitz, 2013). Obviously, as said above, minds/brains are not closed-off, and we are exposed to many forms of classic Pavlonian conditioning (e.g., in advs which pair two stimuli in order to convince us to buy an item). In this sense, we are all commonly "conditioned," and one might ask if that violates mental integrity. I take here mental integrity to be specifically targeted by new (neuro)technologies and in need of special protection, but also an ideal value which can be defended from invasive forms of classic conditioning as well.

My point is that privacy, understood as the secrecy of one's brain data and mental contents, is key to a free conduct, because autonomy is exercised not only in public but also in private. Being spied on through mind-reading inevitably reduces the subject's autonomy in the Kantian sense: the subject is thus limited in self-imposing their own norms of conduct, as they would not be in a condition free of external pressure, which happens when you are being observed without your consent. This can only be avoided by keeping one's thoughts private: what follows from such privacy violation is therefore an inappropriate imposition on the individual, even in the absence of direct harm.

The definition I propose also grasps another aspect: as previously noted, even brain data apparently unrelated to conscious contents or cognitive processes (mental states) can help predict one's behavior. This is, especially relevant in light of the fact that recent interpretations of brain activity are emphasizing its predictive character. In particular, it has been argued that our brains are similar to prediction machines, capable of anticipating the incoming streams of sensory stimulation before they arrive (Clark, 2016; Hohwy, 2013). If the brain is more than a response machine and if actions are more than responses to stimuli, but are a way of selecting the next input, then knowing the present state of the brain (understood as a prediction machine) can say a lot about an individual and their future behavior—much more than one would have thought within a different paradigm of brain functioning.

Now, one may wonder why mental (cerebral) integrity should be granted special value compared to other aspects, and whether the right to integrity is relative or absolute. Mental integrity is the basis for freedom of thought as it was classically conceived, before the era of neurotechnological pervasiveness (Shen, 2013). It is the first and most important freedom that the individual must be granted in order to have all the other freedoms that are usually considered relevant. As already stated, in the Western tradition, the inner life of the individual—the one that no one can see and with which no one can interfere—has always been considered the most precious and intangible resource of the human being. Personal autonomy seems to directly follow from freedom of thought.

In many of its declinations—albeit not all of those proposed in different cultures—human flourishing feeds on freedom of thought understood as a "private repository" where nothing and no one can intrude without the subject's consent. In fact, it has been claimed that "the right and freedom to control one's own consciousness and electrochemical thought processes is the necessary substrate for just about every other freedom" (Sententia, 2004). In this sense, cognitive freedom is an extended form of freedom of thought that "takes into account the power we now have, and increasingly we will have, to monitor and manipulate cognitive function" (*Ibid.*). In other words, cognitive freedom includes and highlights the technological features of the new devices which can exploit and interfere in new ways with our mind/brain functioning.

In his defence of cognitive freedom, Bublitz (2013) distinguishes between three "interrelated but not identical dimensions." These are: (1) the liberty to change one's mind or to choose whether and by which means to change one's mind; (2) protection against interventions into other minds to protect mental integrity; and (3) the ethical and legal obligation to promote cognitive liberty. At this point, it becomes clear that cognitive freedom is a fundamental dimension for the human being, both as a last remaining living space when all the others have been lost for various reasons (think of Milton's Lady and the quotation from Orwell, mentioned in Section "New Threats to Freedom of Thought and Conscience") and as a means of individual flourishing thanks to the possibility of coming into contact with different information and non-standardized or pre-constituted styles of thought.

This also seems to respond to a general feature of our evolutionary history from both a biological and a cultural point of view. In fact, from a biological standpoint, mutations generate new types of individuals and some of these changes offer a selective advantage in terms of survival or reproduction. The increase of mutant types in environments where this competitive advantage appears leads advantageous mutants to eventually replace the previous types. Genetic variability is extremely important for the preservation of the species. In fact, given that adaptation to a changing environment depends on the availability of beneficial mutations that can be useful to cope with the new life conditions of the organism, the more numerous the mutant types, the greater the chances of survival and reproduction of the species in question. And the same can be said to apply to cultural evolution, as Cavalli Sforza has convincingly shown (Cavalli Sforza & Feldman, 1981).

In this sense, we have a strong pragmatic motivation, which we could call consequentialist, to guarantee individual cognitive freedom. In fact, the latter allows everyone to access various and different cognitive contents and styles: this condition permits the development of a greater range of ideas and existential paths, which in turn can increase the chances of well-being and personal flourishing. This is also based on the assumption that knowledge is distributed, and no one can ever master all the best available solutions to the problems that everyone has to face (Hayek, 1945). In addition, this also applies to society understood as a group of individuals who live together and interact. Indeed, societies, as well as their individual members, are always faced with new challenges posed by both the natural environment and the human environment itself. And cognitive diversity is a fundamental resource in order to have more tools and more solutions available, as is the case with genetic diversity in the history of biological evolution (Charlesworth & Charlesworth, 2017).

It therefore seems that cognitive freedom is a decisive tool for individual and social prosperity in the broad sense. The inefficiency and consequent crisis of societies and countries that have sought to severely restrict the freedom of thought of their members illustrates why repressing freedom of thought and standardizing cognitive styles generally lead first to impoverishment and then to a reaction from below aimed to restore that freedom. However, a consequential justification of cognitive freedom is open to exceptions and partial and temporary limitations, which could be introduced precisely in order to achieve—in a more efficient and effective way—the same aims as those to which cognitive freedom is believed to lead. For example, the benefits in terms of the economic development of web platforms could be considered preferable to a greater potential freedom of cognitive styles (see Sect. "New Technologies that Influence Cognitive Processes and Mental Contents"). Alternatively, at a time of strong social turbulence that threatens the stability of a country, the legitimate government might think it better to regulate the dissemination of certain information or ideas, for the good of all.

In this sense, we also need a deontological justification of the right to cognitive freedom. This justification is not subject to exceptions and is based on what has already been said about the mind as the last sanctuary in which the individual can defend their identity and autonomy from external interference. We can also agree on this right if we carry out a rational reflection that takes into account the concrete subject in society. In fact, we know that not everyone enjoys the highest degree of autonomy and self-determination oriented to their own flourishing, whether for obvious cognitive limits or for lack of exposure to suitable external stimuli. This implies that forms of education, instruction, guidance and coercive regulation, as already implemented in our societies, are neither ethically illicit nor violate people's cognitive freedom. Beyond these forms of intervention, however, the mind must remain an inner space that no one can violate.

As I have been illustrating, today, possible uses of neuroscientific advances and the massive spread of digital technology bear unprecedented threats to cognitive freedom. For this reason, as already explained, it seems that we need renewed protection against this massive intrusion into our cognitive freedom and its consequent limitation. This can be achieved first of all by highlighting the potential risks at play and formulating a relevant ethical framework, as has been done so far in this chapter. However, in a completely new technological context, it seems that it is not enough to establish rights. We therefore need an approach that literally incorporates these rights into the artifacts that can potentially interfere with our cognitive freedom.

Using Technology as a Defence Against Technology Itself

Elsewhere (Lavazza, 2018), I have proposed a technical principle for the protection of mental integrity with regards to neuroprostheses. However, this same principle can be extended to all devices that can perform analysis on, or interfere with, the activity of our brains, such as those described in Sect. "Neuroscience Crossing the Final Frontier".

> Def2: The technical principle for the protection of mental integrity is a functional limitation that should be incorporated into any devices capable of interfering with mental integrity (as defined in Def1). Specifically, new neural devices should (a) incorporate systems that can find and signal the unauthorized detection, alteration, and diffusion of brain data (and brain functions as far as possible); (b) be able to stop any unauthorized detection, alteration, and diffusion of brain data (and brain functions as far as possible). This should not

only concern individual devices, but act as a general (technical) operating principle shared by all interconnected systems that deal with decoding brain activity.

For example, neural devices in use could incorporate a mind-reading device detector. This would prevent people from being secretly subjected to, or deceived about, the use of devices capable of threatening the right to mental/brain integrity (Ienca & Haselager, 2016). Think of neural prostheses for cognitive enhancement: in the future, some people might be obliged to use them because of their profession (airplane pilots, surgeons, etc.) and might want to protect themselves against privacy violation or brain manipulation (Santoni de Sio et al., 2014). In addition, brain-altering devices could be equipped with various usage thresholds, each only activated with a different access key, depending on the professional profiles of the users: laboratory technicians, doctors, medical investigators... In this way, the most invasive practices would be only available to a few people that would have been selected over time for their professional skills and ethical integrity, as a safeguard against possible abuse.

As for new neural devices and prostheses (Lebedev et al., 2011)—such as tools for the treatment of consciousness disorders, direct brain-to-brain communication, networks composed of many brains, artificial parts of the brain replacing damaged circuitry and improving the existing one, and even the transfer of brain content and functions to an artificial carrier—it should be made possible to extract brain data only by means of special access keys managed exclusively by the subjects under treatment or by their legal representatives. Some of these devices could be specifically aimed at brain alteration for rehabilitation or other medical purposes, and likewise they should only be made accessible to professionals in charge of their correct use, which should be able to be monitored by a specific authority upon the subject's request. Think, for example, of new portable devices such as the fNIRS instrument for mobile NIRS-based neuroimaging, neuroergonomics, and BCI/BMI applications (von Lühmann et al., 2015).

As for the military or anti-terrorism use of such devices, it would be desirable to stipulate an international treaty like those concerning antipersonnel mines, cluster bombs, or even chemical weapons. These treaties establish the obligation of non-production of such weapons in peacetime. In fact, if such weapons were readily available during a war, countries would likely be tempted to use them. If, however, the production of those weapons had been interrupted long before, it would be much more difficult to resort to them. Similarly, if mind-reading or brain-altering devices were available without mechanisms preventing their improper use, some would likely want to use them for military purposes. This would be very tempting, because the point would not be to threaten anyone's safety, but only to temporarily violate their mental privacy and/or integrity in order to defend the community—say, from a terror attack. However, if unregulated mind-reading devices (that is, without the limits imposed by the technical principle proposed in Def 2) were not available in the first place, it would be much more difficult to resort to them.

Notwithstanding all this, in some medical, legal, and military cases, the right to mental integrity may seem to compete with other fundamental rights, as a result of which the principle of protection of mental integrity may be occasionally bypassed (e.g., in cases of life or death). On the other hand, as already stated, relativizing the right to mental integrity risks weakening it, making it less important and much easier to violate thanks to neurotechnological progress. Consider the positive purposes for which neural devices are generally built in the first place: such goals might be partly hindered by the technical principle of the protection of mental integrity. For example, prostheses able to signal and/or prevent epileptic crises might also be able to do the same for violent outbursts. However, these positive goals should only be pursued if the subject in question has expressed full informed consent on the matter, so that they may explicitly authorize the violation of their mental integrity and, accordingly, of their cognitive freedom (which is based on the former) operated by a device that automatically interferes with their (more or less voluntary and conscious) actions.

Even in this case, though, it might be argued that the individual's freedom, autonomy, and intrinsic value can only be expressed when the subject enacts positive behaviors, without being constrained by some automated device. In other words, if a person suffering from epilepsy may very well give their informed consent to treatment, being fully aware of its pros and cons, the issue seems a lot more complex when it comes to a violent subject. For example, the latter may find themselves forced to choose between staying in prison or accepting a neural prosthesis controlling their violent drives; but it must be kept in mind that violence is sometimes needed to defend oneself or others from a threat, so that having all violent drives automatically controlled at a neural level might prove to be highly dysfunctional and damaging to the individual.

Other cases may involve different levels of consensus and collective security. For example, access to the EEG data of a vehicle driver could allow a built-in tool to detect the neuronal activation pattern that leads to decreased attention while driving (Biondi & Skrypchuk, 2017). The purpose of avoiding serious car accidents could be considered superior to the driver's right not to undergo the constant monitoring of their brain states. Following the same line of thought, protecting the population from terror attacks could result in introducing compulsory "neural" control mechanisms in order to find potential terrorists. In this case, many people's right to mental integrity would be violated, even if most of them would probably have no malicious goals whatsoever. Generally speaking, it might be argued that the best solution would be to let judges decide on a case by case basis whether or not to authorize the violation of the right to integrity. This is a general provision which obviously does not resolve all the legal problems and ethical dilemmas related to freedom violations by the State for crime- or terrorism-fighting purposes. I tend to consider that such cases should be very few, for a pragmatic reason at least. Whereas we know perfectly how mental integrity and freedom are violated by brain-monitoring, we are rarely sure of the results we can obtain by it.

Finally, one may wonder how the need to protect the fundamental right to mental integrity may coexist with the need to violate the related technical principle. Well, in this case, the increasingly sophisticated neurotechnological techniques might help. In fact, if all neural devices were produced according to the technical principle for the protection of mental integrity, it would be more complicated to use them in violation of this fundamental right, even if for a socially positive goal. For example, if the secret codes to control neural devices or prostheses were available to an external authority, this would reduce the risk of potential breaches. Also, the authorization process of any "improper but necessary" use of such prostheses—in cases related to collective security—would be more difficult and rigorous. One can also support a more rigid position, so as to further limit the possibility of mental integrity violations. For example, one can argue that, once implanted, these devices should belong to, and only be "controlled" by, the subject, even in case of maintenance and reprogramming. However, the subject will always need physicians and experts to implant and use the device or prosthesis in question. The problem therefore does not seem to have an easy solution.

The underlying idea is that technology is so pervasive that rules and sanctions are not enough. This is because the technologies described are generally positive in their purpose and are appreciated by users, who seek them out and tend to use them significantly. Moreover, it is very difficult to detect possible abuses from the outside. In this sense, it is necessary to provide users with countermeasures already incorporated in the devices themselves. There is certainly a risk of an "arms race" at play here, which might be expensive and slow down the development and market launch of some devices. But this seems to be an acceptable price for the protection of cognitive freedom as it has been characterized in this chapter.

As to the second type of danger, the one related to the standardization of cognitive styles and to exposure to the same content due to the massive use of centralized web platforms, the forms of protection of cognitive freedom and mental integrity should be different from those described so far. However, the rationale for preventive protection measures should be the same. In this case, automatic and non-removable alerts will be needed to indicate, for example, the amount of time spent on the platform (starting from the first hour). Other alerts should make it explicit that all links or product suggestions result from an algorithm based on our previous choices, and there should also be a button redirecting to a casual selection with respect to the search we have initiated. Furthermore, the functions that exploit brain reward systems, from "likes" and all the appreciation signals on social media to references to the most viewed pages, should be reported as an option to be selected at the beginning of every session and should feature easy alternatives, such as "possibility of receiving negative feedback" and "unguided navigation."

In general, all the default options adopted by the platforms should be highlighted and explained, offering a few simple different settings. In addition, it should be possible to anonymize, at least partially, one's navigation so as not to be profiled and controlled through our actions on the Web—as is the case, for example, with forms of mass control such as the Chinese social credit system (cf. Kobie, 2019). The latter, indeed, also assigns points based on the digital lives of users, rewarding and punishing them according to the strict rules of conduct imposed by the government. The system is inspired by a trade-off between freedom and security: sacrificing a certain amount of privacy and autonomy would achieve greater predictability of interactions and would result in a more orderly society, with less deviant behavior. In reality, it seems that this is primarily the way to encourage conformism, contributing to the standardization of cognitive styles and available information, that is, the denial of freedom of thought and conscience typically imposed by authoritarian or totalitarian regimes. The decentralization of digital systems and platforms could thus also prevent a decrease in the cognitive freedom of users.

Finally, it should be stressed that there is no explicit constraint at play in the digital world, yet this does not seem sufficient to increase the technical literacy of users. On the one hand, general competence seems to be diminishing and a deeper understanding of it seems to require a specialization that only few can achieve; on the other hand, the more complex and refined the machines, the more we like them, the less we know how to understand them in their hidden logic. The solution of using technology against invasive technology thus appears as the most appropriate way to defend cognitive freedom.

Conclusion

As we have seen, technological progress related to neuroscientific knowledge and to the development of digital devices and networks, which generally constitutes a positive advance for both human flourishing and economic growth, may also involve a number of risks in terms of cognitive freedom (Inglese & Lavazza, 2021). This happens, on the one hand, because of possible misuse of thought apprehension devices by misinformed or malicious people. On the other hand, this is due to the massive immersion of users in digital platforms that tend to standardize content and cognitive styles. Faced with these risks, which may well turn into real threats, it is important to affirm the right to cognitive freedom and mental integrity as a fundamental human right.

The justification of this right has both a consequentialist and a deontological component. The novelties brought about by technology and its reasonably foreseeable further advancements lead to the conclusion that the defence of the right to cognitive freedom cannot be achieved only by means of ethical codes, laws, and relative sanctions. The proposal put forward here is that technology itself should be used preventively to provide users, in real time, with information about attempts to restrict their cognitive freedom and thus be able to intervene or at least choose whether and how to protect themselves. Codes and laws should therefore focus on the rules of construction and operation of devices and digital platforms, ensuring that all of them, used for any purpose and under any circumstance, incorporate "defense systems" such as those described.

This approach is certainly not free of difficulties and legal and technical complications, which is why it certainly should be further investigated and better developed in its theoretical and applicative details. However, it seems to be a promising way to protect our cognitive freedom in an era in which technology opens up great opportunities for us to expand the boundaries of that freedom but, for the first time, also proves capable of overcoming the last barrier that protects the deepest and most precious sanctuary of human thought, conscience, and autonomy.[1]

References

Ariely, D. (2008). *Predictably irrational*. HarperCollins.

Abu-Hassan, K., Taylor, J. D., Morris, P. G., Donati, E., Bortolotto, Z. A., Indiveri, G., Paton, J. F. R., & Nogaret, A. (2019). Optimal solid state neurons. Nature. *Communications, 10*(1), 1–13.

Bayne, T., Cleeremans, A., & Wilken, P. (Eds.). (2009). *The oxford companion to consciousness*. Oxford University Press.

Biondi, F., & Skrypchuk, L. (2017). Use your brain (and light) for innovative human-machine interfaces. In I. Nunes (Ed.), *Advances in human factors and system interactions* (pp. 99–105). Springer.

Bublitz, J. C. (2013). "My mind is mine!? cognitive liberty as a legal concept". In E. Hildt & A. G. Franke (Eds.), *Cognitive enhancement. An interdisciplinary perspective* (pp. 233–264). Springer.

Carruthers, P. (2013). *The opacity of mind: An integrative theory of self-knowledge*. Oxford University Press.

Cavalli Sforza, L. L., Feldman, M. D. (1981). *Cultural transmission and evolution: A quantitative approach*. Princeton University Press.

Charlesworth, B., & Charlesworth, D. (2017). *Evolution: A very short introduction*. Oxford University Press.

Clark, A. (2016). *Surfing uncertainty. Prediction, action, and the embodied mind*. Oxford University Press.

Clarke, S. Savulescu, J., Coady, C. A. J., Giubilini, A. Sanyal, S. (Eds.). (2016). *The ethics of human enhancement: Understanding the debate*. Oxford University Press.

Crick, F. (1984). *The astonishing hypothesis: The scientific search for the soul*. Scribner.

[1] I'd like to thank the two editors of the book for their insightful and helpful comments on an early version of this chapter.

Damasio, A. (1994). *Descartes' error: Emotion, reason, and the human brain*. Putnam.

Eickhoff, S. B., & Langner, R. (2019). Neuroimaging-based prediction of mental traits: Road to utopia or Orwell?. *PLoS Biology, 17*(11), e3000497.

Farina, M., & Lavazza, A. (2021). The meaning of Freedom after Covid-19. *History and Philosophy of the Life Sciences, 43*(1). https://doi.org/10.1007/s40656-020-00354-7

Fasoli, M. (2019). Cacciatori (di informazioni) e prede (di trappole cognitive) nel web 2.0: Una lettura cognitivo-evoluzionista dell'attrattività dei social network. *Sistemi Intelligenti, XXXI, 3*, 395–412.

Gazzaniga, M. S. (2009). *Cognitive neurosciences*. The MIT Press.

Geman, D., & Geman, S. (2016). Opinion: Science in the age of selfies. *Proceedings of the National Academy of Sciences, 113*(34), 9384–9387. https://doi.org/10.1073/pnas.1609793113

Hayek, F. (1945). The use of knowledge in society. *The American Economic Review, 35*(4), 519–530.

Hendriks, S., Grady, C., Ramos, K. M., Chiong, W., Fins, J. J., Ford, P., Goering, S., Greely, H. T., Hutchison, K., Kelly, M. L., Kim, S. Y. H., Klein, E., Lisanby, S. H., Mayberg, H., Maslen, H., Miller, F. G., Rommelfanger, K., Sheth, S. A., & Wexler, A. (2019). Ethical challenges of risk, informed consent, and posttrial responsibilities in human research with neural devices: A review. *JAMA Neurology*. https://doi.org/10.1001/jamaneurol.2019.3523

Heersmink, R. (2015). Extended mind and cognitive enhancement: Moral aspects of cognitive artifacts. *Phenomenology and the Cognitive Sciences, 16*(1), 17–32.

Hohwy, J. (2013). *The predictive mind*. Oxford University Press.

Horwitz, J., et al. (2021). The Facebook files. A Wall Street Journal investigation. https://www.wsj.com/articles/the-facebook-files-11631713039?mod=article_inline

Ienca, M., & Andorno, R. (2017). Towards new human rights in the age of neuroscience and neurotechnology. *Life Sciences, Society and Policy, 13*, 5.

Ienca, M., & Haselager, P. (2016). Hacking the brain: Brain–computer interfacing technology and the ethics of neurosecurity. *Ethics and Information Technology, 18*, 117–129.

Ihde, D. (1990). *Technology and the lifeworld: From garden to Earth*. Indiana University Press.

Inglese, S., & Lavazza, A. (2021). What Should We Do With People Who Cannot or Do Not Want to Be Protected From Neurotechnological Threats? *Frontiers in Human Neuroscience*. https://doi.org/1510.3389/fnhum.2021.703092

Just, M. A., Pan, L., Cherkassky, V. L., McMakin, D. L., Cha, C., Nock, M. K., & Brent, D. (2017). Machine learning of neural representations of suicide and emotion concepts identifies suicidal youth. *Nature Human Behaviour, 1*(12), 911.

Khan, S., & Aziz, T. (2019). Transcending the brain: Is there a cost to hacking the nervous system?. *Brain Communications, 1*(1), fcz015.

Klein, E., Goering, S., Gagne, J., Shea, C. V., Franklin, R., Zorowitz, S., Dougherty, D. D., & Widge, A. S. (2016). Brain-computer interface-based control of closed-loop brain stimulation: Attitudes and ethical considerations. *Brain-Computer Interfaces, 3*(3), 140–148.

Kobie, N. (2019). *The complicated truth about China's social credit system.* Wired, https://www.wired.co.uk/article/china-social-credit-system-explained

Koenig-Robert, R., & Pearson, J. (2019). Decoding the contents and strength of imagery before volitional engagement. *Scientific Reports, 9*(1), 3504.

Latour, B. (1994). On technical mediation. *Common Knowledge, 3*(2), 29–64.

Lavazza, A. (2018). Freedom of thought and mental integrity: The moral requirements for any neural prosthesis. *Frontiers in Neuroscience, 12*, 82.

Lavazza, A. (2019a). Thought Apprehension: The "True" Self and The Risks of Mind Reading. *AJOB Neuroscience, 10*(1) 19–21. https://doi.org/10.1080/21507740.2019.1595784

Lavazza, A. (2019b). The Two-Fold Ethical Challenge in the Use of Neural Electrical Modulation. *Frontiers in Neuroscience.* https://doi.org/1310.3389/fnins.2019.00678

Lázaro, G., Yoshor, D., Beauchamp, M. S., Goodman, W. K., & McGuire, A. L. (2018). Continued access to investigational brain implants. *Nature Reviews Neuroscience, 19*(6), 317–318.

Lebedev, M. A., Tate, A. J., Hanson, T. L., Li, Z., O'Doherty, J. E., Winans, J. A., Ifft, P. J., Zhuang, K. Z., Fitzsimmons, N. A., Schwarz, D. A., Fuller, A. M., An, J. H., & Nicolelis, M. A. L. (2011). Future developments in brain-machine interface research. *Clinics, 66*, 25–32.

Lovink, G. (2019). *Sad by design: On platform nihilism.* Pluto Press.

Maffei, L. (2014). *La libertà di essere diversi. Natura e cultura alla prova delle neuroscienze.* il Mulino.

Martin, N. (2019). *Elon Musk is making microchips to link your brain to your smartphone.* Forbes, https://www.forbes.com/sites/nicolemartin1/2019/07/17/elon-musk-is-making-microchips-to-link-your-brain-to-your-smartphone/

Martins, N. R., Angelica, A., Chakravarthy, K., Svidinenko, Y., Boehm, F. J., Opris, I., Lebedev, M. A., Swan, M., Garan, S. A., Rosenfeld, J. V., Hogg, T., & Freitas, R. A., Jr. (2019). Human brain/cloud interface. *Frontiers in Neuroscience, 13*, 112.

Mason, R. A., & Just, M. A. (2016). Neural representations of physics concepts. *Psychological Science, 27*(6), 904–913.

Meynen, G. (2019). Ethical issues to consider before introducing neurotechnological thought apprehension in psychiatry. *AJOB Neuroscience, 10*(1), 5–14.

Monti, M. M., Vanhaudenhuyse, A., Coleman, M. R., Boly, M., Pickard, J. D., Tshibanda, L., Owen, A., & Laureys, S. (2010). Willful modulation of brain activity in disorders of consciousness. *New England Journal of Medicine, 362*(7), 579–589.

Pitt, J. C. (2014). "Guns Don't Kill, People Kill"; Values in and/or Around Technologies. In P. Kroes & P.-P. Verbeek (Eds.), *The moral status of technical artefacts* (pp. 89–101). Springer.

Ryberg, J. (2019). *Neurointerventions, crime, and punishment: Ethical considerations*. Oxford University Press.

Ryle, G. (1949). *The concept of mind*. University of Chicago Press.

Samuel, S. (2019). *Brain-reading tech is coming. The law is not ready to protect us*. Vox, https://www.vox.com/2019/8/30/20835137/facebook-zuckerberg-elon-musk-brain-mind-reading-neuroethics

Santoni de Sio, F., Faulmüller, N., & Vincent, N. A. (2014). How cognitive enhancement can change our duties. *Fro ntiers in Systems Neuroscience, 8*, 131.

Schneider, S. (2019). *Artificial you: Ai and the future of your mind*. Princeton University Press.

Sententia, W. (2004). Neuroethical considerations: Cognitive liberty and converging technologies for improving human cognition. *Annals of the New York Academy of Sciences, 1013*, 221–228.

Shen, F. X. (2013). Neuroscience, mental privacy, and the law. *Harvard Journal of Law and Public Policy, 36*, 653–713.

Tangermann, V. (2019). *Darpa wants soldiers to control drones with their thoughts*. Futurism, https://futurism.com/the-byte/darpa-brain-machine-interface-drone

Verbeek, P. P. (2006). Materializing morality: Design ethics and technological mediation. *Science, Technology, & Human Values, 31*(3), 361–380.

Verbeek, P. P. (2015). Beyond interaction: A short introduction to mediation theory. *Interactions, 22*(3), 26–31.

von Lühmann, A., Herff, C., Heger, D., & Schultz, T. (2015). Toward a wireless open source instrument: Functional near-infrared spectroscopy in mobile neuroergonomics and BCI applications. *Frontiers in Human Neuroscience, 9*, 617.

Wakefield, J. (2019). Brain implants used to fight drug addiction in US. *BBC News*, https://www.bbc.com/news/technology-50347421

Wansink, B., Painter, J. E., & North, J. (2005). Bottomless bowls: Why visual cues of portion size may influence intake. *Obesity Research, 13*(1), 93–100.

Zuboff, S. (2018). *The age of surveillance capitalism: The fight for a human future at the new frontier of power*. Public Affairs.

CHAPTER 10

Varieties of (Extended) Thought Manipulation

J. Adam Carter

Introduction

It is uncontroversial that the rise of the cognitive sciences, broadly construed, has had a significant impact on how we understand how humans think and behave. Robust sets of neurobiological and psychological findings concerning human cognitive processes have both challenged orthodox positions in, and raised new questions for, the disciplines of economics, philosophy, politics, and beyond.

A central platitude in legal and political philosophy, and which lies at the heart of many democratic constitutional systems, is that all individuals enjoy—in slogan form—the *freedom of thought*. Even if your actions are constrained by laws, your capacity to exercise your own mind as you wish is not equally constrained.

J. A. Carter (✉)
University of Glasgow, Glasgow, UK
e-mail: Adam.Carter@glasgow.ac.uk

M. J. Blitz et al. (eds.), *The Law and Ethics of Freedom of Thought, Volume 1*, Palgrave Studies in Law, Neuroscience, and Human Behavior,
https://doi.org/10.1007/978-3-030-84494-3_10

Kant ([1797] 1991) famously committed himself to this idea by defining the scope of juridical laws so as to exclude them from applying to the mind, insisting that juridical laws apply only to 'external actions.'[1] Other philosophers, like Mill ([1859] 1998), have defended the freedom of thought by pointing to the disutility of its absence: the suppression of opinion thwarts a community's capacity to discover and maintain the truth.[2]

Outside of philosophy, a defence of the freedom of thought is enshrined explicitly in Article 18 of the Universal Declaration of Human Rights, which ensures that 'everyone has the right to freedom of thought, conscience and religion.' Elsewhere, in U.S. Constitutional legal scholarship, it is lauded by Supreme Court Justice Oliver Wendel Holmes as the principle that 'most imperatively calls for attachment.'

But even if the existence of a freedom so described is not controversial, things get thorny quickly when we zero in on what constitutes a plausible *violation* of it. This is especially so when we distinguish what is involved in violating one's freedom as pertains to (i) *expression of thought*; versus (ii) *the thought itself*. We can easily conceive of what it takes to violate (i) by looking to egregious examples of such violations—e.g., political persecution of minority opinions as expressed through religious and political demonstration and speech.

Question: But what would it be, exactly, to *violate* one's *freedom to simply form and possess her own thoughts* as opposed to express them, and to violate this freedom non-trivially? (A trivial way to violate *any* kind of freedom in thinking, categorically, would be to cause injury to the physical brain, injury to which is already legislated against as a paradigmatic *physical* harm.) Is the freedom to (in short) think as one wishes—at least on those matters on which it is possible when functioning normally to control thought[3]—something that could be violated any *other* way? And

[1] For discussion, see Bublitz (2013, 241).

[2] See, e.g., *On Liberty*, ([1859] 1998, Chap 2).

[3] Even when paradigmatically free, our thinking is not entirely in our control—as philosophers have recognized in denying *doxastic voluntarism*, the view that (in short) we can believe what we desire to believe, and to do so directly without any intermediate steps in thinking. A simple kind of counterexample to doxastic voluntarism concerns perception. If there is a red table in front of you, and you desire to see a blue table and to immediately form the belief < *There is a blue table* > , you will not be able to do it. The denial of doxastic voluntarism is compatible with the thought that you have a kind of indirect control over (some) beliefs about what is true, which can be brought

if not, then did we even need to make this freedom explicit in the first place?

2.

It is tempting to think the answer to these questions is 'no,'[4] given how pervasive the Cartesian picture of the mind, as a kind of private 'inner theatre,' remains in ordinary thought and talk, as well as, implicitly, in legal and political thinking.[5] On the Cartesian view, according to which a thinker alone has privileged and exclusive *access* to the content of her own thoughts, thought itself is *in principle* unregulatable (apart from regulating against physical injury to the brain) and so there would seem to be no point to legislating it in a way that goes beyond regulating physical harm. We could at most, on the Cartesian view, attempt to regulate a thinker's thoughts *indirectly* by regulating (e.g., punishing) the *behavior we take to be evidence of thought*.[6] However, and in line with Kant's thinking, these regulations themselves would be de facto regulations of (e.g., verbal and physical) *behavior*, and not regulations of anything like the shape and character of thought *as such*.

But—as contemporary thinking in the philosophy of mind and cognitive science suggest—Descartes was wrong in (at least) two important ways about the 'inner' nature of the mind. First—as Putnam (1975), Kripke (1980), and Burge (1986) showed in the 1970s and 80s, it is mistaken to think that the content of our thoughts is either (i) transparent to us or (ii) determined solely by the inner workings of the mind. *Content internalism* has since been rejected almost universally for *content externalism*, which holds that the content of our thoughts—viz., what

about by intentionally taking steps to acquire certain kinds of evidence. For some notable discussions of doxastic voluntarism and the philosophical issues surrounding it, see, e.g., Audi (2001), Clarke (1986), and Steup (2000).

[4] Perhaps one exception though is found in debates surrounding indoctrination in the philosophy of education. It's beyond the scope of what I can do to cover this here, but some relevant stances are found in Hand (2002), Gardner (2004), Hansson (2018), and Siegel (2004).

[5] For some discussion on this point, see Carter and Palermos (2016). See also Blitz (2010).

[6] Alternatively, one might indirectly regulate thought by depriving another of information (or tools to generate information). For example, one might indirectly regulate a mathematician's ability to discover a certain result—and thus, to believe that result is true—by depriving her of a pencil and paper. Thanks to John Tillson for noting this other indirect form of thought regulation.

our thoughts are *about*—is at least partly determined by facts about our physical and socio-linguistic environments that might be inaccessible to us on reflection.[7] For example, on this view, when you think about the wet, blue stuff that you see in oceans, whether you are thinking about *water* (which is type identical with H_2O) or about something *else* (as Putnam imagined: 'XYX'—viz., something which is very similar to water but which is not identical to H_2O), depends on what the physical environment you are interacting with is *actually like*, and this is something you might not have reflective access to while entertaining the image of the blue, wet stuff.[8]

More importantly for our purposes, though, a *second* kind of Cartesian doctrine about the mind's inner nature—*cognitive internalism*—has also fallen into disrepute and has increasingly done so over the past 10 years.[9] Whereas *content* internalism concerned the *content* of thoughts—viz.., what your thought counts as being a thought *about*—*cognitive* internalism is a thesis about the kinds of things that *materially realize* cognition, viz., about the kinds of physical processes on which cognition supervenes.

Prior to Clark and Chalmers' landmark paper 'The Extended Mind' (1998), even most *content* externalists in the philosophy of mind were still *cognitive* internalists. They held that although one's physical and socio-linguistic[10] environment can partially determine the content of one's thoughts, only *intracranial* processes—i.e., biological processes that play out in brain—are the sorts of things that can materially 'bring about' cognitive processes like memory, reasoning, perception and the like.

But even this more basic kind of internalism about the mind is falling to the wayside. According to the *hypothesis of extended cognition* (HEC), our assessments of what kinds of things can feature in 'cognitive' process should be guided by common-sense functionalist thinking, rather than by considerations to do with physical make-up or special location. For

[7] See Carter et al. (2014).

[8] The denial of content externalism is closely related to a range of puzzles in the contemporary literature on self-knowledge. For discussion, see, e.g., Gertler (2000, 2010), Parent (2017), McKinsey (1991), and Pritchard (2002).

[9] See, e.g., Clark (2008), Menary (2007), Palermos (2011, 2014b), Wilson (2000, 2004). For criticism, see, e.g., Adams and Aizawa (2008).

[10] See, e.g., Burge's (1986) arthritis/tharthritis example.

example, according to the HEC proponent, if you are using a well-integrated smartphone to *do* what biomemory does—viz., to play the role biomemory normally plays in storing and retrieving information—then to the extent you have dispositional beliefs (i.e., which become occurrent beliefs when retrieved and brought to conscious awareness) stored in biomemory, you also have 'extended' dispositional beliefs stored in your phone's memory, or in the cloud.[11] This might seem radical, but to say otherwise, on this line of thought, commits one to an unprincipled kind of 'bioprejudice'[12] that gives arbitrary weight to material constitution and special location when demarcating the bounds of the cognitive.

3.

Against this background, it should be obvious that the question of what it would be to violate one's freedom of thought—and not *just* her expression of thought in speech and action—can hardly be set aside as moot or purely theoretical. And this is because the latest cognitive science allows beliefs, memories, perceptions,[13] and the like, to be materially realized by processes that *include parts of the world* which themselves are not in principle 'hidden away' in some Cartesian theatre, but subject publicly to various kinds of manipulation by other parties.[14]

And the picture is complicated further when we include recent advancements in, and the potential future of, *brain-computer interface* (BCI) technologies.[15] To make concrete the kinds of possibilities generated by BCIs, consider that in October 2019, a French dentist who had fallen 15 feet walked for the first time in two years using his mind to control an exoskeleton suit. The man who goes by the first name 'Thibault' has implants on his brain that read its activity and send this to a

[11] For discussion, see, e.g., Clark (2010, 2008), Carter and Kallestrup (2016, 2017), Carter and Pritchard (2020), Menary (2010), Pritchard (2010, 2018), and Palermos (2014a).

[12] The use of this term is due to Chalmers (2008).

[13] While HEC is often explained in terms of extended memory processes, the thesis also applies to perception, and this can be illustrated with reference to tactile visual substitution systems (TVSS). See, e.g., Bach-y-Rita (1983), Bach-y-Rita and Kercel (2003), and Palermos (2016).

[14] For a discussion of such violations and their potential legal ramifications, see Carter and Palermos (2016).

[15] For some representative recent developments in BCI technologies for use in cognitive enhancement, see, e.g., Ghafoor et al. (2019), Wang et al. (2019), and Pisarchik et al. (2019).

nearby computer, which in turn uses this information to send instructions to the exoskeleton. The result is that Thibault can, simply by thinking, control the limbs of the exoskeleton in three-dimensional space.[16]

The reasoning of Elon Musk, who has launched BCI start-up Neuralink (and other BCI startups such as BrainCo, Emotiv, Kernel, Mindmaze, NeuroSky, NeuroPro, Neurable, and Pandromics) if that we can use neural implants to communicate with computers in the *therapeutic* case—viz., where the aim is to restore an individual to normal, healthy levels of human functioning in order to correct disease and pathology[17]—why not use it to take already healthy individuals *beyond* normal levels of functioning, especially when it comes to *cognitive* functioning, where more sophisticated BCIs can in principle allow us to not only send but also *receive* information immediately through thought commands.[18]

To make this idea a bit more concrete, think about what you do when you say 'Hey Google/Siri what's the weather today?' Moments later, Google/Siri tells you the answer. Now just imagine streamlining this process. You *think*, rather than verbalize 'What's the weather?' And soon after, perhaps immediately, your brain receives the information[19] from the computer you've just communicated with via a thought command.

I am going to assume from here on in that these kinds of BCI enhancement technologies are worth taking seriously, even if they have not yet arrived fully functional. What is important for philosophical and legal thinking about the freedom of thought is whether we have a clear way to think about how to protect freedom of thought in connection with them when (if) they arrive.

4.

Current international law frameworks recognize three key elements to one's freedom of thought which can be threatened in different ways by

[16] See Carter (2020b, Chap 1) for a recent discussion of this case.

[17] For discussion on the distinction between cognitive enhancement and mere therapeutic cognitive improvements, see, e.g., Bostrom and Sandberg (2009) and Carter and Pritchard (2019).

[18] See Musk (2019).

[19] Different possible BCIs might realize this operation differently, e.g., by prompting content representation and regulating attention via the implant; the assumption should be that these ways will trend in the direction of being increasingly seamless and non-obtrusive as BCI technologies continue to improve in the more distant future.

new technologies.[20] These are, as Susan Alegre (2017, 225) summarizes: (i) the right not to reveal one's thoughts or opinions; (ii) the right not to have one's thoughts or opinions manipulated; (iii) the right not to be penalised for one's thoughts.

The advent of HEC bears directly on (i) and by extension (iii). HEC implies that, to the extent that your mind is partly located (in certain circumstances) in external memory storage, the right you have not to *reveal* your thoughts is a right that extends also to certain protections from inspection of such external storage.[21]

The rise of BCI technologies, by contrast, poses special challenges for (ii), and better understanding these challenges helps to equip us for future thinking about the freedom of thought. Or so I want to argue. Here is the plan for what follows. In Sect. 5, I propose by using illustrative BCI-style cases, a sufficient condition for freedom-of-thought violating 'extended' thought manipulation, viz., thought manipulation that involves some kind of distortion of a thinker's non-biological mental faculties.[22] Once this condition is set out and defended, I will, in the remaining sections, taxonomize four distinct varieties of freedom-of-thought-violating extended thought manipulation which have interestingly different structures, but which all satisfy the proposed sufficient condition.

5.

Let's distinguish two kinds of cases where a thinker might be fitted with a BCI: *pre-arranged* cases and *non-pre-arranged* cases, e.g., where in the latter kind of case, one's being fitted with a BCI is not in accordance with one's past autonomous decisions.[23] For example: a person is unwillingly 'experimented on.'

[20] For discussion, see Alegre (2017).

[21] For an interesting recent take on this idea, see Riley v. California (2014), and in particular, John Roberts' majority opinion on the case, in which he draws comparisons between cell phones and human biological anatomy. http://www.supremecourt.gov/opinions/13pdf/13-132_8l9c.pdf For an overview in the context of the extended cognition debate, see Carter and Palermos (2016).

[22] Note that manipulation is distinct from coercion. For discussion on this difference, see, e.g., Baron (2003), cf., Ghafoor et al. (2019).

[23] I am setting aside for the purposes of discussion here issues to do with thought manipulation via *genetic* enhancement, or by testing and selecting for certain embryos; these cases, while interesting and important, are difficult to address without a foray into questions of personal identity that go beyond what I can cover here.

Non-pre-arranged BCI cases constitute *trivial* violations of one's freedom against thought manipulation, however, such freedom may be plausibly construed; but such cases are covered under the wider class of protections against physical harm and injury. What I want to suggest in what follows is that even *pre-arranged* BCI implementation cases can very easily serve as ones where a thinker's freedom against thought manipulation is violated. Appreciating this has potential practical import in a very possible future in which consenting to BCI fitting is a typical and common form of cognitive enhancement.[24]

More specifically, what I want to propose and then sharpen is the following *sufficient condition* on freedom-of-thought violating thought manipulation:

Thought Manipulation (Sufficiency) (TMS): The right not to have one's thoughts or opinions manipulated is violated if one is (i) caused to acquire non-autonomous propositional attitudes (*acquisition manipulation*) or (ii) caused to have otherwise autonomous propositional attitudes non-autonomously eradicated (*eradication manipulation*).

Regarding the *acquisition manipulation* component of (TMS): a term that needs clarified is that of a *non-autonomous propositional attitude*.[25] Examples of propositional attitudes are beliefs and desires, e.g., your belief that Paris is the capital of France, your desire that you not eat liver for dinner this evening. Following influential work on autonomous attitudes by Al Mele (2001), I am going to assume that, sufficient for a propositional attitude's *not* being autonomous, and thus, not being such that it is properly attributable[26] to the agent, is the conjunction of two conditions:

[24] For an influential defence of the idea that we can expect to increasingly incorporate BCIs, see Clark (2003).

[25] Note that, on the proposed account—which states just a sufficiency condition and not a necessary condition—it's entirely possible that the right not to have one's thoughts or opinions manipulated could be violated *non-propositionally* as well, e.g., via the compromise of faculties or dispositions in such a way as to leave all representational content as is. For the purpose of this paper, I'm keeping my focus on propositional manipulation; an interesting and relevant question for further work concerns the matter of freedom-of-thought manipulation via one's dispositions directly. Thanks to John Tillson for discussion on this point.

[26] I am using attributability here in the sense of Watson (1996) as denoting 'character revealing.' Your striking someone as a result of being pushed into that person is, for example, is not properly attributable to you, as it in no way reveals your character—viz., your stable dispositions of mind.

(i) a *bypass condition*—viz., a condition pertaining to whether the attitude in question was acquired in a way that 'bypassed' the subject's relevant (e.g., cognitive and conative) faculties[27]; and (ii) an *unsheddability condition*—viz., a condition pertaining to whether the subject is able to (easily enough) give up, or at least attenuate the strength of, the relevant attitude.[28] Regarding the *eradication manipulation* component of (TMS). To unpack this further, say that an otherwise autonomous propositional attitude is caused to be *non-autonomously eradicated* if it is caused to be either (a) *shed* (e.g., to go out of existence, or to decrease in severity) or (b) blocked from manifesting in ways that relevantly bypass a thinker's cognitive and conative faculties. The core idea of TMS is, in sum, that your freedom of thought is violated if you're caused to either acquire an unsheddable attitude that your own faculties played no role in acquiring or are caused to shed (or block) an attitude that your own faculties played no role in your shedding.

A simple and egregious case of *acquisition manipulation* is having beliefs or desires 'implanted' in a clandestine fashion. A simple and egregious case of *eradication manipulation* is having beliefs or desires 'wiped' in a clandestine fashion. But these are just 'limit' cases; what's more interesting (as we'll see in Sect. 6) are the less egregious but nonetheless morally and epistemically significant violations.

A final point of clarification: (TMS), it should be emphasized, does not imply that if a subject had an implanted belief or desire that *was* sheddable, then it would thereby *not* constitute a violation of her freedom against acquisition manipulation. This is because (TMS)—both its acquisition and eradication clauses—offers sufficiency conditions but not necessity conditions on freedom-of-thought violating thought manipulation. However, even as a (disjunctive) sufficiency condition for attitudinal acquisition and eradication manipulation, TMS is of philosophical interest. As we'll see in the next section—using some BCI-based

[27] See Carter (2020b, Chap 2) for a detailed discussion of different ways to interpret this condition.

[28] For alternative ways of thinking about attitudinal autonomy, see, e.g., Dworkin (1981) and Frankfurt (1988). For developments of a Mele-style approach to attitudinal autonomy—an approach which denies that attitudinal autonomy is entirely a matter of one's present psychological structure and can also include such things as the attitude's history—see, e.g., Weimer (2009) and Carter (2020b, Chap 2, 2020a).

thought experiments—any plausible right we have against thought manipulation can be violated (with reference to the clauses in TMS) in crucially different *kinds* of ways, which map on to (at least) four interestingly different 'varieties' of freedom-of-thought violating thought manipulation.

6.

Let's now consider the following cases:

> **Case 1:** Otto, due to gradually failing biomemory, asks to be fitted with a sophisticated 'Neuralink Memory-Pro' brain-computer interface that will help 'pick up the slack' where his memory is failing, when it comes to scheduling and organizing his life.[29] The BCI is designed so that when Otto learns something he wants to include in his calendar, the information goes, via a thought command, rather than to his biomemory, straight to the BCI's cloud storage (e.g., much like a Google Calendar). When he attempts to recall old information from the Memory-Pro, he receives the information that is stored. For Otto, the Neuralink Memory-Pro plays the role that biomemory, pen-and-paper, as well as manually operated computers used to play for structuring his day. Unbeknownst to Otto, the Memory Pro's software update has now automatically 'auto-integrated' national and bank holiday dates into Otto's cloud storage.
>
> *Assessment*: Otto's initial and consensual fitting of the Neuralink Memory-Pro violated no thought-based right of his. However, the software update *did*. The reason, with reference to TMS, is that the update causes him to acquire non-autonomous (extended) propositional attitudes (e.g., national and bank holiday dates). For classification purposes, let's call Case 1 a **Type 1 case** for the following reason: freedom against thought manipulation is violated due to the acquisition of a non-autonomous belief in such a way that *no faculty (cognitive or conative) was exercised whatsoever* in the acquisition of the (extended) belief; in other words, his faculties have

[29] This case is a twist on Clark and Chalmers' (1998) case of 'Otto,' which they use to motivate the extended mind.

been *fully bypassed* in the course of propositional attitude acquisition.[30] As we'll see in Case 2, freedom against thought manipulation can be violated (with reference to the acquisition manipulation clause in TMS) even when faculties are only ***partially*** bypassed.

Case 2: Everything is the same as with Case 1, except for some of the details about the nature of the Neuralink Memory-Pro's update and Otto's knowledge about it. First, the update is much more extensive, in that it inserts (along with bank holidays) various other kinds of information, which will continuously be added, including on the basis of algorithmic suggestions synced from his other devices (e.g. 'Stores open late in your area tonight for Black Friday shopping' … 'This Sunday, new Netflix WWI documentary available', etc.). Due to high demand for the product, Otto is given an impossibly brief period of time to decide whether to opt-in or out-out of the update, not long enough for him to understand what kinds of things it will include (and how they're included), and he's provided no further information from the company. On good faith, Otto opts in, and is soon after baffled by what seem to be his own beliefs (and by extension, plans), and he begins losing his grip on what he had intentionally stored and what was prompted by the Memory-Pro's algorithms.[31]

Assessment: Unlike Case 1, it's not the case that Otto's acquisition of the algorithmically generated information inserted in his

[30] This includes no such exercise of a cognitive faculty in the past, as would be the case, for example, if one prearranged to have bank dates auto-inserted via the update at a future date.

[31] It's worth registering an important difference between the kind of situation described in Case 2 (a genuine acquisition manipulation case) with a superficially similar situation depicted in the science fiction show *Almost Human*. In that show—set in a cyborg future—it is common for individuals to see ***personalised*** hologram advertising, which targets the individual user. For example, while walking to the store, you might see a hologram on the side of a building which appears there (keyed to your GPS) to target you specifically, on the basis of sophisticated algorithms. In such a case, you would be—like Otto in Case 2—'bombarded' with content it would be very easy to 'uptake,' and further, in both cases, you are cognitively influenced. But the *Almost Human* situation is not a genuine case of acquisition manipulation (of either Type 1 or Type 2) because, in this case, your autonomy is being respected; you are ***nudged***, but not ***caused*** to uptake or endorse anything that features in the aggressive hologram-style advertising. However, the situation is different in Case 2 (as well as in Case 1) where acquisition manipulation is present.

Memory-Pro via the update *completely* bypassed his faculties. (He was after all informed that there would be some updates; he understood this much and consented to the update in so far as he understood it, which was limited given his unusually restricted opportunities). Nonetheless, this is a case where, with reference to TMS, the update causes him to acquire non-autonomous (extended) propositional attitudes and in doing so violates his freedom against acquisition manipulation. For classification purposes, let's call this a **Type 2 case** for the following reason: freedom against thought manipulation is violated due to the acquisition of non-autonomous (extended) beliefs (like Case 1), where these extended beliefs are non-autonomous *not* because (as in Case 1) their acquisition bypasses faculties altogether, but because it bypasses suitable *opportunities to exercise* those faculties.[32]

In sum: whereas Type 1 acquisition manipulation involves acquiring attitudes in ways that *completely* bypass the thinker's faculties, Type 2 acquisition manipulation involves acquiring attitudes in ways that bypass suitable *opportunities* to exercise those faculties, even if not bypassing the faculties wholesale.

Case 3: Everything is the same as with Case 1, with a few important exceptions. The Neuralink Memory-Pro's creators, inspired by the efficacy in 'strategic forgetting'[33] demonstrated by deep neural networks and reinforcement learning techniques at Google DeepMind, have introduced an algorithm in the latest update that *deletes* information stored in the Memory-Pro deemed to be 'clogging' up the system. This includes, for example, information about plans that have been canceled or superseded by other plans. It also includes information stored in the Memory-Pro that is both flagged by the algorithm as 'unimportant details' (e.g., the weather back on May 5, 2024) and which has gone long enough without

[32] Plausibly, after all, your faculties are 'bypassed' in the acquisition of an attitude in a way that is relevant to whether the attitude is autonomous if you acquire it without suitable opportunity to exercise those faculties. (By way of comparison: An otherwise non-autonomous attitude whose acquisition bypasses one's faculties wouldn't be 'converted' into an autonomous attitude simply were it the cases that one was able to exercise one's faculties in unsuitable circumstances in coming to acquire the attitude.) For a detailed discussion of this issue, framed in terms of competences rather than faculties, see Carter (2020b, Chap 2).

[33] See, e.g., Beierle and Timm (2019) and Silver et al. (2017).

being retrieved. The update's functions, including how the algorithm targets information for deletion, are not made suitably explicit to users.

Assessment: With reference to (TMS), Case 3 is a case of *eradication manipulation* rather than *acquisition manipulation*. Recall that, on (TMS), an otherwise autonomous propositional attitude is caused to be *non-autonomously eradicated* if it is caused to be *shed* (e.g., to go out of existence, or to decrease in intensity) in ways that relevantly bypass your cognitive and conative faculties. In this case, Otto's faculties have been bypassed precisely because he lacks an explanation for how the algorithm is targeting stored information. Call this kind of case—where the mechanisms of memory eradication (as opposed to acquisition) are opaque to one—a **Type 3 case**. The difficulty of legislating Type 3 cases, it is worth noting, is already evidenced in recent debates following the 2018 GDPR (Art. 22, 13–15, Recital 71) about a data subject's 'right to an explanation', when purely algorithmic decisions are used to make decisions that affect someone's interests.[34]

Case 4: After years of enjoying his Neuralink Memory-Pro BCI, Otto wants 'the next big thing,' which is Neuralink's 'i-Connect' BCI device, which promises to help a thinker better 'organise one's mind.' The device's key trick is to use semantic tagging to sort information committed via thought command to information storage into compartments. Algorithms are then run on specific compartments in order to 'connect' information a thinker might not have connected themselves, which is then 'suggested' to the user on the basis of retrieval cues. The i-Connect promises, for example, to help users make better decisions on issues ranging from whom to trust (e.g., by storing track-record information) to which things to do to best relax. The suggestions made by the i-Connect, however, interfere with a thinker's own natural capacities for insight and

[34] For discussion, see Goodman and Flaxman (2017) and Selbst and Powles (2017). For criticism that the GDPR can reasonably be interpreted as insuring a 'right to an explanation' on the part of data subjects when purely algorithmic decisions are made that affect their interests, see Wachter et al. (2017).

creativity. In particular, the i-Connect does this by (albeit, inadvertently) blocking the efficacy of 'incubation' in insight problem solving tasks.[35]

Assessment: Let's assume, *ex hypothesi*, that Otto is fully aware of what kind of information the i-Connect enables him to acquire and even how it does this, such that the case is not, with reference to (TMS), a case of *acquisition manipulation*; in short, in Case 4, the 'bypass' condition on acquisition manipulation is not met *ex hypothesis*. That said, with reference to (TMS), Case 4 *is* a case of *eradication manipulation*. But this is *not* due (as in Case 3) to the 'shedding' proviso on eradication manipulation but due to the 'blocking' proviso. In Case 4, various insights Otto would have had have been effectively, even if not by intentional design, 'blocked.' Having insights blocked needn't violate a plausible freedom against thought manipulation (with reference to TMS's eradication manipulation component) if such blocking *itself* did not relevantly bypass cognitive and conative faculties. In Case 4, though, it does. Call this, accordingly, a **Type 4 case**: a case of eradication manipulation that qualifies as such, with reference to TMS, via 'blocking' rather than via 'shedding.'

Extended thought manipulation of Types 1–4[36] hardly exhaust possible categories. In fact, we can imagine subcategories of several of these, which map on to, e.g., partial or total bypassing, partial or total shedding, partial or total blocking, etc.

7.

The aim of the above taxonomy is to reveal a few of the salient contrast points when it comes to *violations* of a plausible freedom against thought manipulation—viz., one that is framed (as TMS is) in terms of a freedom against (at least) the caused acquisition of *non-autonomous attitudes*

[35] Sternberg and Davidson (1995), Metcalfe and Wiebe (1987), and Carter (2017).

[36] I've intentionally illustrated the varieties of thought manipulation I have by using BCIs. This is because BCIs—even if less practically applicable today than smartphones, through which thought manipulation is also in principle possible—allow us to frame these kinds of manipulation in an particularly sharp way. It is worth noting though that BCIs aren't *necessary* for thought manipulation. If the extended mind and cognition theses (see Sects. 2–3) hold water, and one's mind supervenes partly on a thinker's extracranial environment, the ingredients are present to manipulate 'extended' thought. See Carter and Palermos (2016) for discussion on this point.

and against the *non-autonomous eradication* of (would-be) autonomous attitudes.

As we continue to develop new technologies that make thought manipulation possible in new ways—including (and in addition to BCIs) various kinds of brain 'implants'[37] along with potential new breakthroughs in research on artificial neurons[38] and deep brain stimulation[39]—it becomes more important to anticipate and understand varieties of thought manipulation that such technologies enable. The above is an attempt at engaging in this kind of anticipation.

Further work in (extended) thought manipulation will go beyond the kind of sufficient condition (TMS) advanced here in order to make progress *vis-à-vis* the articulation of conditions *necessary* as well as sufficient for extended thought manipulation, a project more ambitious than what I've set out to do here.[40]

References

Adams, F., & Aizawa, K. (2008). *The bounds of cognition.* Blackwell.

Alegre, S. (2017). Freedom of thought, belief and religion and freedom of expression and opinion. In *Human Rights of Migrants in the 21st Century* (pp. 72–77). Routledge.

Audi, R. (2001). Doxastic voluntarism and the ethics of belief. *Knowledge, Truth, and Duty*, 93–111.

Bach-y-Rita, P. (1983). Tactile vision substitution: Past and future. *International Journal of Neuroscience, 19*(1–4), 29–36.

Bach-y-Rita, P., & Kercel, S. W. (2003). Sensory substitution and the human-machine interface. *Trends in Cognitive Sciences, 7*(12), 541–546.

Baron, M. (2003). Manipulativeness. *In Proceedings and Addresses of the American Philosophical Association, 77*(37–54), 2.

Beierle, C., & Timm, I. J. (2019). Intentional forgetting: An emerging field in Ai and beyond. *KI-Künstliche Intelligenz, 33*(1).

Blitz, M. J. (2010). Freedom of thought for the extended mind: Cognitive enhancement and the constitution. *Wisconsin Law Review*, 1049.

[37] For discussion of research on the implantation of 'false memories,' see, e.g., Ramirez et al. (2013). See also Carter (2020b, Chap 1).

[38] See, e.g., Simon et al. (2015).

[39] See, e.g., Suthana and Fried (2014) and Flöel et al. (2008).

[40] Thanks to Marc Blitz, Christoph Bublitz, John Tillson, and Ruaridh Gilmartin for helpful comments on a draft of this paper.

Bostrom, N., & Sandberg, A. (2009). Cognitive enhancement: Methods, ethics, regulatory challenges. *Science and Engineering Ethics, 15*(3), 311–341.
Bublitz, J. (2013). My mind is mine!? Cognitive liberty as a legal concept. In *Cognitive Enhancement*, 233–264.
Burge, T. (1986). Individualism and psychology. *Philosophical Review, 95*(January), 3–45.
Carter, J. A. (2017). Virtuous insightfulness. *Episteme, 14*(4), 539–554.
Carter, J. A. (2020a). Epistemic autonomy and externalism. In J. Matheson and K. Lougheed (Eds.), *Epistemic Autionomy.*
Carter, J. A. (2020b). *The future of knowing: Knowledge, radical enhancement, and epistemic autonomy.*
Carter, J. A., & Kallestrup, J. (2016). Extended cognition and propositional memory. *Philosophy and Phenomenological Research, 92*(3), 691–714.
Carter, J. A., & Kallestrup J. (2017). Extended circularity. In J. Adam Carter, J. K. Andy Clark, S. O. Palermos & D. Pritchard (Eds.), *Extended epistemology.* Oxford University Press.
Carter, J. A., Jesper, S., Palermos, O., & Pritchard, D. (2014). Varieties of externalism. *Philosophical Issues, 24*(1), 63–109.
Carter, J. A., Orestis, S., & Palermos. (2016). Is having your computer compromised a personal assault? The ethics of extended cognition. *Journal of the American Philosophical Association, 2*(4), 542–560.
Carter, J. A., & Pritchard, D. (2019). The epistemology of cognitive enhancement. *The Journal of Medicine and Philosophy, 44*(220–42), 2.
Carter, J. Adam, & Duncan, P. 2020. Extended entitlement. In Peter Graham & Nikolaj J. L. L. Pederson (Eds.), *New essays on entitlement.* Oxford University Press.
Chalmers, David. 2008. Foreword to Andy Clark's supersizing the mind. *A. Clark, supersizing the mind: Embodiment, action, and cognitive extension*, 000–000.
Clark, A. (2003). *Natural-born cyborgs: Minds, technologies, and the future of human intelligence.* Oxford University Press.
Clark, A. (2008). *Supersizing the mind: Embodiment, action, and cognitive extension: Embodiment, action, and cognitive extension.* Oxford University Press.
Clark, A. (2010). Memento's Revenge: The Extended Mind, Extended. In R. Menary (Ed.), *The Extended Mind* (pp. 43–66). MIT Press.
Clark, A., & Chalmers, D. (1998). The extended mind. *Analysis, 58*(1), 7–19.
Clarke, M. (1986). Doxastic voluntarism and forced belief. *Philosophical Studies, 50*(1), 39–51.
Dworkin, G. (1981). The concept of autonomy. *Grazer Philosophische Studien, 12*, 203–213.

Flöel, A., Rösser, N., Michka, O., Knecht, S., & Breitenstein, C. (2008). Noninvasive brain stimulation improves language learning. *Journal of Cognitive Neuroscience, 20*(8), 1415–1422.

Frankfurt, Harry G. 1988. *The importance of what we care about: Philosophical essays.* Cambridge University Press.

Gardner, P. (2004). Hand on religious upbringing. *Journal of Philosophy of Education, 38*(1), 121–128.

Gertler, B. (2000). The mechanics of self-knowledge. *Philosophical Topics, 28*(2), 125–146.

Gertler, B. (2010). *Self-knowledge.* Routledge.

Ghafoor, U., Zafar, A. Atif Yaqub, M., and Keum-Shik, H. (2019). Enhancement in Classification Accuracy of Motor Imagery Signals with Visual Aid: An fNIRS-Bci Study. *International Conference on Control Robot System Society*, 1201–1206.

Goodman, B., & Flaxman, S. (2017). European Union regulations on algorithmic decision-making and a 'right to explanation.' *AI Magazine, 38*(3), 50–57.

Hand, M. (2002). Religious upbringing reconsidered. *Journal of Philosophy of Education, 36*(4), 545–557.

Hand, M., Mackenzie, J., Gardner, P., & Tan, C. (2004). Religious upbringing: A rejoinder and responses. *Journal of Philosophy of Education, 38*(4), 639–662.

Hansson, L. (2018). Science education, indoctrination, and the hidden curriculum. In *History, philosophy and science teaching*, 283–306. Springer.

Kant, I. ((1797) 1991). *The metaphysics of morals* (Trans. M. Gregor). Cambridge University Press.

Kripke, S. (1980). *Naming and necessity.* Harvard University Press.

McKinsey, M. (1991). Anti-individualism and privileged access. *Analysis, 51*(1), 9–16.

Mele, A. R. (2001). *Autonomous agents: From self-control to autonomy.* Oxford University Press on Demand.

Menary, R. (2007). *Cognitive integration: Mind and cognition unbounded.* Springer.

Menary, R. (2010). *The extended mind.*

Metcalfe, J., & Wiebe, D. (1987). Intuition in insight and noninsight problem solving. *Memory & Cognition, 15*(3), 238–246.

Mill, J.S. (1859 [1998]). *On liberty and other essays.* Oxford University Press USA.

Musk, E. (2019). An integrated brain-machine interface platform with thousands of channels. *Journal of Medical Internet Research, 21*(10), e16194.

Palermos, S. O. (2011). Belief-forming processes, extended. *Review of Philosophy and Psychology, 2*(4), 741–765.

Palermos, S. O. (2014a). Knowledge and cognitive integration. *Synthese, 191*(8), 1931–1951.

Palermos, S. O. (2014b). Loops, constitution, and cognitive extension. *Cognitive Systems Research, 27*, 25–41.

Palermos, S. O. (2016). The dynamics of group cognition. *Minds and Machines, 26*(4), 409–440.

Parent, T. (2017). Externalism and self-knowledge. In E. N. Zalta (Ed.), *The stanford encyclopedia of philosophy* (Fall). https://plato.stanford.edu/archives/fall2017/entries/self-knowledge-externalism/; Metaphysics Research Lab, Stanford University.

Pisarchik, A. N., Maksimenko, V. A., & Hramov. A. E. (2019). From novel technology to novel applications: Comment on 'an integrated brain-machine interface platform with thousands of channels' by Elon Musk and Neuralink. *Journal of Medical Internet Research. 21*(10), e16356.

Pritchard, D. (2002). McKinsey paradoxes, radical scepticism, and the transmission of knowledge across known entailments. *Synthese, 130*(2), 279–302.

Pritchard, D. (2010). Cognitive ability and the extended cognition thesis. *Synthese, 175*(1), 133–151.

Pritchard, D. (2018). Extended epistemology. *Extended epistemology*, 90–104.

Putnam, H. (1975). The meaning of 'meaning'. *Minnesota Studies in the Philosophy of Science, 7*, 131–193.

Ramirez, S., Liu, X., Lin, P. A., Suh, J., Pignatelli, M., Redondo, R. L., Ryan, T. J., & Tonegawa, S. (2013). Creating a false memory in the Hippocampus. *Science, 341*(6144), 387–391.

Schwitzgebel, E. (2008). The unreliability of naive introspection. *Philosophical Review, 117*(2), 245–273.

Selbst, A. D., & Powles, J. (2017). Meaningful information and the right to explanation. *International Data Privacy Law, 7*(4), 233–242.

Siegel, H. (2004). Faith, knowledge and indoctrination: A friendly response to hand. *Theory and Research in Education, 2*(1), 75–83.

Silver, D, Hubert, T., Schrittwieser, J., Antonoglou, I., Lai, M., Guez, A., Lanctot, M., et al. (2017). Mastering Chess and Shogi by Self-Play with a General Reinforcement Learning Algorithm. *arXiv Preprint* arXiv:1712.01815.

Simon, D. T., Larsson, K. C., Nilsson, D., Burström, G., Galter, D., Berggren, M., & Richter-, A. (2015). An organic electronic biomimetic neuron enables auto-regulated neuromodulation. *Biosensors and Bioelectronics, 71*, 359–364.

Sternberg, R. J., & Davidso, J. E. (1995). *The nature of insight*. The MIT Press.

Steup, M. (2000). Doxastic voluntarism and epistemic deontology. *Acta Analytica, 24*, 25–56.

Suthana, N., & Fried, I. (2014). Deep brain stimulation for enhancement of learning and memory. *NeuroImage, 85*, 996–1002.

Wachter, S., Mittelstadt, B., & Floridi, L. (2017). Why a right to explanation of automated decision-making does not exist in the general data protection regulation. *International Data Privacy Law, 7*(2), 76–99.

Wang, W., Liu, Y., Li, Z., Wang, Z., He, F., Ming, D., & Yang, D. (2019). Building multi-modal sensory feedback pathways for Srl with the aim of sensory enhancement via Bci. In *2019 Ieee International Conference on Robotics and Biomimetics (Robio)*, 2439–44. IEEE.

Watson, G. (1996). Two faces of responsibility. *Philosophical Topics, 24*(2), 227–248.

Weimer, S. (2009). Externalist autonomy and availability of alternatives. *Social Theory and Practice, 35*(2), 169–200.

Wilson, R. A. (2000). *The mind beyond itself*. Oxford University Press.

Wilson, R. A. (2004). *Boundaries of the mind: The individual in the fragile sciences-cognition*. Cambridge University Press.

Index

M. J. Blitz et al. (eds.), *The Law and Ethics of Freedom of Thought, Volume 1*, Palgrave Studies in Law, Neuroscience, and Human Behavior,
https://doi.org/10.1007/978-3-030-84494-3

The manufacturer's authorised representative in the EU is Springer Nature Customer Service Centre GmbH, Europaplatz 3, 69115 Heidelberg, Germany. If you have any concerns regarding our products, please contact ProductSafety@springernature.com

Printed and bound by CPI Group (UK) Ltd, Croydon, CR0 4YY
30/09/2026
02225507-0004